# Mutations in Man

Edited by Günter Obe

With 73 Figures

Springer-Verlag
Berlin Heidelberg New York Tokyo 1984

Professor Dr. GÜNTER OBE
Institut für Genetik, Freie Universität Berlin
Arnimallee 5–7, 1000 Berlin 33, FRG

ISBN 3-540-13113-2 Springer-Verlag Berlin Heidelberg New York Tokyo
ISBN 0-387-13113-2 Springer-Verlag New York Heidelberg Berlin Tokyo

Library of Congress Cataloging in Publication Data. Main entry under title: Mutations in man.
Includes index. 1. Chemical mutagenesis. 2. Human chromosome abnormalities. 3. Human
genetics. 4. Medical genetics. I. Obe, G. QH465.C5M88  1984  573.2'292  84-1213

Typesetting, offsetprinting and bookbinding: Brühlsche Universitätsdruckerei, Giessen.
2131/3130-543210

# Preface

This year we remember the 39th anniversary of the atomic bomb explosions in Hiroshima and Nagasaki, which led to the exposure of thousands of people to high doses of ionizing radiations. Nearly 18 years earlier, on the 15th of September, 1927, H.J. Muller presented his paper *The Problem of Genic Modification* at the Fifth International Congress of Genetics in Berlin, in which he brilliantly demonstrated the mutagenic activity of X-rays. In 1928, K.H. Bauer formulated his mutation theory of the origin of cancer, and already in 1914, Th. Boveri speculated that tumor cells originate from an abnormal chromosomal complement. In the meantime we have learned that also nonionizing radiation and an immense number of environmental chemicals, both, man-made and naturally occurring, are mutagenic in a variety of test systems, including human cells. In no case has it been shown unequivocally that physical or chemical mutagens have led to an elevation of the mutation rate in the germ cells of man, but in view of the huge body of experimental data this seems to be a problem of detection. It can be expected that germ cell mutations are induced as a consequence of exposure to mutagens in man, as yet undetectable with the methods at hand. An uncontrolled addition of mutations to the human gene pool may well have unforeseen and catastrophic consequences in future generations for whom we should feel responsible. The scenario which can be imagined is a rising frequency of various types of genetic ill-health including heritable types of cancer. The induction of mutations in somatic cells of man has been detected without any doubt, and there is ample evidence that some of these mutations are initiators of cancer. Therefore it is of central importance to understand the mechanisms of mutation induction, to recognize mutagenic agents, to limit the exposure to such mutagens, and to prevent that new mutagens are introduced carelessly into our environment. The book *Mutations in Man* may be helpful in a better understanding of these problems and their solution. The following aspects are discussed:

1. Chemical mutagens in the human environment and their detection (in Chapter Sobels, p. 1ff.), their reactions with cellular DNA (in Chapter Doerjer, Bedell and Oesch, p. 20ff.), and the repair of DNA lesions (in Chapter van Zeeland, p. 35ff.).

2. Structure and organization of the human genome (in Chapter Evans, p. 58ff.).
3. Frequencies and origin of gene or point mutations (in Chapter Vogel, p. 101ff.), and of chromosomal abnormalities (in Chapters Sperling, p. 128ff., and Hansmann, p. 147ff.).
4. Origin and significance of chromosomal alterations (in Chapter Natarajan, p. 156ff.).
5. The human lymphocyte test system (in Chapter Obe and Beek, p. 177ff.), chromosomal alterations in lymphocytes of patients under chemotherapy (in Chapter Gebhart, p. 198ff.), and of cigarette smokers (in Chapter Obe, Heller and Vogt, p. 223ff.).
6. Sperm anomalies in smokers and nonsmokers (in Chapter Vogt, Heller and Obe, p. 247ff.).
7. Estimation of the genetic risk (in Chapter Ehling, p. 292ff.).

I thank the authors of this book for having taken their time to write chapters and the staff of the Springer-Verlag, especially Dr. Dieter Czeschlik, Miss Antonella Cerri, and Mrs Marion Kreisel for their help.

Berlin-Dahlem, January 1984                          GÜNTER OBE

# Contents

# Contributors

You will find the addresses at the beginning of the respective contributions

Bedell, M.A.   20
Beek, B.   177
Doerjer, G.   20
Ehling, U.H.   292
Evans, H.J.   58
Gebhart, E.   198
Hansmann, I.   147
Heller, W.-D.   223, 247

Natarajan, A.T.   156
Obe, G.   177, 223, 247
Oesch, F.   20
Sobels, F.H.   1
Sperling, K.   128
Van Zeeland, A.A.   35
Vogel, F.   101
Vogt, H.-J.   223, 247

# Problems and Perspectives in Genetic Toxicology

F.H. SOBELS[1]

## 1 Introduction

The past 12 years have witnessed the rapid development of the field of chemical mutagenesis which has now attained a new dimension as Environmental Mutagenesis or Genetic Toxicology. This growth is a direct consequence of the concern for human ill-health resulting from damage to the genetic material. Mutagenic chemicals occur in the food we eat, the water we drink, and the air we breathe. Moreover, there is a striking parallelism between the mutagenic and carcinogenic potential of a large number of these chemicals.

The problems of adequately protecting the human population against mutagenic and carcinogenic hazards are enormous and extend far beyond national boundaries. Intensive collaboration on a world-wide scale was urgently needed. Thus, ICPEMC, the International Commission for Protection against Environmental Mutagens and Carcinogens was founded in January 1977. ICPEMC is dedicated to identifying and evaluating genotoxic chemicals in the environment and to suggesting approaches to regulatory and legal decision-making.

In the following, a brief review is given of what has been accomplished in the evaluation of genotoxic chemicals and some reflections on what should perhaps be done now. I have been closely associated with ICPEMC since its foundation. Consequently, ongoing discussions in the Commission have greatly contributed to the development of my ideas on the subject. This paper, however, should not be construed as reflecting the views of ICPEMC.

## 2 Identification of Mutagenic Activity

The formulation of adequate measures to restrict exposure to mutagenic chemicals first of all requires the recognition of mutagenic activity. For this purpose many useful test systems are now in operation and various approaches to evaluate mutagenic activity have been envisaged. In discussions with my colleagues of the Dutch Advisory

---

1 Department of Radiation Genetics and Chemical Mutagenesis, State University of Leiden, Wassenaarseweg 72, 2333 AL Leiden, The Netherlands

Mutations in Man, ed. by G. Obe
© Springer-Verlag Berlin Heidelberg 1984

Committee on Chemical Mutagenesis, a simple scheme for evaluating genotoxic activity was proposed. It consists of a qualitative and a quantitative phase (Sobels 1980). In the qualitative phase the major aim is to establish whether a particular compound with or without metabolic activation will exhibit mutagenic properties. For this purpose a battery consisting of one prokaryotic and two eukaryotic test systems is selected to minimize the chances of scoring false negatives. The use of the quantitative phase is indicated when an assessment of genetic risks is required; for this purpose, mammalian in vivo systems are employed.

Selection of the most appropriate test systems should be dictated by the sensitivity of their detection capacity, which in turn requires careful studies in comparative chemical mutagenesis (Sobels 1977). It was with this purpose in mind that a coordinated project for a number of different chemicals was initiated under the Environmental Research Programme of the European Community. Extensive data are now available for MMS, DEN, mitomycin C, procarbazine, atrazine, and benz(a)pyrene in some 20 different assay systems (Adler 1980; Loprieno and Adler 1980).

Ideally such studies should be based on comparison of the induction frequencies of different kinds of genetic damage and the reaction products of the chemical with the DNA in the target cells at different concentrations. Results of my colleagues from the Leiden laboratory on the quantitative comparison of mutation induction in *E. coli*, V79 Chinese hamster cells and L5178Y mouse lymphoma cells with EMS (ethyl methane sulfonate) show a remarkably good correlation between exposure concentration and the dose in DNA expressed as ethylations per nucleotide in very different in vitro systems (Fig. 1, left). However, under identical treatment conditions, large differences were observed in the mutation frequencies between *E. coli*, V79 and L5178Y cells, even if the levels of alkylation in the DNA were similar (Fig. 1, right). Such a quantitative calibration of the systems would not have been possible without DNA-dose measurements (Aaron et al. 1980; Mohn et al. 1982; van Zeeland et al. 1983). A further study was directed at clarifying which of the various ethylation products may be relevant for mutation induction. Thus, the genetic effects of two different ethylating agents, ENU (ethyl nitrosourea) and EMS were compared in V79 Chinese hamster cells (van Zeeland et al., to be published). From the data in Table 1

Table 1. A comparison of the mutation frequencies induced by ENU and EMS in V79 Chinese hamster cells shows that only at equal levels of $O^6$-ethylguanine the same frequencies of mutation were observed

|  | Ratio of ENU and EMS induced mutant frequency measured at |
| --- | --- |
| Equal exposure concentration | 1.8 |
| Equal level of total alkylations | 4.2 |
| Equal level of 7-ethylguanine | 23.8 |
| Equal level of $O^6$-ethylguanine | 1.04 |

Treatments were for 1 h with ENU and for 2 h with EMS. Mutations were determined at the HGPRT locus. DNA dose measurements were carried out immediately after treatment of the cells (Zeeland A.A. van, unpublished). ENU: ethylnitrosourea; EMS: ethylmethanesulphonate

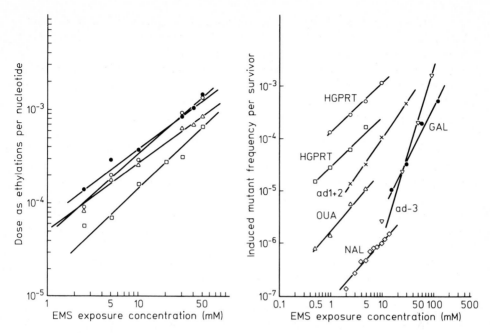

**Fig. 1.** *Left* correlation between exposure concentration and the dose in DNA after a 2-h treatment in buffer with various concentrations of EMS. ○ Chinese hamster cells; ● *E. coli*; △ Neuro-
*Right* induced mutation frequencies under identical treatment conditions in a variety of cells, ranging from bacteria, fungi, and yeast to mammalian cells in culture. It can be seen that at similar doses of EMS, mutation frequencies show striking variation both between assay systems and the genetic loci employed. ○ Chinese hamster cells (*HGPRT⁻*); □ Mouse lymphoma cells (*HGPRT⁻*); △ Mouse lymphoma cells (*OUAᴿ*); ● *E. coli* (*GAL⁺*); ◇ *E. coli* (*NALᴿ*); ▽ Neurospora (*ad-3*); ✕ Yeast (*ad 1+2*). (Aaron et al. 1980; van Zeeland et al. 1983 and unpublished data of Knaap AGAC)

it can be seen that at equal levels of $O^6$-ethylguanine the same mutation frequencies were obtained for the two agents. When the mutation frequencies were compared on the basis of other parameters, such as exposure concentration, total level of alkylations, or the amount of 7-ethylguanine, there are no clear correlations. These data illustrate that the degree of alkylation at the $O^6$ position of guanine constitutes a reliable indicator for mutation induction by ethylating agents in this system (van Zeeland et al. unpublished work).

In this context it is appropriate to say a few words about ICPEMC's Committee 1 (1983) report which has just appeared. Committee 1, chaired by Dr. D. Brusick and Dr. B.J. Kilbey, was directed at the evaluation of short-term screening tests for predicting heritable damage in humans. The whole-mammal tests that the Committee chose to use as standard are the specific locus test, the heritable translocation test and the dominant lethal test. Test performance in 12 different assays ranging from microbial systems to mammalian cell cultures, were compared with those of the above whole-animal tests. As a consequence of the limited data-base which is skewed toward positive chemicals, no significant correlation could be established. ICPEMC believes however, that this exercise has been useful as a framework for future activities.

In the interpretation of test results difficulties may be encountered in decision-making when both positive and negative data are involved. Towards this objective of producing criteria to evaluate mixed data from a battery of tests, ICPEMC Committee 1 will continue its activities. In addition to the qualitative judgment, a quantitative component focusing on in vivo potency may be established.

## 3 The Advantage of Short-Term Tests for Identifying Carcinogens

The development and refinement of the bacterial test systems have played a major role in the identification of substances with carcinogenic properties present in the human environment. Since many carcinogens require activation in the mammalian body, major improvement in the bacterial test systems was made when representative parts of mammalian metabolism were included as part of the test protocol, either in vitro or in the host-mediated assay (Clayson 1980; Mohn 1981).

Collating data on about 300 substances whose carcinogenicity or non-carcinogenicity had been established in animals, Ames and colleagues were able to show a high correlation between the capacity to produce cancer in animals and mutations in the *Salmonella* test, that is the ability to damage DNA (Bridges 1979; McCann et al. 1975; McCann and Ames 1976; Ames 1979). These findings reflected, as Bridges (1979) puts it: „almost a quantum jump in our progress towards understanding the difficult area of carcinogenesis". At present one-fourth to one-fifth of the human population will develop cancer and clearly the identification of possible causes underlying this high tumor incidence deserves high priority. In this context it is important to remember that in vitro mutagenicity monitors only the early stages of carcinogenesis, that is the interaction of the electrophile with electron-rich groups such as DNA. This is the initiation stage and for that reason it is the group of genotoxic carcinogens that can be detected (Clayson 1980).

Of particular interest is the identification of many mutagenic components in food for human consumption. This raises the question of the extent to which these mutagens are causative agents in the large dietary component of human cancer revealed by epidemiological studies. According to Weisburger's (1979) estimate 50% of current cancer incidence in the USA could be due to their daily food. Similar estimates have been made by Higginson and Muir (1979). The most convincing epidemiological evidence is provided by (1) differences in mortality rate of specific cancer between different countries, (2) the observation that immigrants tend to show the cancer mortality incidence of their host country, and (3) differences in cancer incidence in well-defined religious groups, such as 7th Day Adventists or Mormons.

A particularly high incidence of stomach and esophagus cancer was found among Japanese fishermen. Sugimura and colleagues then investigated their lifestyle and identified what particular food habits were possibly different from the rest of the population (for review see Sugimura 1982a,b,c; Sugimura et al. 1977). Thus, it was observed that the fishermen roast all their fish on charcoal braisures. By testing the various components of the roasted fish in an Ames test, they found that the charred surface of the broiled fish and also of meat contained mutagens. The data in Fig. 2

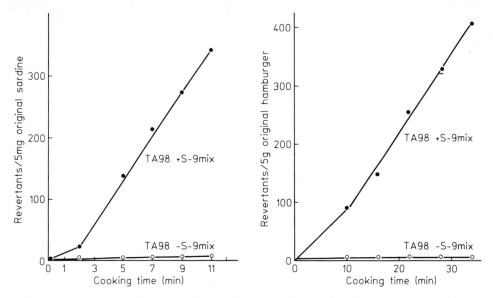

**Fig. 2.** Mutagenicity of broiled sardines (*left*) and hamburgers (*right*) depending on cooking time. The stronger response from the fish than from the meat results from the fact that the sardines were dried and the meat was fresh. The average cooking time for sardines would be about 5 min and for the hamburgers about 10 min. (Sugimura 1982a)

show how mutagenicity depends on cooking time. The high temperature in the char-coal fires leads to the formation of pyrolysates of components of protein (Fig. 2). Pyrolysates of particular amino acids, tryptophan and glutamic acid which lead to Trp–P-2 and Glu–P-1 are highly mutagenic and the same was found for the more recently isolated quinolines IQ, MeIQ, and MeIQX (see Table 2); these are substances which are formed already at lower temperatures of frying meat or normal boiling. All these substances are heterocyclic amines. Their structure and interaction with DNA was determined (Fig. 3) and animal cancer bioassays were performed, and a high incidence of liver tumors and interscapular tumors was observed after the application of pyrolysates of tryptophan and glutamic acid (Table 3).

Another important group of substances in food and beverages, such as coffee and tea, that merits consideration, are the flavonoids. Quercetin occurs in all vegetables and fruits. It is mutagenic in *Salmonella* and produces chromosome aberrations in vitro, micronuclei in vivo, and recessive lethals in *Drosophila*. The data on tumor induction are not consistent. In one study employing Norwegian rats, a clear carcino-genic effect could be demonstrated, but this could not be repeated in extensive Japanese studies using very high doses of quercetin. The ubiquitous occurrence of flavonoids warrants intensive studies of whether these substances reach mammalian germ cells and can act as mutagens; the same holds, in fact, for the pyrolytic products cited above. Comparative studies on the mutagenic activity, as indicated by the *Salmonella* test, of different breeds of vegetables show great variation in the amount of quercetin (see Table 4); selection directed at low quercetin contents could well be successful (van der Hoeven et al., to be published). (These data, like those in Fig. 2,

**Table 2.** Mutagens isolated from pyrolysates. (Sugimura 1982a)

| Full name (abbreviation) | Structure | Source of isolation | Spec. Mut. Act. Revertants/µg TA98, +S9 mix | Present in |
|---|---|---|---|---|
| 2-Amino-3,4-dimethyl-imidazo[4,5-f]quinoline (MeIQ) | | Broiled sardine (Kasai et a. 1980b,c) | 661,000 | Fried beef (K. Wakabayashi, unpublished observation), heated beef extract (Spingarn et al. 1980) |
| 2-Amino-3-methylimidazo-[4,5-f]quinoline (IQ) | | Broiled sardine (Kasai et al. 1980a,c) | 433,000 | |
| 2-Amino-3,8-dimethyl-imidazo[4,5-f]quinoxaline (MeIQx) | | Fried beef (Kasai et al. 1981) | 145,000 | |
| 3-Amino-1-methyl-5 H-pyrido[4,3-b]indole (Trp-P-2) | | Tryptophan pyrolysate (Sugimura et al. 1977) | 104,000 | Broiled sardine (Yamaizumi et al. 1980) |
| 2-Amino-6-methyldipyrido-[1,2-a:3′,2′-d]imidazole (Glu-P-1) | | Glutamic acid pyrolysate (Yamamoto et al. 1978) | 49,000 | |
| 3-Amino-1,4-dimethyl-5H-pyrido[4,3-b]indole (Trp-P-1) | | Tryptophan pyrolysate (Sugimura et al. 1977) | 39,000 | Broiled sardine (Yamaizumi et al. 1980), broiled beef (Yamaguchi et al. 1980b) |
| 2-Aminodipyrido[1,2-a:3′,2′-d]imidazole (Glu-P-2) | | Glutamic acid pyrolysate (Yamamoto et al. 1978) | 1,900 | Broiled dried-cuttlefish (Yamaguchi et al. 1980a) |

**Fig. 3.** The adducts of the pyrolytic products Trp-P-2 and Glu-P-1 with guanine. (Sugimura 1982a)

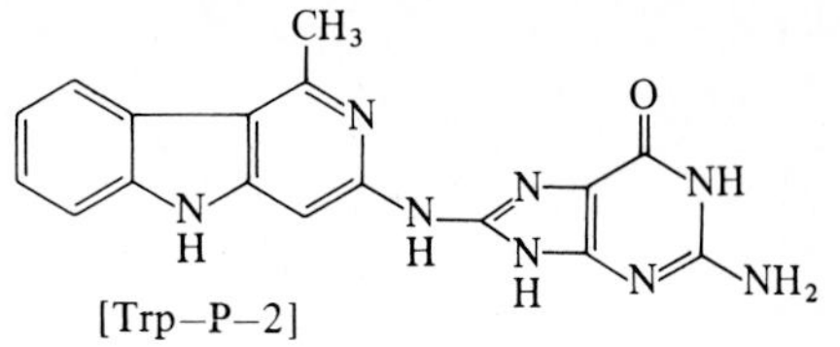

[Trp–P–2]

3–(C$^8$–guanyl) amino–1–methyl–5H–pyrido [4,3–b]indole

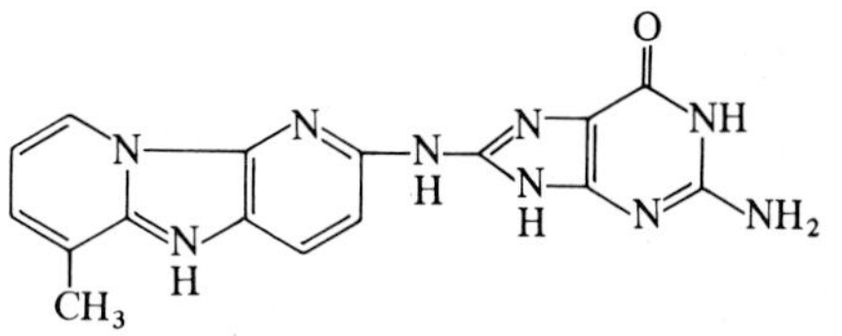

2–(C$^8$–guanyl)amino–6–methyldipyrido [1,2–a:3′,2′–d]imidazole

illustrate how the Ames test can be used as an analytical tool.) In this context I will repeat Sugimura's (1978) plea for modesty rather than alarmist statements: "Scientists now have so great a responsibility that they should be very cautious about mak-

---

Literature cited in the table 2:

Kasai H, Yamaizumi Z, Wakabayashi K, Nagao M, Sugimura T, Yokoyama S, Miyazawa T, Spingarn NE, Weisburger JH, Nishimura S (1980a) Potent novel mutagens produced by broiling fish under normal conditions. Proc Jpn Acad 56(B):278–283

Kasai H, Yamaizumi Z, Wakabayashi K, Nagao M, Sugimura T, Yokoyama S, Miyazawa T, Nishimura S (1980b) Structure and chemical synthesis of Me-IQ, a potent mutagen isolated from broiled fish. Chem Lett 11:1391–1394

Kasai H, Nishimura S, Wakabayashi K, Nagao M, Sugimura T (1980c) Chemical synthesis of 2-amino-3-methylimidazo[4,5-f]quinoline (IQ), a potent mutagen isolated from broiled fish. Proc Jpn Acad 56(B):382–384

Kasai H, Yamaizumi Z, Shiomi T, Yokoyama S, Miyazawa T, Wakabayashi K, Nagao M, Sugimura T, Nishimura S (1981) Structure of a potent mutagen isolated from fried beef. Chem Lett 4: 485–488

Spingarn NE, Kasai H, Vuolo LL, Nishimura S, Yamaizumi Z, Sugimura T, Matsushima T, Weisburger JH (1980) Formation of mutagens in cooked foods. III. Isolation of a potent mutagen from beef. Cancer Lett 9:177–183

Sugimura T, Kawachi T, Nagao M, Yahagi T, Seino Y, Okamoto T, Shudo K, Kosuge T, Tsuji K, Wakabayashi K, Iitaka Y, Itai A (1977) Mutagenic principle(s) in tryptophan and phenylalanine pyrolysis products. Proc Jpn Acad 53(1):58–61

Yamaguchi K, Shudo K, Okamoto T, Sugimura T, Kosuge T (1980b) Presence of 3-amino-1,4-dimethyl-5H-pyrido[4,3-b]indole in broiled beef. Gan 71:745–746

Yamaguchi K, Shudo K, Okamoto T, Sugimura T, Kosuge T (1980a) Presence of 2-aminodipyrido-[1,2-a:3′,2′-d]imidazole in broiled cuttle-fish. Gan 71:743–744

Yamaizumi Z, Shiomi T, Kasai H, Nishimura S, Takahashi Y, Nagao M, Sugimura T (1980) Detection of potent mutagens, Trp-P-1 and Trp-P-2, in broiled fish. Cancer Lett 9:75–83

Yamamoto T, Sutji K, Konsuge T, Okamoto T, Shudo K, Takeda K, Iitaka Y, Yamaguchi K, Seino Y, Yahagi T, Nagao M, Sugimura T (1978) Isolation and structure determination of mutagenic substances in L-glutamic acid pyrolysate. Proc Jpn Acad 54(B):248–250

**Table 3.** Induction of tumors in CDF$_1$ mice by oral administration of Glu-P-1 and Glu-P-2. (Sugimura 1982a)

| Group | Sex | Initial no. | No. of mice examined | No. of mice with (%) | | | |
|---|---|---|---|---|---|---|---|
| | | | | Tumor | Interscapular tumor | Liver tumor | Other tumor |
| Glu-P-1 | M | 40 | 40 | 34 (85) | 33(83) | 8 (20) | 9[2] |
| (0.05%) | F | 40 | 40 | 40(100) | 33(83) | 40(100) | 7 |
| Glu-P-2 | M | 40 | 37 | 29 (78) | 21(57) | 7 (19) | 15[6] |
| (0.05%) | F | 40 | 37 | 35 (95) | 20(54) | 35 (95) | 14[5] |
| Control | M | 40 | 13 | 0 | 0 | 0 | 3 |
| | F | 40 | 2 | 2 | 0 | 1 | 2 |

**Table 4.** Mutagenicity of methanol extracts of 5 cultivars of lettuce towards TA98 *Salmonella typhimurium*. The striking difference between, for example, cultivars Ravel and Renate may be noted. (van der Hoeven et al. to be published)

| Cultivar | Amount of lettuce extract (mg/plates) | Revertants per plate | |
|---|---|---|---|
| | | – GFE | + GFE |
| Sonate | 197 | 12 | 173(132)[a] |
| Ravel | 146 | 6 | 41 (42) |
| Renate | 121 | 10 | 279(346) |
| Dandy | 164 | 6 | 185(169) |
| Deci minor | 167 | 11 | 109 (98) |

[a]   The numbers between brackets correspond with the calculated numbers expressed as revertants per 150 mg lettuce per plate
GFE: Gutflora extract from rats was applied as a metabolizing system

ing statements that will alarm the general public until their data have been carefully evaluated. They must also realize and confess frankly that sometimes they know too little to give the correct estimate of risk that the public is anxious to hear". In a later paper he continues: "Otherwise the general population will come to distrust scientists, and this will be a great loss to our society" (Sugimura 1982b).

The problem of establishing carcinogenic activity is one of unusual dimension. There are at present some 60 to 100,000 synthetic or naturally occurring chemicals in our environment and many new bacterial mutagens have been defined by the about 2,000 laboratories employing microbial test systems and the 100 or so others using in vitro assays. Whole animal cancer tests are too expensive (approximately $ 500,000 per compound) and too time-consuming (about 3 years per compound) to be employed for testing all the thousands of suspicious chemicals. Since the resources required to conduct animal carcinogenesis bioassays are too great for routine use, John Ashby in a recent position paper to ICPEMC proposed an interesting approach, namely the use of short-term cytogenetic assays in rodents; and Ashby then discusses the various factors that may prevent or modulate the expression of latent genotoxic damage in these short-term assays (Ashby 1983).

## 4 The Nitroarenes

Recently, some very interesting data have been reported by Rosenkranz and co-workers on a new group of highly mutagenic substances, the nitroarenes, nitrated polycyclic aromatic hydrocarbons. They appear to have widespread distribution in the environment, and were first detected as trace impurities ($\sim$1 ppm) in xerographic toners (Rosenkranz et al. 1980). Subsequently their presence was demonstrated in diesel emissions and in fly ash (Rosenkranz and Mermelstein 1983). They are among the most potent microbial mutagens and some members of the group have been shown to be carcinogenic to laboratory animals. In excess of 200 nitroarenes have been detected in diesel emissions and in the ambient environment. However, only about 50 nitroarenes have been tested for mutagenicity in *Salmonella* and only 9 or 10 have been assayed in mammalian cells for gene mutations, genotoxicity or transforming activity.

A number of studies have established that nitroreduction is essential for mutagenicity, this is catalyzed by a series of newly described enzymes: nitroreductases, nitrosoreductases, and transacetylases. In mammalian cells, ring oxidation also occurs, but it differs drastically from that of the non-nitrated parent polycyclic aromatic hydrocarbon with the nitro function exercising a strong directing effect. This oxidation may be followed by nitroreduction (Rosenkranz and Mermelstein 1983).

As it is quite obvious that the synthesis as well as the determination of the mutagenic properties of the various nitroarenes that have been identified in the environment are beyond our current resources, it would seem that the most profitable approaches would be those that use available structure-activity data to predict the activity of nitroarenes that have as yet not been tested. Two such approaches are described here:

In view of the fact that nitroreduction appears to be essential for mutagenicity, it was thought that if a relationship between ease of reduction and mutagenicity existed, it could be used in a predictive manner if it was amenable to theoretical treatment. The ease of electrochemical nitroreduction can be related to the energy of the lowest unoccupied molecular orbital (LUMO), a value that can be determined theoretically. Indeed, the voltage of the half-wave potential ($E_{1/2}$) was found to be related to the mutagenic potency (Fig. 4). $E_{1/2}$ in turn is linearly related to the calculated LUMO energies of the nitroarenes and hence mutagenicity is also related to the LUMO energy (Fig. 5). Since LUMO energies can be calculated, the net result is that mutagenic potency can be predicted from theoretical consideration (Klopman et al. 1983).

The second approach involves an entirely new and unique artificial intelligence system to analyze the structural determinants of the mutagenicity of nitroarenes. The system identified two fragments whose presence is required for mutagenicity (Fig. 6, structures 1 and 2); in addition, deactivating fragments were also identified; molecules containing none of these fragments are also inactive (Fig. 6, structures 3 and 4) (Klopman and Rosenkranz 1983).

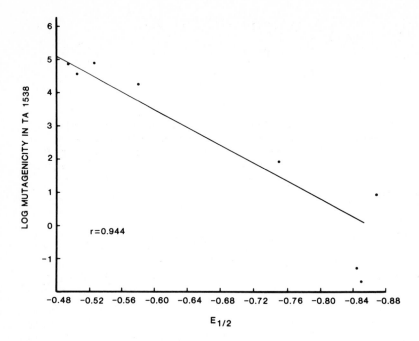

**Fig. 4.** Plot of the logarithm of the mutagenicities of nitroarenes in strain TA1538 versus the voltage of the half-wave potential ($E_{1/2}$) for the reduction of the first nitro group. (Klopman et al. 1983)

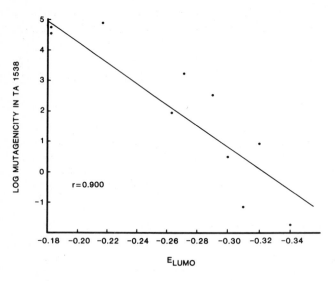

**Fig. 5.** Plot of the logarithm of the mutagenicity of nitroarenes in TA1538 against $E_{LUMO}$, the energy of the lowest unoccupied orbital. (Klopman et al. 1983)

This approach to the study of nitroarenes is not unique and restricted to that class of chemicals; it exemplifies the results that are achievable when an interdisciplinary approach is used to elucidate the basis of the biological activity of chemicals of environmental importance.

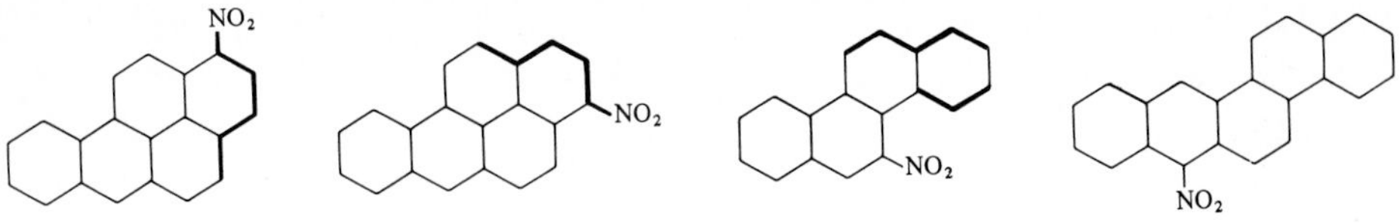

**Fig. 6.** Examples of the predictivity of CASE ("Computer Automated Structure Evaluation"). *1* (1-nitrobenzo(a)pyrene) and *2* (3-nitrobenzo(a)pyrene) are reported to be mutagenic. *Heavy lines* indicate active fragments. *3* (5-nitrochrysene) and *4* (6-nitrobenzo(a)pyrene) are reported as non-mutagenic (Rosenkranz and Mermelstein 1983). The heavy line for *3* indicates a deactivating fragment. *4* has neither active nor deactivating fragment and is, therefore, predicted to be inactive. (Rosenkranz et al. 1983)

## 5 Lesions Leading to Mutations and Chromosome Aberrations

One of the first comprehensive studies in comparative mutagenesis was carried out by my colleague Ekkehart Vogel with *Drosophila* (Vogel 1975, 1976, 1977; Vogel and Leigh 1975; Vogel and Sobels 1976). *Drosophila* offers the opportunity to score for point mutations and chromosome aberrations in progeny of the same flies. For a great variety of different mutagens, Vogel observed that chromosome aberrations, if induced at all can only be recovered at considerably higher concentrations than the ones inducing recessive lethal mutations. For alkylating agents with one alkylating group, the monofunctional ones, the molecular basis for this "Two-level Effect" has been clarified in an extensive study by Vogel and Natarajan (1979a,b). They were able to show that in *Drosophila* agents with a low s (the Swain-Scott factor), like ethylnitroso-urea (ENU), predominantly alkylating the O atoms in the DNA, mainly induce gene mutations and very little chromosome breakage effects and then only so at extremely high concentrations. Agents with a high s factor, mainly alkylating N-7 guanine, like methylmethane sulphonate (MMS), are much more effective in producing chromo-some aberrations (see Table 5). With the latter class of agents that are effective in

**Table 5.** These data, derived from Vogel and Natarajan (1979a), show that agents with a low Swain Scott factor s, predominantly alkylating oxygen, like ethylnitrosourea (ENU), produce gene mutations at high frequency, but are much less effective in producing chromosome aberrations than agents, like methylmethane sulphonate (MMS), with a high s value, mainly alkylating N7 guanine and proteins

| Agent | s factor | Alkylation pattern | | Mutations | Chromosome |
|---|---|---|---|---|---|
| | | Guanine $O_6$ vs. $N_7$ | Proteins | | Aberrations |
| ENU | 0.26 | 0.4 | Low | 30–40% | < 0.1% |
| MMS | 0.86 | 0.004 | High | Below 1% | None |
| | | | | 1–3% | Some |
| | | | | > 15% | > 15% |

Conclusion: O-alkylation is highly mutagenic, but does not lead to chromosome aberrations

inducing translocations, there is an increase in translocation frequencies with sperm storage, and high frequencies of mosaic lethals; in somatic cells, the frequencies of induced twin spots are also higher with those agents (Vogel et al. 1982). These findings are thus instructive in that they help to delineate some specific aspects in which the effects of chemical mutagens differ from those produced by ionizing radiation: that is, a relative shortage of translocations – whose frequency may increase with storage of sperm in the inseminated female – and a high frequency of mutations with mosaic expression, as was noted many years ago by Auerbach (1949, 1967).

A striking difference was also observed between the reparability of lesions induced by MMS (high s) and ENU (low s). Thus, in matings of MMS-treated males to repair-deficient mei-9 females the recessive lethal frequency was increased in comparison to those with repair-proficient females. In contrast, ENU-induced mutational lesions were unaffected by such a repair deficiency, presumably, as Vogel (1983) suggests, because they arise from direct miscoding. All these data illustrate how the specific activity of the various alkylating agents may vary as a function of their chemical reaction pattern. The implication is that in tests for chromosome aberrations in vivo (and most routine mammalian in vivo assay systems score for chromosome aberrations only), the critical dose in the target cells may not be reached, resulting in false negatives.

Van Zeeland's recent studies relating chemical dose to mutation induction in mammalian cells tend to support Vogel's findings in *Drosophila*. Van Zeeland observed that the number of $O^6$ alkylations needed to produce a point mutation is the same for agents differing in s values by a factor of 2, like ENU and EMS. The induction frequencies of chromosome aberrations and sister chromatid exchanges do not correlate well with the number of $O^6$ alkylations, indicating that different lesions must be responsible for these other genetic end points (see Table 1).

In considering the kind of genetic damage that is likely to be most deleterious in man, small multi-locus deletions would deserve high priority. This is a consequence of the fact that deletions will exhibit a certain degree of dominance and thus will become expressed in the immediate progeny of exposed parents. In an exploratory study, Shukla and Auerbach (1980) recently demonstrated that the capacity to produce deletions vastly differs between chemicals. An important question therefore is which agents can be expected to be most effective in producing deletions. If one assumes that deletions as a rule are brought about by chromosome breakage, I would expect agents with a high s factor that predominantly alkylate the N7 position in guanine and proteins to be more effective than agents with a low s, which predominantly induce point mutations; $O^6$ alkylations, however, might well be more important for the induction of tumor-initiating events (Rajewsky 1980).

The above observations are applicable only to alkylating agents. After treatment with hexamethylphosphamide (HMPA), Vogel (1983) did not observe an increase of the translocation frequency following storage of sperm. Chemical dosimetry studies revealed no binding with $O^6$ or N-7 sites of guanine, but only with unidentified products, possibly the pyrimidines. That the induction of a high frequency of deletions is not necessarily accompanied by gross chromosomal rearrangements is exemplified by the studies of Kramers et al. (1983) with the antischistosomal agent hycanthone.

## 6 Problems Involved in Quantification, Extrapolation, and Risk Assessment

In the near future, it is to be expected that testing for mutagenic activity will increase considerably. Consequently, both industry and public health authorities will be more often confronted with the problem of how to handle mutagenic compounds with extensive population exposure. For compounds that have no societal benefits and those for which nonmutagenic substitutes are available, the decision is not difficult, because in these cases exposure should, obviously, be avoided or, at least, minimized.

The question becomes a pressing one when mutagenic compounds cannot be avoided or easily missed. We are then entering into a phase of weighing risks against benefits. In fact this is more easily said than done, as may be illustrated by the case of diagnostic X-rays. Here we are dealing with the frequent application of a mutagenic agent, while the beneficial health effects are recognized by the whole medical world. Thanks to some 30 years of intensive studies with mammalian assay systems, genetic risk estimates with sufficient precision have now been developed. The estimate is about 10–40 cases per million of serious genetic disease or malformations in the direct progeny after exposure to an average dose of 20 mrad of low LET irradiation per year over a 30-year period (UNSCEAR 1977). Ironically, however, no reliable quantitative estimates are available on the number of lives saved, or the number of sick cured, by the application of appropriate radiodiagnostic procedures.

Risk assessment for chemical mutagens will pose even greater problems than has been the case with radiation. Relevant studies with mammalian assay systems may not be possible in each specific case, since the great majority of environmental chemicals may only exhibit weak mutagenic activity. For such chemicals specific locus studies will have to remain restricted to high concentrations. Whether extrapolation from data obtained at high concentrations to what can be expected at low realistic ones is permissible has to remain uncertain, since such factors as repair, activation or inactivation of enzyme systems may well lead to different yields of mutation per unit dose at different concentrations or with different modes of exposure (see Chap. 14, this Vol.).

In cases where a numerical estimate is required and specific locus or other mutation data with the mouse cannot easily be generated, "the parallelogram" approach may be used (Sobels 1977, 1980, 1982a). This approach is based on the principle that information can be obtained on genetic damage that cannot be assessed directly, for example, mutations in mouse germ cells, by comparison of endpoints that can be determined experimentally, such as mutations in somatic cells in vitro, and specific DNA adducts in the somatic mutation system and in the mouse germ cells (Fig. 7, left).

At present, Dr. van Zeeland is engaged in a collaborative effort with Dr. U. Ehling from the G.S.F. in Munich, to investigate the various parameters involved in the parallelogram. Dr. Ehling measures the induction of specific locus mutations in mice treated with EMS. Dr. van Zeeland uses mice of the same strain to measure the number of ethylations per nucleotide at the $O^6$ position of guanine. The data in Table 1 show that $O^6$ alkylation of guanine is a reliable parameter for the prediction of mutation

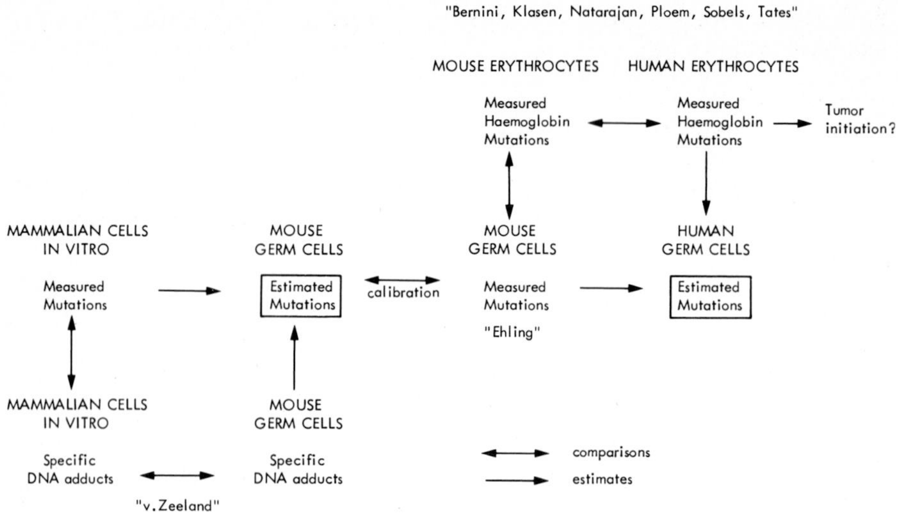

Fig. 7. The parallelogram. (After Sobels and Delehanty 1982)

induction. Van Zeeland observed that the level of $O^6$ alkylguanine in the mouse testis is in the same range as that in cultured mammalian cells. The mutation frequencies in cultured mammalian cells at these levels of $O^6$ alkylation are not outside the range of what can be expected in the mouse testis.

A future extension of the parallelogram makes use of measured somatic mutation frequencies in man and in the mouse, to estimate mutation frequencies in human germ cells (Sobels and Delehanty 1982; Fig. 7, right). These somatic mutations can now be detected in human blood samples thanks to new technical developments. The system makes use of an automatic scanning system designed by Professor J.S. Ploem and his group and of fluorescent mono-specific antibodies against hemoglobin variants developed by Professor L.F. Bernini. The Leytas optical scanning machine has a detection capacity of finding one mutated cell in ten million. By measuring comparable hemoglobin mutations in mouse erythrocytes and specific locus mutations in mouse germ cells, estimates of mutations in germ cells of exposed human populations can be obtained by appropriate comparisons. This new methodology offers important perspectives for monitoring exposed populations and for comparing mutation frequencies with chromosome aberrations in the same individuals, that is, comparative mutagenesis in man himself will now become a realistic possibility.

Viewed against the problems intrinsic to quantitative risk estimation of mutagenic chemicals, we may again pose the question, what kind of approach could be followed in the case of indispensible chemicals showing undisputed mutagenic activity. In cases of certain drugs, as for example, the antischistosomal agent, hycanthone, non-mutagenic compounds have been successfully synthesized retaining their useful therapeutic function (Bueding and Batzinger 1977). Another example of ingenious chemical engineering towards loss of mutagenic activity is presented below (Sobels 1982b).

2,4-Diaminoanisole is a meta-diamine that cannot be missed in the manufacture of hair dyes. It acts as a so-called coupling agent in the production of red and blue colors.

**Fig. 8.** Substitution of the methoxy group by a hydroxyethoxy group leads to considerable reduction of the mutagenic activity. (Sobels 1982b)

2,4 Diaminoanisole

MUTAGENIC

2–(2',4' diaminophenoxy) ethanol

NON–MUTAGNIC

2,4-Diaminoanisole has produced mutations in a variety of assay systems ranging from *Salmonella*, and yeast to *Drosophila* and mammalian cells.

In a search for non-mutagenic substitutes, one particular compound in which the methoxy group of 2,4-diaminoanisole had been replaced by a β-hydroxy-ethoxy group, showed most promising properties from the technological point of hair-dyeing (Fig. 8). This observation led to extensive testing of the new compound, 2-(2',4'-diaminophenoxy) ethanol, in a large variety of different assay systems. These tests consistently produced negative results [see Mutation Research (1982) 102:309–372]. [Recently, Venitt et al. (1983) reported weak mutagenic effectiveness with *Salmonella* TA1538 and TA98.]

An important message to be learned from a study of this kind is that fundamental research on the mechanisms of how a particular chemical structure produces mutations can be of great practical interest. As studies in this particular case clearly show, a simple substitution in part of the molecule may result in a great reduction of mutagenic effectiveness with retention of the useful function.


# 7 Mutation Epidemiology


A major handicap in evaluating the effects of chemical exposure on genetic disease is the paucity of data on genetic damage in human germ cells. ICPEMC's Committee 5 (1983) under the chairmanship of Dr. J. Miller elaborated the need for mutation epidemiology, because there is no assurance that the various test-systems are fully effective with regard to man and thus, substances cleared through all screening tests may not in fact be really harmless. The experimental methodology in mutation tests will ultimately require the confirmation or modification in the light of human experience. The magnitude of the difficulty, may be illustrated by the experience from studies of the offspring of survivors of the atomic bombs at Hiroshima and Nagasaki. Despite the 35 years of intensive work on offspring of parents who received radiation doses large enough to produce genetic effects in animal germ cells, no clear-cut significant effects could be demonstrated. During 5 meetings, Committee 5 has discussed a variety of topics relevant to epidemiology. An outline has now been made of the necessary conditions required for such studies and one can only hope that the resources can be found to make mutation epidemiology a reality.

## 8 The Need for Understanding Mutation Mechanisms

Finally, in considering the most pressing research needs now, I would reply that this question appeared a lot easier to answer some 5 to 7 years ago. The whole field of genetic toxicology has developed so rapidly and its scope touches so many different disciplines that simple generalisations are no longer possible.

At present a wide range of short-term tests are in operation, and extensive evaluation efforts, like EPA's Gene-Tox Programme are underway. These will facilitate qualitative decision-making regarding mutagenic activy of both individual compounds and complex mixtures. Quantitative evaluations to permit extrapolation from experimental systems to man and epidemiological data still leave a lot more to be desired. The important question of the extent to which mutagenic potency in short-term tests can be translated into genetic and carcinogenic risks in man obviously needs to be pursued.

In setting priorities for future research, I would attach great importance to studies directed at elucidating fundamental mechanisms of mutagenesis. Thus more precise insight should be obtained on the major classes of mutagens, to which the majority of environmental chemicals belong. With this knowledge one could then establish priorities for more detailed investigations of the most frequently occurring compounds.

Molecular dosimetry serves a central function both for sharpening the tools for the detection of mutagens and for quantitative evaluation directed at risk assessment via the paralellogram method. The relationships between chemical structure, mutagenic effectiveness, interaction products with DNA and the types of genetic changes induced are of paramount importance. Studies on metabolic and repair pathways and the recent discoveries concerning the role of oxidizing radicals, anti- and desmutagens should be mentioned in this context. More insight into the mechanisms of mutagenesis would contribute to a better understanding of the shape of the dose-response relationships. This is particularly important at environmental exposure levels where actual measurement of mutational response will be impracticable. In fact, one of the most fundamental issues in the practical assessment of genetic risk concerns the question of whether at low-dose exposures, threshold-like responses exist or whether the only correct procedure is one of linear extrapolation from high doses. Evidence for the existence of both kinds of response has been recently reviewed in an ICPEMC paper by Ehling et al. (1983). I would believe that progress in answering this question can be expected primarily from studies on repair and other cellular processes involved in the production of mutations, in addition to those on dose-effect relationships.

For geneticists the clarification of the causes underlying spontaneous mutation is an important task. It may be noted that so far induced mutations, whether by physical or chemical agents, have not yet been observed in man. All assessments of induced genetic damage are based on the notion that the spontaneous load of genetic disease and malformation would increase in a proportional manner. In various organisms ranging from bacteria to higher plants and *Drosophila*, evidence is now accumulating that transposition events by insertion of mobile DNA sequences contribute in a significant manner to the total incidence of spontaneous mutation (CSHSB 1981), and the possible involvement of these events in the initiation of carcinogenesis has been postulated by some authors. A recent study by Eeken and Sobels (1983) suggests that, as in bacteria, in *Drosophila* mutation involving removal or transposition of an insertion element is not affected by mutagenic treatments. This finding may have consequences for the evaluation of induced genetic damage on the basis of the spontaneous load of genetic disease in man. Since we are still largely ignorant about the steps involved in these mysterious transposition or jumping gene events, their further elucidation presents an interesting challenge for future research.

*Acknowledgements*. I would like to thank my friend and colleague, Professor K. Sankaranarayanan for critically reading the manuscript and Dr. A.A. van Zeeland for providing me with unpublished data of his work on chemical dosimetry. I am grateful to Professor H.S. Rosenkranz for his help with the section on nitroarenes.

# References

Aaron CS, Zeeland AA van, Mohn GR, Natarajan AT, Knaap GAC, Tates AD, Glickman BW (1980) Molecular dosimetry of the chemical mutagen ethyl methanesulfonate. Quantitative comparison of the mutation induction in *Escherichia coli*, V79 Chinese hamster cells and L5178Y mouse lymphoma cells and some cytological results in vivo and in vitro. Mutat Res 69:201–216

Adler I-D (1980) A review of the coordinated research effort on the comparison of test systems for the detection of mutagenic effects, sponsored by the E.E.C. Mutat Res 74:77–93

Ames BN (1979) Identifying environmental chemicals causing mutation and cancer. Science 204:587–593

Ashby J (to be published 1983) The unique role of rodents in the detection of possible human carcinogens and mutagens. Mutat Res 115:177–213

Auerbach C (1949) Chemical mutagenesis. Biol Rev 24:355–391

Auerbach C (1967) The chemical production of mutations. Science 158:1141–1147

Bridges BA (1979) Short-term tests and human health – the central role of DNA repair. In: Emmelot P, Kriek E (eds) Environmental carcinogenesis occurrence, risk evaluation and mechanisms. Elsevier/North-Holland Biomedical, Amsterdam, p 319

Bueding E, Batzinger RP (1977) Hycanthone and other schistosomicidal drugs: lack of obligatory association between chemotherapeutic effect and mutagenic activity. In: Hiatt HH, Watson JD, Winston JA (eds) Origins of human cancer, A. Cold Spring Harbor Laboratory, pp 445–463

Clayson DB (1980) Comparison between in vitro and in vivo tests for carcinogenicity. An overview. Mutat Res 75:205–213

Cold Spring Harbor Symposia on Quantitative Biology (1981) Movable genetic elements, parts 1 and 2, vol 45. Cold Spring Harbor Laboratory, New York, p 1025

Eeken JCJ, Sobels FH (to be published 1983) The effect of two chemical mutagens ENU and MMS on MR-mediated reversion of an insertion sequence mutation in *Drosophila melanogaster*. Mutat Res 110:279–310

Ehling UH, Averbeck D, Cerutti PA, Friedman J, Greim H, Kolbey AC, Mendelsohn ML (to be published 1983) Review of the evidence for the presence or absence of thresholds in the induction of genetic effects by genotoxic chemicals. ICPEMC Publication No. 10. Mutat Res 123:231–341

Higginson J, Muir CS (1979) Environmental carcinogenesis: misconceptions and limitations to cancer control. J Natl Cancer Inst 63:1291–1298

Hoeven JC van der, Lagerweij WJ, Bruggeman IM, Voragen FG, Koeman JH (to be published 1983) Mutagenicity of extracts of some vegetables commonly consumed in the Netherlands. J Agricultural and Food Chemistry

ICPEMC Committee 1 Final Report (1983) Screening strategy for chemicals that are potential germ-cell mutagens in mammals. Mutat Res 114:117−177

ICPEMC Committee 5 Final Report (to be published 1983) Mutation epidemiology: review and recommendation. Mutat Res 123:1−11

Klopman G, Tonucci DA, Holloway M, Rosenkranz HS (1983) Relationship between polarographic reduction potential and mutagenicity of nitroarenes (submitted)

Kramers PGN, Schalet AP, Paradi E, Huiser-Hoogteyling L (1983) High proportion of multi-locus deletions among hycanthoneinduced X-linked recessive lethals in *Drosophila melanogaster*. Mutat Res 107:187−201

Loprieno N, Adler I-D (1980) Cooperative programme of the European Economic Community on short-term assays for mutagenicity. In: Montesano R, Bartsch H, Tomatis L (eds) Molecular and cellular aspects of carcinogen screening tests. IARC Sci Publ, No. 27, Lyon, pp 323−331

McCann J, Choi E, Yamasaki E, Ames BN (1975) Detection of carcinogens in the *Salmonella*/ microsome test: assay of 300 chemicals. Proc Natl Acad Sci USA 72:5135−5139

McCann J, Ames BN (1976) Detection of carcinogens in the *Salmonella*/microsome test: assay of 300 chemicals. Discussion. Proc Natl Acad Sci USA 73:950−954

Mohn GR (1981) Bacterial systems for carcinogenicity testing. ICPEMC working paper 2/7. Mutat Res 87:191−210

Mohn GR, Zeeland AA van, Glickman BW (1982) Influence of experimental conditions and DNA repair ability on EMS-induced mutagenesis and DNA binding in *Escherichia coli* K12. Comparison with mammalian cell mutagenesis. Mutat Res 92:15−27

Rajewsky MF (1980) Specificity of DNA damage in chemical carcinogens. In: Montesano R, Bartsch H, Tomatis L (eds) Molecular and cellular aspects of carcinogen screening tests. IARC Sci Publ, No. 27, Lyon, pp 41−55

Rosenkranz HS, Haimes YY, Klopman G (1984) Prediction of environmental cancers: a strategy for the mid-1980's, health matrix, in press

Rosenkranz HS, McCoy EC, Sanders DR, Butler M, Kiriazides DK, Mermelstein R (1980) Nitropyrenes: isolation, identification, and reduction of mutagenic impurities in a carbon black and toners. Science 209:1039−1043

Rosenkranz HS, Mermelstein R (1983) Mutagenicity and genotoxicity of nitroarenes: all nitrocontaining chemicals were not created equal. Mutat Res 114:217−267

Shukla PT, Auerbach C (1980) Genetic tests for the detection of chemically induced small deletions in *Drosophila* chromosomes. Mutat Res 72:231−243

Sobels FH (1977) Some problems associated with the testing for environmental mutagens and a perspective for studies in "Comparative Mutagenesis". Mutat Res 46:245−260

Sobels FH (1980) Evaluating the mutagenic potential of chemicals: the minimal battery and extrapolation problems. Arch of Toxicol 46:21−30

Sobels FH (1982a) The parallelogram: an indirect approach for the assessment of genetic risks from chemical mutagens. In: Bora KC, Douglas GR, Nestmann ER (eds) Progress in mutation research, vol 3. Elsevier Biomedical, pp 323−329

Sobels FH (1982b) Chemical modification toward loss of mutagenic activity as an alternative to problems involved in risk assessment. Mutat Res 102:305−307

Sobels FH, Delehanty J (1982) The first five years of ICPEMC: the International Commission for Protection against Environmental Mutagens and Carcinogens. In: Sugimura T, Kondo S, Takebe H (eds) Environmental mutagens and carcinogens, Proc 3rd Intern Conference on Environmental Mutagens. Univ of Tokyo Press, Tokyo/New York, pp 81−90

Sobels FH, Kramers PGN (to be published 1983) The assessment of chemomutagenic activity as envisaged in The Netherlands, presented at GUM Conference on Legislative Aspects of Mutagenicity Testing, December 4, 1980, Munich, FRG

Sugimura T (1978) Let's be scientific about the problem of mutagens in cooked food. Mutat Res 55:149−152

Sugimura T (1982a) The Trust W Bertner Memorial Award Lecture: Tumor initiators and promotors associated with ordinary foods. In: Arnott MS, Eys J van, Wang Y-M (eds) Molecular interrelations of nutrition and cancer. Raven, New York, p 3−24

Sugimura T (1982b) A view of a cancer researcher on environmental mutagens. In: Sugimura T, Kondo S, Takebe H (eds) Environmental mutagens and carcinogens. Proc 3rd Intern Conference on Environmental Mutagens. Univ of Tokyo Press, Tokyo/Liss, New York, pp 3–21

Sugimura T (1982c) Mutagens in cooked food. In: Fleck RA, Hollaender A (eds) Genetic toxicology. An agricultural perspective, vol 21. Plenum, New York London, pp 243–271

Sugimura T, Kawachi T, Matsushima T, Nagao M, Sato S, Yahagi T (1977) A critical review of submammalian systems for mutagen detection. In: Scott D, Bridges BA, Sobels FH (eds) Progress in genetic toxicology. Elsevier/North-Holland Biomedical, Amsterdam, pp 125–140

UNSCEAR (1977) Sources and effects of ionizing radiation. United Nations Scientific Committee on the Effects of Atomic Radiation, New York

Venitt S, Crofton-Sleigh, Osborne MR (1983) The hair dye reagent 2-(2′,4′-diaminophenoxy)-ethanol (2,4-DAPE) is mutagenic to *Salmonella typhimurium*. U.K. Environmental Mutagen Society, annual meeting, Edinburgh, February 1983 (abstract in Environmental Mutagenesis 5:953)

Vogel E (1975) Mutagenic activity of cyclophosphamide, trofosfamide, and ifosfamide in *Drosophila melanogaster*. Specific induction of recessive lethals in the absence of detectable chromosome breakage. Mutat Res 33:383–396

Vogel E (1976) The relation between mutational pattern and concentration by chemical mutagens in *Drosophila*. In: Montesano R, Bartsch H, Tomatis L (eds) Screening tests in chemical carcinogenesis. IARC Sci Publ, No. 12, Lyon, pp 117–132

Vogel E (1977) Identification of carcinogens by mutagen testing in *Drosophila*: the relative reliability for the kinds of genetic damage measured. In: Hiatt HH, Watson JD, Winston JA (eds) Origins of human cancer, A. Cold Spring Harbor Laboratory, pp 1483–1498

Vogel EW (to be published 1983) Approaches to comparative mutagenesis in higher eukaryotes: significance of DNA modifications with alkylations in *Drosophila melanogaster*. In: Williams GM, Dunkel V, Ray LA (eds) Cellular systems of toxicity testing. The New York Academy of Sciences 407:200–220

Vogel E, Leigh B (1975) Concentration-effect studies with MMS, TEB, 2,4,6-triCl-PDMT, and DEN on the induction of dominant lethals, recessive lethals, chromosome loss and translocations in *Drosophila* sperm. Mutat Res 29:221–228

Vogel E, Sobels FH (1976) The function of *Drosophila* in genetic toxicology testing. In: Hollaender (ed) Chemical mutagens, vol IV. Plenum, New York London, pp 93–142

Vogel E, Natarajan AT (1979a) The relation between reaction kinetics and mutagenic action of monofunctional alkylating agents in higher eukaryotic systems. I. Recessive lethal mutations and translocations in Drosophila. Mutat Res 62:51–100

Vogel E, Natarajan AT (1979b) The relation between reaction kinetics and mutagenic action of monofunctional alkylating agents in higher eukaryotic systems. II. Total and partial sex-chromosome loss in Drosophila. Mutat Res 62:101–123

Vogel EW, Blijleven WGH, Kortselius MJH, Zijlstra JA (1982) A search for some common characteristics of the effects of chemical mutagens in *Drosophila*. Mutat Res 92:69–87

Weisburger JH (1979) On the etiology of gastro-intestinal tract cancers, with emphasis on dietary factors. In: Emmelot P, Kriek E (eds) Environmental carcinogenesis. Elsevier/North-Holland Biochemical, Amsterdam, pp 215–184

Zeeland AA van, Mohn GR, Aaron CS, Glickman BW, Brendel M, de Serres FJ, Hung CY, Brockman HE (1983) Molecular dosimetry of the chemical mutagen ethyl methanesulfonate. Quantitative comparison of the mutagenic potency in *Neurospora crassa* and *Saccharomyces cerevisiae*. Mutat Res 119:45–54

# DNA Adducts and Their Biological Relevance

G. DOERJER, M.A. BEDELL, and F. OESCH[1]

## 1 Introduction

Epidemiologic studies of human carcinogenesis have demonstrated that exposure to certain chemicals significantly increases cancer risks. Additionally, the carcinogenic effects of chemicals have been observed in many animal experiments. As a result of these studies, strong regulations have been imposed worldwide to limit exposure to chemical carcinogens, primarily during occupational, and industrial exposure as well as restricting the use of potentially hazardous pharmaceutical products and dangerous food additives. Despite the efforts to minimize human exposure to chemical carcinogens, the abundance of these carcinogens in our natural environment implies that exposure to chemical carcinogens is not only a problem of industrialized civilization but is in fact inevitably associated with normal food consumption etc. The ubiquitous distribution of cancer-causing chemicals makes imperative the study of the biochemical mechanisms of chemical carcinogenesis.

Interdisciplinary studies indicate that somatic mutation of normal cells, as one crucial step during a multistep process, is the most convincing concept to explain the development of cancer after exposure to chemical carcinogens. The transformation of normal cells into neoplastic cells can be observed in tissue culture experiments and has been utilized as a method to detect potential chemical carcinogens. The somatic mutation theory and the heritability of cancerogenic information in cancer cells suggested that changes in the genetic information in DNA were responsible for cancerogenic events (Fig. 1). The interest of many research groups was therefore focussed on interactions of chemical carcinogens with DNA. DNA adducts of a variety of chemical carcinogens have been identified and their abilities to alter the genetic information have been observed in various test systems. The validity of the somatic mutation theory and the importance of DNA adducts to the development of mutations have recently gained further support by the analysis of relevant DNA sequences before and after transformation of cells. In one of these studies, the alteration of a single base pair in a proto-oncogene, which was not able to transform other cells after transfection, resulted in the formation of an oncogene which was then able to transform cells (Reddy et al. 1982).

As indicated in Fig. 1, the formation of DNA adducts of chemical carcinogens can result from very complex systems for both activation and detoxification. The identi-

---

1  Institut für Toxikologie der Universität, Obere Zahlbacher Str. 67, 6500 Mainz, FRG

Mutations in Man, ed. by G. Obe
© Springer-Verlag Berlin Heidelberg 1984

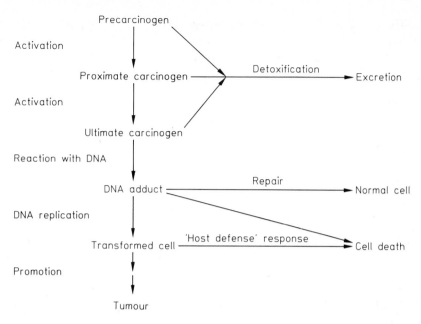

**Fig. 1.** Proposed scheme of chemical carcinogenesis

fication of metabolites which are not responsible for the formation of relevant DNA adducts is as important as identifying the ultimate carcinogenic metabolite. The miscoding potential of specific DNA adducts during DNA replication is used to estimate the biological importance of the different DNA adducts. Another important factor indicated in Fig. 1 is the repair of DNA adducts. The alteration of the genetic information by DNA adducts becomes irreversible after the reduplication of DNA. During the time between the formation of adducts and their miscoding action DNA adducts can be detoxified by repair processes. Information pertaining to mechanisms by which DNA adducts are repaired and to factors influencing the rate of repair is a substantial aspect in the evaluation of their biological relevance to chemical carcinogenesis.

## 2  Polycyclic Aromatic Hydrocarbons (PAH)

PAH are ubiquitously distributed carcinogens to which human exposure is inevitable because of their presence in our environment and to the formation of these compounds from many organic compounds by incomplete combustion processes. There is substantial evidence that PAH require metabolic activation to exert their biological effects. The main pathway of activation is by cellular monooxygenases resulting in the formation of highly reactive epoxides which are able to alkylate nucleophilic centers of molecules. Besides the detoxification of the epoxides to water-soluble conjugates, two reactions seem to limit the interaction of epoxides with DNA. First, the non-enzymatic rearrangement of epoxides to phenols and second, the enzymatic formation of dihydrodiol by epoxide hydrolase (Oesch 1979). The complex enzyme systems which are responsible for the activation and detoxification renders it difficult to compare metabolite patterns resulting after the activation of PAH in cell free systems, in tissue culture and in vivo. Because qualitative and quantitative differences in the formation of DNA adducts can be observed in various systems (Ashurst and Cohen 1982, Bigger et al. 1980), single reactive epoxides must be tested for their mutagenic and carcinogenic activity to obtain more information about the biological relevance of their respective DNA adducts.

The activation of PAH and the DNA binding of reactive metabolites between PAH of different structure show similarities related to analogous reactive centers and to structurally identical subunits of the molecule. The formal double bond in PAH with the highest electron density is the K-region (K is derived from Krebs, the German word for cancer). DNA adducts from K-region epoxides usually represent the major portion of the total amount of DNA adducts when PAH are activated by cell-free systems and are often also detectable in animal tissues. A second region has been named by its topologic characteristics, the bay-region. DNA adducts from epoxides activated in the bay-region always have prior to the ultimate activation an initial activation and detoxification step in the same ring of the molecule leading to the formation of a diol epoxide. The DNA adducts of bay-region diol epoxides represent the major part of DNA adducts formed in vivo, however, in vitro modulation of relevant enzymes also leads almost exclusively to the formation of DNA adducts resulting from bay-region diol epoxides (Oesch and Guenthner 1983). In Fig. 2 the K- and the bay-regions are indicated for the carcinogens B(a)P and 7,12-dimethyl-benz(a)anthracene. The formation of DNA adducts of benzo(a)pyrene [B(a)P] will be discussed in more detail as a typical example of PAH.

From all theoretically possible epoxides which can be formed by a single metabolic step of B(a)P, DNA adducts have only been detected from the K-region epoxide, the B(a)P 4,5-oxide. Another epoxide of the K-region, the 9-hydroxy-B(a)P 4,5-oxide is derived from the 9-phenol metabolite which can be formed by the nonenzymatic re-arrangement of B(a)P 9,10-oxide. The bay-region diol epoxide of B(a)P results after a second activation of B(a)P-7,8-diol which is formed by the hydrolysis of B(a)P 7,8-oxide mediated by the action of the microsomal epoxide hydrolase. The impor-tance of this enzyme has been shown by the alteration of the relative amounts of DNA adducts derived from the three epoxides by addition or inhibition of microsomal epoxide hydrolase. In vitro, modulation of the enzyme activity achieved the almost exclusive formation of DNA adducts derived from B(a)P-7,8-diol-9,10-oxide (BPDE) (Oesch and Guenthner 1983). Additionally, the formation of DNA adducts derived from other phenols, diols or other epoxides of B(a)P-7,8-diol can be prevented completely. For example, B(a)P-9,10-diol is a major metabolite of B(a)P in many

Fig. 2. K-regions and bay-regions of PAH

activation systems and can be activated to B(a)P-9,10-diol-7,8-oxide. In hamster embryo cells, this epoxide binds in relatively high levels to nuclear protein but not to DNA (Macleod et al. 1980). The formation and the structures of the B(a)P epoxides mentioned are shown in Fig. 3.

B(a)P 4,5-oxide, BPDE and 9-hydroxy-B(a)P (the epoxide has not yet been synthesized) have been studied extensively for their mutagenic and carcinogenic effects. Because BPDE is the most powerful carcinogenic metabolite of B(a)P, all of its stereoisomeric forms have been tested individually. From the four possible isomers of BPDE, the (+)-7β,8α-dihydroxy-9α,10α-epoxy-7,8,9,10-tetrahydro-B(a)P [(+)-anti-BPDE] is the most potent carcinogen. The other isomers of BPDE, the (−)-anti-BPDE and the both enatiomers of racemic syn-BPDE, are at most only weakly carcinogenic as are the K-region epoxide and the 9-hydroxy-B(a)P (Levin et al. 1978; Slaga et al. 1977, 1978, 1979). Because of the high activity of (+)-anti-BPDE and the preferential formation of its DNA adducts in vivo, this epoxide is considered by many investigators to be the ultimate carcinogen of B(a)P. However, a minor contribution to the carcinogenicity of B(a)P by reactive metabolites other than (+)-anti-BPDE is still possible. Even 9-hydroxy-B(a)P was found to be active as a tumor initiator in mouse skin (Slaga et al. 1978). The potential genotoxicity of all epoxides has also been demonstrated by their potent mutagenicity in *Salmonella typhimurium* strains (Malaveille et al. 1977; Wood et al. 1976). In these cells syn-BPDE shows a greater mutagenicity than anti-BPDE in contrast to the carcinogenic potential of the isomers. Better correlations are obtained in V79 cells where only (+)-anti-BPDE was mutagenic

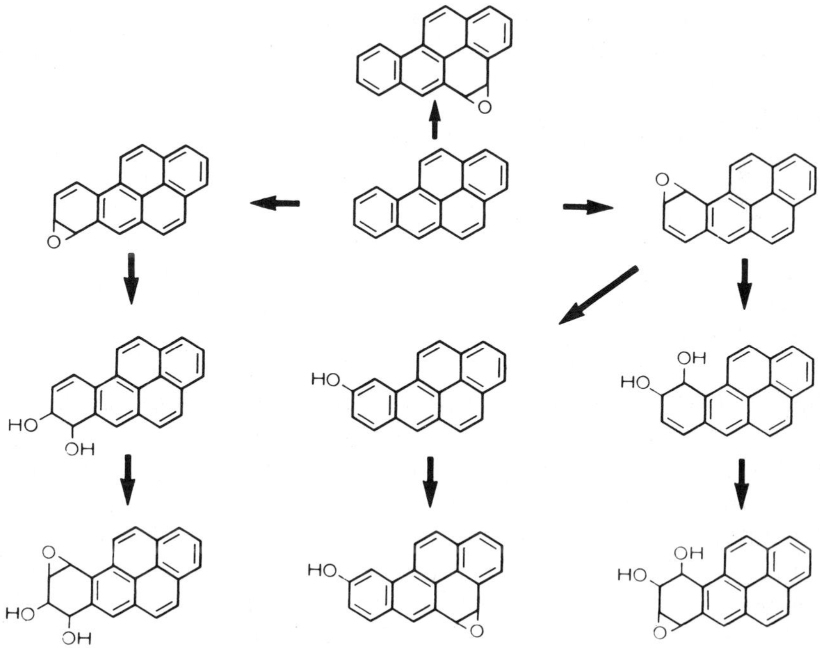

**Fig. 3.** Formation and structure of B(a)P epoxides

(Wood et al. 1977). In a more extended study with (+)-anti- and (−)-anti-BPDE, both enantiomers bound to DNA to a similar extent in V79 cells (Brookes and Osborne 1982). In this study, the total amount of DNA binding correlated only with the cytotoxicity of the epoxides. The marked contrast in their mutagenicity could not be explained by a slightly different pattern of DNA adducts or by repair processes. From these findings, it has been concluded that the biological effects of (+)-anti-BPDE must be caused by the specific spatial orientation of its DNA adducts.

(+)-anti-BPDE reacts exclusively at its 10-position with many nucleophilic centers in the DNA. Although over 90% of the alkylation occurs at the exocyclic amino group of guanine (G), other sites of alkylation include the $O^6$- and the 7-position of G and the exocyclic amino groups of adenine (A) and cytosine (C) (Straub et al. 1977; Osborne et al. 1978). In comparison, only 60% of alkylation by the less active (−)-anti-BPDE occurs at the $N^2$-positions of G whereas 20% and 15% of alkylation could be found at the $O^6$-position of G and the $N^6$-position of A (Osborne et al. 1981). Despite the possibility that minor DNA adducts can contribute to the biological effects of (+)-anti-BPDE its $N^2$-G adduct is considered the primary promutagenic DNA lesion of B(a)P.

The biochemical action of the $N^2$-G adduct of (+)-anti-BPDE is not fully understood at the present time. Stereochemical studies indicate that the pyrene moiety of the adduct is intercalated in the helix causing a distortion of the helix at this position. This substantial lesion is not limited to the complementary base but involves an extended region of the helix (Hogan et al. 1981). A genetic alteration may then originate from the introduction of a noncomplementary base somewhere in the distorted area to stabilize the DNA helix. Other alkylated bases can cause a mutation in a different manner such as the potential miscoding BPDE adduct in the $O^6$-position of G. DNA adducts from other epoxides of B(a)P should lead to another type of DNA distortion because they differ in their topology from (+)-anti-BPDE. Further investigations are required to demonstrate how these stereochemical characteristics explain the biological activities.

The biochemical mechanisms for the repair of all DNA adducts derived from B(a)P has yet to be examined. There are several possibilities which differ in the mechanism of the repair and also indicate whether or not a secondary DNA lesion can be introduced by the removal of the DNA adducts. First, an incision-excision repair, like that for the repair of pyrimidine dimers, leads to the excision of mono- or oligonucleotides and creates single strand breaks. Second, the excision of an altered base by an N-glycosylase leads to an apurinic/apyridimic site (AP) in the DNA. Third, the enzymatic hydrolysis of the adduct-base bond excludes secondary alterations of the DNA. Kinetics of the persistence of anti-BPDE DNA adducts correspond to the cytotoxic and mutagenic effects of this epoxide and suggest that the repair of this lesion is error-free in normal human cells (Yang et al. 1980). The persistence of anti-BPDE DNA adducts in tissue culture and in vivo demonstrates that the DNA adducts are repaired slowly and can persist in the DNA for a long period. There seems to be a limitation of the repair since after a fast initial repair within the first 20 h, some cultured cells still contain about 50% of the initial adduct concentration after several days with a very slowly declining rate of adduct concentration thereafter (Feldman et al. 1978; Kaneko and Cerutti 1982). After a single dose of B(a)P to rats, a relatively fast repair rate was observed during the first 8 days, whilst the adduct concentrations in all organs examined remained almost constant between 30 and 90 days (Kleihues et al. 1980). The persistent adducts and the adducts removed during the first period of repair were chemically and biochemically indistinguishable. One possible explanation for this finding is that the distorted DNA becomes stabilized by repair events which recognize only the distortion of the DNA and not the adducts. The remaining adducts then become part of the DNA and are no longer recognized by enzymes responsible for the removal of the adducts. Investigations of a possible preferential repair of certain adducts of B(a)P consistently show that all major adducts are removed at the same rate (Feldman et al. 1978; Kleihues et al. 1980). There are some reports concerning a faster removal of an A adduct (Feldman et al. 1978; Alexandrov et al. 1982). However, this adduct is difficult to quantitate because its initial concentration is close to the detection limit. With the present results, the repair of the DNA adducts derived from various B(a)P epoxides does not correlate with the respective carcinogenic potency of the epoxides. The available data therefore indicate that the specific distortion of the DNA by adducts of different epoxides is responsible for the respective carcinogenicity of B(a)P metabolites.

Similar results have been obtained with other PAH in comparison to B(a)P. The importance of the bay-region diol epoxides and the specific stereochemistry of the most reactive stereoisomer has been demonstrated for many compounds including benz(a)anthracene (Hemminki et al. 1980) and chrysene (Hodsgon et al. 1982). These studies support the concept of using B(a)P as a valuable model compound for other PAH. To evaluate the importance of the DNA adducts of PAH to human carcinogenesis, epidemiologic studies have been already initiated to quantitate the formation and persistence of the anti-BPDE guanine adduct using monoclonal antibodies or fluorometric techniques for the detection of this DNA adduct (Perera et al. 1982; Doerjer et al. 1982).

## 3 Methylating and Ethylating Carcinogens

Chemical carcinogens which alkylate DNA by the transfer of a methyl or ethyl group include widely distributed chemicals such as N-nitrosamines, nitrosoureas, hydrazines, halogenides, and reactive esters. Some carcinogens, such as N-methyl-N-nitrosourea (MNU) or methylmethanesulfonate (MMS) do not require metabolic activation and can directly react with DNA. Other alkylating carcinogens need multiple steps of enzymatic activation before a reactive alkyl group is formed. An example for this is 1,2-dimethylhydrazine. The pathways of activation of this compound are given in Fig. 4.

The pattern of alkylated sites in DNA can vary to a considerable extent depending on the electrophilicity of the ultimate carcinogen and the nucleophilicity of reactive centers in DNA (Lawley 1974). Typical amounts of alkylated bases, A, G, C, and thymine (T), resulting after in vitro reaction of DNA with four different carcinogens are given in Table 1. There are remarkable differences in the ratios of O- and N-alkylations between the four alkylating agents. For example, ratios of alkylation in the

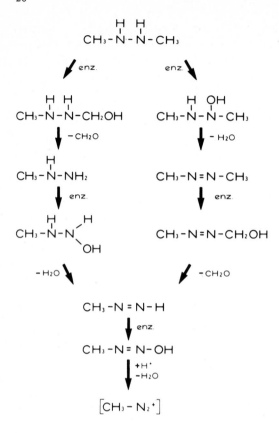

**Fig. 4.** Metabolic activation of 1,2-dimethylhydrazine

$O^6$- to $N^7$-position of G are highest for N-ethyl-N-nitrosourea and lowest for MMS (Table 1). Moreover, the ratios between bases with the same type of alkylation (O- or N-alkylation) can be also changed such as the ratios of 3-alkyl A (3-alkA) to 7-alkG which is highest for N-ethylnitrosoureas and lowest for ethylmethanesulfonate. The indicated range for the relative amounts of alkylation products is not only due to normal experimental deviations but rather caused by using DNA of different origin. In contrast to the reaction sites of PAH, alkylation of DNA by methylating or ethylating agents is usually not detectable at any exocyclic amino group. At the 1-position of G, alkylation is only observed under non-physiological conditions, e.g., at alkaline pH or in non-aqueous solutions (Singer 1972; Farmer et al. 1973).

The principal N-alkylated base, 7-methyl G (7-meG), was believed to miscode with T during DNA replication due to its ionization (Lawley and Brookes 1963). However, this could never be proved experimentally. On the contrary, it has been shown that the presence of 7-meG in DNA leads to a normal incorporation of C (Abbott and Saffhill 1979) and that 7-meG is incorporated as G into DNA (Hendler et al. 1970). Similar results have been obtained for 3-meA, another observable N-alkylation product (Abbott and Saffhill 1977). Therefore, 7-meG and 3-meA are not considered as miscoding DNA lesions. Also, if the N-alkylation involves the hydrogen bridges of normal base pairing, no significant miscoding effects have been observed during DNA replica-

**Table 1.** Distribution of DNA adducts after alkylation with four different carcinogens. (Data from Beranek et al. 1980)

| Alkylation of DNA by | | Methyl methane-sulfonate | Ethyl methane-sulfonate | N-methyl-N-nitrosourea | N-ethyl-N-nitrosourea |
|---|---|---|---|---|---|
| Site of alkylation | Method of digestion | Percent of total DNA alkylated | | Percent of total DNA alkylated | |
| Adenine | | | | | |
| 1- | B | 1.1– 2.0 | 0.2–2.0 | 0.4– 2.7 | 0.1–0.5 |
| 3- | A | 9.8–11.7 | 3.8–5.6 | 7.9–11.0 | 2.7–4.3 |
| $N^6$- | B | n.d. | n.d. | n.d. | n.d. |
| 7- | A | 0.3– 1.9 | 1.4–1.9 | 1.2– 2.6 | 0.2–0.4 |
| Cytosine | | | | | |
| $O^2$- | A | n.d. | 0.3 | n.d. | 2.8–4.0 |
| 3- | A | n.d. | 0.3–0.8 | 0.2–3.2 | 0.2–0.4 |
| Guanine | | | | | |
| 1- | C | n.d. | n.d. | n.d. | n.d. |
| $N^2$- | C | n.d. | n.d. | n.d. | n.d. |
| 3- | A | 0.6– 0.7 | 0.2– 1.9 | 0.6– 1.1 | 0.6 |
| $O^6$- | B | 0.3 | 0.2– 2.2 | 4.4– 7.7 | 7.0– 8.1 |
| | C | 0.2 | 1.7– 2.3 | 5.3 | 9.4– 9.6 |
| 7- | A, B | 78.8–84.0 | 57.0–75.0 | 60.1–78.0 | 10.6–14.0 |
| Thymidine | | | | | |
| $O^2$- | C | n.d. | n.d. | 0.1 | 6.0–8.4 |
| 3- | C | 0.1 | n.d. | n.d. | 0.8 |
| $O^4$- | C | n.d. | n.d. | 0.1–0.7 | 0.8–4.0 |
| Phosphodiesters | | | | | |
| (dTpT) | C | 0.06 | 0.9–1.0 | 0.9–1.2 | 3.8–4.3 |

Method of digestion: A = thermal hydrolysis (pH 7.0, 100 °C, 30 min)
B = acid hydrolysis (0.1 N HCl, 70 °C, 20 min)
C = enzymic hydrolysis
n.d. – not detectable
The percentage of phosphotriesters is obtained from phosphodiesters (dTpT) $\times$ 13.7 (Swenson and Lawley 1978)

tion. In an in vitro system using poly(dG–dC,3-meC) as template for DNA polymerase I, no misincorporation was detectable resulting from 3-meC (Abbott and Saffhill 1979).

The loss of 7-meG and 3-meA from DNA occurs either by spontaneous hydrolysis or by enzymatic excision of this lesion by N-glycosylases. Whereas in vitro under physiological conditions the half-lives are 150 h for 7-meA and 30 h for 3-meA, the half-lives in intact cells are about 25 h for 7-meG and 3 h for 3-meA (Margison et al. 1976). One form of 3-meA N-glycosylase is specific for this DNA lesion whereas another form also possesses some activity towards other N-alkylation products (Lindahl 1982). 7-meG N-glycosylase has been detected in human and rodent tissues as well as an N-glycosylase responsible for the removal of imidazole ring-opened derivatives which result after alkali treatment of 7-meG (Margison and Pegg 1981). These enzymes are active against ethylated bases as well. The spontaneous loss of thermally unstable adducts and the action of N-glycosylases create AP sites in DNA. These secondary DNA lesions generally do not lead to observable misincorporations, however, in SOS-induced cells, the preferential incorporation of A opposite to AP sites has been reported (Schaaper et al. 1983). AP sites labilize the remaining sugar-phosphate bond and can create single strand breaks. With a half-time of 500 h for AP sites under physiological conditions, the spontaneously formed single strand breaks are probably less important than the enzymatic cleavage by AP endonucleases (Lindahl 1982). These tertiary DNA lesions are considered to be responsible for cyto-toxic events. Single strand breaks are repaired by the action of exonucleases, which remove several nucleotides including the AP sites, followed by the resynthesis of the missing segment by DNA polymerases and connection with the DNA by a DNA ligase (Lindahl 1982). In contrast to this mechanism, the repair of AP sites by the insertion of a single base is still uncertain (Lindahl 1982).

A second type of DNA adduct represents the DNA alkylphosphotriesters which account in some cases for over 50% of total DNA alkylation (Table 1). These DNA adducts are determined by quantitating the yield of thymidine phosphodiesters fol-lowed by estimation of the total amount of phosphotriester present in the DNA (Swenson and Lawley 1978). These DNA adducts are responsible for many alkali-labile lesions, but in intact cells they do not appear to contribute to cytotoxicity and mutagenicity.

The miscoding potential of O-alkylated bases is illustrated in Fig. 5. Since the original proposal that $O^6$-meG causes C–T transitions (Loveless 1969), experiments have supported the biological relevance of this suggestion. In a system with poly-$(dG,O^6$-meG–dC) as template, quantitative analyses show that DNA polymerase I leads to the incorporation of T but not A. However, the levels of incorporated T were always less than the amount of $O^6$-meG present and varied with the concentra-tions of other nucleotides in the system. With equal amounts of all relevant nucleo-tides, three molecules of $O^6$-meG in the template led to the incorporation of only one T. Thus, in this system, $O^6$-meG must be able to pair also with C like G (Abbott

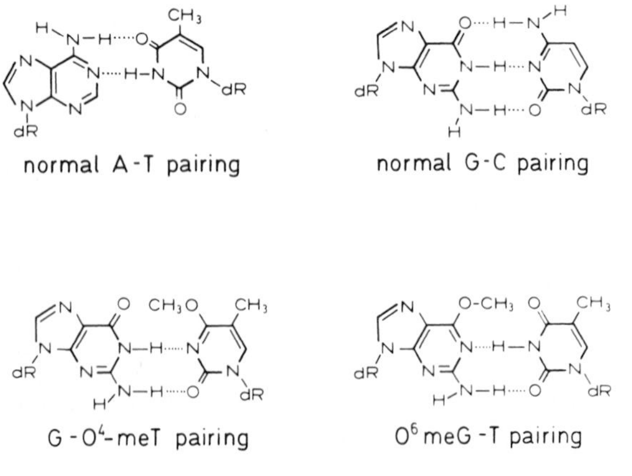

normal A-T pairing                    normal G-C pairing

G - $O^4$-meT pairing                    $O^6$meG -T pairing

Fig. 5. Miscoding potential of O-alkylated DNA bases

and Saffhill 1979). In Fig. 5, the miscoding ability is also given for $O^4$-meT. Due to low amounts after DNA methylation, the contribution of O-alkylated pyrimidines to the biological effects of methylating carcinogens is not yet clear. In V79 cells, $O^4$-meT gets incorporated as C, whereas $O^2$-meT is not incorporated (Brennard et al. 1982). The thermal instability of C alkylated at the $O^2$-position suggests that this modified DNA base is of minor importance to mutagenesis or carcinogenesis (Beranek et al. 1980).

$O^6$-alkG is repaired by transfer of the alkyl group to a cysteine residue in a repair protein. Because of lack of regeneration, the protein reacts only by an 1:1 stoichiometry with the DNA lesion (Lindahl 1982). Due to this biochemical action, the repair of $O^6$-alkG is related to initial concentrations of the adduct and to the resynthesis of the repair protein. In vivo, such a mechanism leads to a longer persistence of $O^6$-alkG in DNA when greater initial concentrations of this DNA lesion are present (Kleihues et al. 1979). In an attempt to describe the characteristics of the repair protein, it has been named $O^6$-meG-DNA methyltransferase because it is exclusively active towards DNA containing $O^6$-meG and not active for $O^6$-meG as a monomer or for other methylated bases in DNA. To indicate the stoichiometrical reaction and the activity towards $O^6$-etG the term $O^6$-alkG alkyl acceptor protein is also used.

The biological importance of unrepaired $O^6$-meG to somatic mutation has been supported by numerous investigations in cultured cells and in animal carcinogenesis models. The mutagenicity of MNU and MMS to Chinese hamster cells correlated only with the concentrations of $O^6$-meG, and not with any other detectable methylation product (Beranek et al. 1983). The initial amount and the persistence of $O^6$-meG in different organs (Kleihues et al. 1979) or specific cell populations (Bedell et al. 1982) of animals exposed to methylating agents correlated well with susceptibility to carcinogenesis. Less information is available for in vivo repair or persistence of O-alkyl pyrimidines. There are suggestions that $O^4$-etT is repaired at a much slower rate than $O^6$-etG (Scherer et al. 1980; Singer et al. 1981). Recent data indicate that the levels of $O^4$-etT, but not those of $O^6$-etG, correlate with liver cell specificity of carcinogenesis by diethylnitrosamine (Swenberg et al. 1983). Therefore, it can be suggested that the miscoding effect of $O^6$-alkG dominates for methylating agents, whereas the miscoding of $O^4$-alkT is more important for ethylating carcinogens.

Since the first report that endogeneous DNA alkylation occurs during hydrazine intoxication (Quintero-Ruiz et al. 1981) a continuous formation of methylated DNA adducts by S-adenosylmethionine can be expected (Barrows and Magee 1982). The consequences of this pathway for chemical carcinogenesis have to be further explored. However, they already suggest an exceptional importance of premutagenic methylated DNA adducts. To determine the relevance of $O^6$-alkG and its repair for human carcinogenesis, epidemiologic studies have been initiated (Waldstein et al. 1982; Doerjer et al. 1983).

# 4 Bifunctional Carcinogens

Chemical carcinogens with two reactive centers are able to react with DNA at more than one position. This class of compounds includes those with reactive functions of the same type and compounds with chemically different groups. Important examples are anticancer drugs, vinyl chloride and dihalogenides. The possibility that only one reactive site of bifunctional agents interacts with DNA and the other is detoxified leads to the formation of DNA adducts comparable to those formed by monofunctional carcinogens. A bifunctional reaction may involve reactive centers in the same DNA strand, interstrand cross-links of DNA, cross-linking of DNA with protein and anellation of a new ring to a DNA base if two positions in the same base are involved.

Due to the exponentially increasing number of possible DNA adducts from bifunctional compared to monofunctional carcinogens, only few defined adducts have been identified which demonstrate the alkylation of two DNA bases. However, there are indirect methods available to observe the presence of interstrand cross-links in DNA. Two of the most commonly applied systems are the renaturability of alkali denatured DNA (Geiduschek 1961) and alkaline elution techniques (Kohn 1979).

The formation of interstrand cross-links is inherently associated with additional monofunctional alkylation. The ratio of both types of DNA lesions depends on reaction parameters such as dose and temperature, and detoxification processes (Rutman et al. 1969; Wolpert and Ruddon 1969). In comparing the cytotoxic effects of mono- and bifunctional sulfur mustards, it has been concluded that one cross-link is equivalent in lethality to 200 DNA lesions of monofunctional alkylations (Lawley and Brookes 1967). Moreover, without repair a single cross-link in a genome seems to be a lethal DNA lesion (Verly and Brakier 1969). In addition to these DNA lesions, cross-links from DNA to proteins can contribute to the cytotoxic effects of bifunctional agents (Grunike et al. 1973).

The exact mechanism of repair of DNA cross-links is unknown, but an excision-incision mechanism with or without a prior action of a N-glycosylase is widely accepted (Wunder et al. 1981). Whereas an efficient repair has been observed in most cells, patients with the heritable, cancer-prone disease Fanconi's anemia show a high sensitivity to bifunctional agents which relates to the deficient repair of DNA cross-links in all cell types of these patients. Recent studies indicate that the repair deficiency may originate from the misfunction of a DNA topoisomerase (Wunder et al. 1981).

More information is available for the structure of DNA adducts which result after reaction of bifunctional agents with two positions in the same DNA base. These include the formation of etheno bridges from N-(2-haloethyl)-N-nitrosoureas (Gombar et al. 1980) and the formation of adducts resulting from reaction of DNA with glyoxal (Yuki et al. 1972). Major interest has recently focussed on the importance of etheno adducts to biological effects. These adducts occur after treatment of DNA with chloroacetaldehyde, a reactive metabolite of vinyl chloride. In Fig. 6, the structures are given for all etheno adducts which have been isolated at the present time (Oesch and Doerjer 1982). Recent studies show that these adducts lead to miscoding in DNA (Hall et al. 1981; Barbin et al. 1981; Kusmierek and Singer 1982). Whereas it is difficult to explain miscoding effects of etheno A and etheno C with the alteration of hydrogen bridges, the miscoding of etheno G can be formulated in analogy to $O^6$-meG (Fig. 7). The repair or persistence of etheno adducts have still to be explored. However, their miscoding potential renders them valuable models for the biological relevance of DNA adducts resulting from reaction of both groups of a bifunctional agent with the same DNA base.

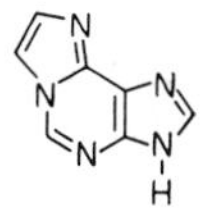

O$^6$meG-T pairing

EtoA
1,N$^6$-etheno adenine

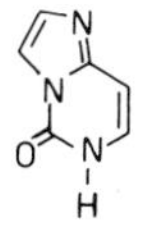

EtoC
3,N$^4$-etheno cytosine

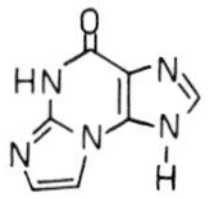

EtoG
N$^2$,3-etheno guanine

EtoG-T pairing

**Fig. 6.** Anellation products derived from reaction of chloroacetaldehyde with DNA

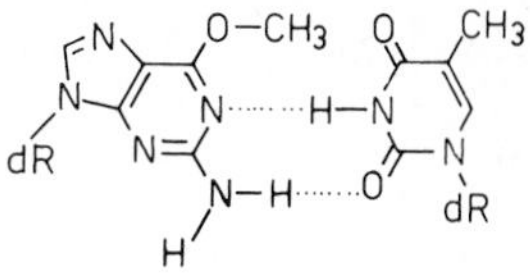

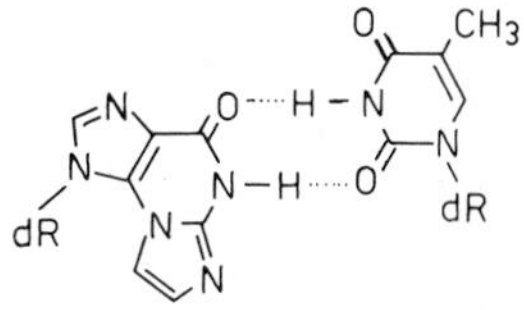

**Fig. 7.** Miscoding potential of ethenoguanine (EtoG)

## 5 Conclusions

The review of present data demonstrates that DNA adducts of chemical carcinogens exert their biological activity (1) by distortion of the DNA helix by bulky adducts, (2) by miscoding of complementary bases when an alteration of relevant hydrogen bridges is involved, and (3) by persistent DNA cross-links. Whereas all kinds of DNA cross-links seem to be responsible for the cytotoxic effects of these DNA lesions, the other types of DNA lesions show a high specificity. The marked biological differences of B(a)P diol epoxide isomers are not reflected in different rates of formation or persistence of their DNA adducts. Thus, the overall formation of these DNA adducts does not correlate with mutagenicity or carcinogenicity of B(a)P diol epoxides. Moreover, not all DNA adducts of the ultimate carcinogen (+)-anti-BPDE are responsible for its biological action. Similar results have been obtained with methylating and ethylating carcinogens. From all formed DNA adducts, only the potentially miscoding alkylations at the O$^6$-position of G and the O$^4$-position of T gave positive correlations with biological effects. Additionally, the relative importance of both promutagenic DNA lesions seems to differ between methylating and ethylating carcinogens.

The present results indicate that only distinct DNA adducts are important for biological effects, particularly for carcinogenesis. The promutagenic potential of these DNA lesions supports the theory that somatic mutation is important to chemical carcinogenesis. The examples of DNA adducts presented here also demonstrate that a broad data set is necessary to evaluate the biological relevance of specific DNA adducts.

# References

Abbott PJ, Saffhill R (1977) DNA synthesis with methylated poly(dA-dT); possible role of $O^4$-methylthymidine as a promutagenic base. Nucleic Acids Res 4:761–769

Abbott PJ, Saffhill R (1979) DNA synthesis with methylated poly(dG-dC) templates, evidence for a competitive nature to miscoding by $O^6$-methylguanine. Biochem Biophys Acta 562:51–61

Alexandrov K, Becker M, Frayssinet C, Dubowska W, Gerry R (1982) Persistence of benzo(a)pyrene diol epoxide DNA adduct in mouse skin. Cancer Lett 16:247–251

Ashurst SW, Cohen GM (1982) The formation of benzo(a)pyrene-diolepoxide adducts in vivo and in vitro. Carcinogenesis 3:267–273

Barbin A, Bartsch H, Leconte P, Radman U (1981) Studies on the miscoding properties of $1,N^6$-ethenoadenine and $3,N^4$-ethenocytosine, DNA reaction products of vinyl chloride metabolites, during in vitro DNA synthesis. Nucleic Acids Res 9:375–387

Barrows LR, Magee PN (1982) Nonenzymatic methylation of DNA by S-adenosylmethionine in vitro. Carcinogenesis 3:349–351

Bedell MA, Lewis JG, Billings KC, Swenberg JA (1982) Cell specificity in hepatocarcinogenesis: $O^6$-methylguanine preferentially accumulates in target cell during continuous exposure of rats to 1,2-dimethylhydrazine. Cancer Res 42:3079–3083

Beranek DT, Weis CC, Swenson DH (1980) A comprehensive quantitative analysis of methylated and ethylated DNA using high pressure liquid chromatography. Carcinogenesis 1:595–606

Beranek DT, Heflich RH, Kodell RL, Morris SM, Casciano DA (1983) Correlation between specific DNA-methylation products and mutation induction at the HGPRT locus in Chinese hamster ovary cells. Mutat Res 110:171–180

Bigger CAH, Tomaszewski JE, Dipple A (1980) Limitations of metabolic activation systems used with in vitro tests for carcinogens. Science 209:503–505

Brennard J, Saffhill R, Fox M (1982) The effects of methylated thymidines upon cultures of V79 cells and the mechanism of incorporation of $O^4$-methylthymidine into their DNA. Carcinogenesis 3:219–222

Brookes P, Osborne MR (1982) Mutation in mammalian cells by stereoisomers of anti-benzo(a)pyrene-diol-epoxide in relation to the extent and nature of the DNA reaction products. Carcinogenesis 3:1223–1226

Doerjer G, Platt KL, Schneckenburger A, Oesch F (1982) Quantitation of DNA adducts of benzo(a)pyrene-7,8-diol-9,10-oxide in DNA hydrolysates with fluorometric methods. Mutat Res 113:334

Doerjer G, Aulmann W, Oesch F (1983) Intra- and interindividual differences in the repair of $O^6$-methylguanine in human cell homogenates. J Cancer Res Clin Oncol 105:27

Farmer PB, Forster AB, Jarman M, Tisdale MJ (1973) The alkylation of 2'-deoxyguanosine and of thymidine with diazoalkanes. Biochem J 135:203–213

Feldman G, Remsen J, Shinohara K, Cerutti P (1978) Excisability and persistence of benzo(a)pyrene DNA adducts in epitheliod human lung cells. Nature 274:796–798

Geiduschek EP (1961) "Reversible" DNA. Proc Natl Amer Soc USA 47:950–955

Gombar CT, Tong WP, Ludlum DB (1980) Mechanism of action of nitrosoureas – IV, reactions of bis-chloroethyl nitrosourea and chloroethyl cyclohexyl nitrosourea with deoxyribonucleic acid. Biochem Pharmacol 29:2639–2643

Grunike H, Bock KW, Becher H, Gäng V, Schnierda J, Puschendorf B (1973) Effect of alkylating antitumor agents on the binding of DNA to protein. Cancer Res 33:1048–1053

Hall JA, Saffhill R, Green T, Hathaway DE (1981) The induction of errors during in vitro synthesis following chloroacetaldehyde treatment of poly(dA-dT) and poly(dG-dC) templates. Carcinogenesis 2:141–145

Hemminki K, Cooper CS, Ribeiro O, Grover PL, Sims P (1980) Reactions of "bay-regions" and non-"bay-region" diol epoxides of benz(a)anthracene with DNA: evidence that the major products are hydrocarbon-$N^2$-guanine adducts. Carcinogenesis 1:1051–1056

Hendler S, Fürer E, Srinivasan PR (1970) Synthesis and chemical properties of monomers and polymers containing 7-methylguanine and an investigation of their substrate or template properties for bacterial deoxyribonucleic acid or ribonucleic acid polymerases. Biochemistry 9:4141–4153

Hodgson RM, Pal K, Grover PL, Sims P (1982) The metabolic activation of chrysene by hamster embryo cells. Carcinogenesis 3:1051–1056

Hogan ME, Dattagupta N, Whitlock JP (1981) Carcinogen-induced alteration of DNA structure. J Biol Chem 256:1504–1513

Kaneko M, Cerutti PA (1982) Excision of benzo(a)pyrene diol epoxide I adducts from nucleosomal DNA of confluent normal human fibroblasts. Chem Biol Interact 38:261–274

Kleihues P, Doerjer G, Keefer LK, Rice JM, Roller PP, Hodgson RM (1979) Correlation of DNA methylation by methyl(acetoxymethyl)nitrosamine with organ-specific carcinogenicity in rats. Cancer Res 39:5136–5140

Kleihues P, Doerjer G, Ehret M, Guzman J (1980) Reaction of benzo(a)pyrene and 7,12-dimethylbenz(a)anthracene with DNA of various rat tissues in vivo. Arch Toxicol [Suppl] 3: 237–246

Kohn KW (1979) DNA as a target in cancer chemotherapy: measurement of macromolecular DNA damage produced in mammalian cells by anticancer agents and carcinogens. Methods Cancer Res 16:291–345

Kusmierek JT, Singer B (1982) Chloroacetaldehyde treated ribo- and deoxyribonucleotides. 2. Errors in transcription by different polymerases resulting from ethenocytosine and its hydrated intermediate. Biochemistry 21:5723–5728

Lawley PD, Brookes P (1963) Further studies on the alkylation of nucleic acids and their constituent nucleotides. Biochem J 89:127–138

Lawley PD, Brookes P (1967) Interstrand cross-linking of DNA by bifunctional alkylating agents on T7 coliphage. Biochem Biophys Acta 174:674–685

Lawley PD (1974) Some chemical aspects of dose response relationships in alkylation mutagenesis. Mutat Res 23:283–295

Levin W, Wood AW, Wisloki PG, Chang RL, Kapitulnik J, Mah MD, Yagi H, Jerina DM (1978) Mutagenicity and carcinogenicity of benzo(a)pyrene and benzo(a)pyrene metabolites. In: Gelboin HV, Ts'o POP (eds) Polycyclic hydrocarbons and cancer. Academic, New York, pp 189–202

Lindahl T (1982) DNA repair enzymes. Ann Rev Biochem 51:61–87

Loveless A (1969) Possible relevance of $O^6$-alkylation of deoxyguanosine to the mutagenicity and carcinogenicity of nitrosamines and nitrosamides. Nature 223:206–207

Macleod MC, Koostra A, Mansfield BK, Slaga TJ, Selkirk JK (1980) Specificity in interaction of benzo(a)pyrene with nuclear macromolecules: implication of derivatives of two dihydrodiols in protein binding. Proc Natl Acad Sci USA 77:6396–6400

Malaveille C, Kuroki T, Sims P, Grover PL, Bartsch H (1977) Mutagenicity of isomeric diol epoxides of benzo(a)pyrene and benz(a)anthracene in S. typhimurium TA98 and TA100 and in V79 Chinese hamster cells. Mutat Res 44:313–326

Margison GP, Margison JM, Montesano R (1976) Methylated purines in the deoxyribonucleic acid of various Syrian-golden-hamster tissues after administration of a hepatocarcinogenic dose of dimethylnitrosamine. Biochem J 157:627–634

Margison GP, Pegg AE (1981) Enzymatic release of 7-methylguanine from methylated DNA by rodent liver extracts. Proc Natl Acad Sci USA 78:861–865

Oesch F (1979) Epoxide hydratase. In: Bridges JW, Chasseaud LF (eds) Progress in drug metabolism, vol 3. Wiley, Chichester (England), pp 253–301

Oesch F, Doerjer G (1982) Detection of $N^2$,3-ethenoguanine after reaction of DNA with chloroacetaldehyde in vitro. Carcinogenesis 3:663–665

Oesch F, Guenthner TM (1983) Effects of the modulation of epoxide hydrolase activity on the binding of benzo(a)pyrene metabolites to DNA in the intact nuclei. Carcinogenesis 4:57–65

Osborne MR, Harvey RG, Brookes P (1978) The reaction of trans-7,8-dihydroxy-anti-9,10-epoxy-7,8,9,10-tetrahydrobenzo(a)pyrene with DNA involves attack at the N7-position of guanine moieties. Chem Biol Interact 20:123–130

Osborne MR, Jacobs S, Harvey RG, Brookes P (1981) Minor products from the reaction of (+) and (–) benzo(a)pyrene-anti-diol epoxide with DNA. Carcinogenesis 2:553–558

Perera FP, Poirier MC, Yuspa SH, Nakayama J, Jaretzki A, Curnen MM, Knowles DM, Weinstein IB (1982) A pilot project in molecular cancer epidemiology: determination of benzo(a)-pyrene-DNA adducts in animal and human tissues by immunoassays. Carcinogenesis 3:1405 to 1410

Quintero-Ruiz A, Paz-Neri LL, Villa-Trevino S (1981) Indirect alkylation of CBA mouse liver DNA and RNA by hydrazine *in vivo*. A possible mechanism of action as a carcinogen. J Nat Cancer Inst 67:613–618

Reddy EP, Reynolds RK, Santos E, Barbacid M (1982) A point mutation is responsible for the acquisition of transforming properties by the T24 human bladder carcinoma oncogene. Nature 300:149–152

Rutman RJ, Chun EHL, Jones J (1969) Observations on the mechanism of the alkylation reaction between nitrogen mustard and DNA. Biochem Biophys Acta 174:663–673

Schaaper RM, Kunkel TA, Loeb LA (1983) Infidelity of DNA synthesis associated with bypass of apurinic sites. Proc Natl Acad Sci USA 80:487–491

Scherer E, Timmer AP, Emmelot P (1980) Formation by diethylnitrosamine and persistence of $O^4$-methylthymidine in rat liver DNA in vivo. Cancer Lett 10:1–6

Singer B (1972) Reaction of guanosine with ethylating agents. Biochemistry 52:655–693

Singer B, Spengler S, Bodell WJ (1981) Tissue-dependent enzyme-mediated repair or removal of O-ethylpyrimidines and ethylpurines in carcinogen-treated rats. Carcinogenesis 2:1069–1073

Slaga TJ, Braken WM, Viaje A, Levin W, Yagi H, Jerina DM, Conney AH (1977) Comparison of the tumour-initiating activities of benzo(a)pyrene arene oxides and diol epoxides. Cancer Res 37:4130–4133

Slaga TJ, Braken WM, Dresner S, Levin W, Yagi H, Jerina DM, Conney AH (1978) Skin tumour-initiating activities of twelve isomeric phenols of benzo(a)pyrene. Cancer Res 38:678–651

Slaga TJ, Braken WJ, Gleason G, Levin W, Yagi H, Jerina DM, Conney AH (1979) Marked differences in the skin tumour-initiating activities of the optical enantiomers of the diastereomeric benzo(a)pyrene-7,8-diol-9,10-epoxides. Cancer Res 39:67–71

Straub KM, Meehan T, Burlingame AL, Calvin M (1977) Identification of the major adducts formed by reation of benzo(a)pyrene diol epoxide with DNA in vitro. Proc Natl Acad Sci USA 74:5285–5289

Swenberg JA, Bedell MA, Billings KC, Huh N, Kirstein U, Rajewsky MF (1983) $O^4$-ethylthymidine accumulates in DNA of hepatocytes from rats exposed continuously to diethylnitrosamine. Proc Amer Assoc Cancer Res 24:270

Swenson DH, Lawley PD (1978) Alkylation of deoxyribonucleic acid by carcinogens dimethyl sulfate, ethyl methanesulfonate, N-ethyl-N-nitrosourea, and N-methyl-N-nitrosourea. Biochem J 171:575–587

Verly WG, Brakier L (1969) The lethal action of monofunctional and bifunctional alkylating agents on T7 coliphage. Biochem Biophys Acta 174:674–685

Waldstein EA, Cao EH, Bender MA, Setlow RB (1982) Abilities of extracts of human lymphocytes to remove $O^6$-methylguanine from DNA. Mutat Res 95:405–416

Wolpert MK, Ruddon RW (1969) A study on the mechanism of resistence to nitrogen mustard (HN2) in Ehrlich ascites tumour cells: comparison of uptake of HN2-[14]C into sensitive and resistant cells. Cancer Res 29:873–879

Wood AW, Wislocki PG, Chang RL, Levin W, Lu AY, Yagi H, Hernandez O, Jerina DM, Conney AH (1976) Mutagenicity and cytotoxicity of benzo(a)pyrene benzo-ring epoxides. Cancer Res 36:3358–3366

Wood AW, Chang RL, Levin W, Yagi H, Thakker DR, Jerina DM, Conney AH (1977) Differences in the mutagenicity of the optical enantiomers of the diastereomeric benzo(a)pyrene-7,8-diol-9,10-epoxides. Biochem Biophys Res Comm 77:1389–1396

Wunder E, Burghardt U, Lang B, Hamilton L (1981) Fanconi's anemia: anomaly of enzyme passage through the nuclear membrane. Hum Genet 58:149–155

Yang LL, Maher VM, McCormick JJ (1980) Error-free excision of the cytotoxic, mutagenic N2-deoxyguanosine DNA adduct formed in human fibroblasts by (+)-anti-7,8-dihydroxy-9,10-epoxy-7,8,9,10-tetrahydrobenzo(a)pyrene. Proc Natl Acad Sci USA 77:5933–5937

Yuki H, Sempuku C, Park M, Takiura K (1972) Fluorometric determination of adenine and its derivatives by reaction of glyoxal hydrate trimer. Anal Biochem 46:123–128

# DNA Repair

A.A. VAN ZEELAND[1]

## 1 Introduction

It has been known for some time that genetic changes in living organisms are caused by changes in the cellular DNA. In order to conserve the genetic information from one generation to another, an organism has to make sure that during DNA replication as well as during cell division the genetic information, located in the DNA, is properly transferred to both daughter cells. Damage in DNA caused by chemical or physical agents can cause changes in the genetic information contained in the DNA. However, this does not necessarily mean that at each damaged site a genetic change will occur. The cell has usually several possibilities to cope with the problem of damaged DNA.

This is illustrated in Fig. 1. If for instance the DNA of a cell is damaged by a mutagen while the cell is in the G1 phase of the cell cycle, part of the damage might be removed from the DNA before the cell enters the DNA-synthetic phase of the cell cycle (S-phase). Usually fixation of a genetic alteration takes place during the S-phase. Therefore, damage removed from DNA before replication will not cause genetic changes. However this concept is not true for all types of DNA damage. For instance, chromosome aberrations such as dicentrics or reciprocal translocations induced by exposure to ionizing radiation or treatment with the chemical bleomycin can be visualized before the cells enter the S-phase (for review see Kihlman 1977). These types of genetic alterations are probably caused by double strand breaks in DNA which are misrepaired (Natarajan et al. 1980a).

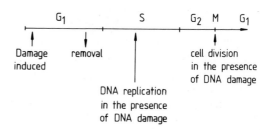

**Fig. 1.** Scheme in which the possible effects of DNA damage in different stages of the cell cycle is illustrated

1 Department of Radiation Genetics and Chemical Mutagenesis, State University of Leiden, Wassenaarseweg 72, 2333 AL Leiden, The Netherlands

Mutations in Man, ed. by G. Obe
© Springer-Verlag Berlin Heidelberg 1984

## 2 DNA Repair Mechanisms

Most cells have several possibilities to cope with damage introduced in their DNA. A list of different DNA repair pathways is given in Table 1. Most of the mechanisms listed in this table lead to the removal of certain types of damage from DNA. However, this is not true in the case of post-replication repair, which rather is a mechanism that allows the cell to replicate its DNA despite the presence of DNA damage. In the following sections of this chapter, the different repair mechanisms will be illustrated and the possible consequences for genetic effects will be discussed.

**Table 1.** DNA repair pathways

|                                                      | Type of DNA-lesions repaired                                           |
| ---------------------------------------------------- | --------------------------------------------------------------------- |
| 1. Repair of DNA breaks                              | Single strand breaks <br> Double strand breaks                        |
| 2. Photoreactivation                                 | Ultraviolet light induced pyrimidine dimers                           |
| 3. Excision repair                                   | Base damage induced by ionizing radiation <br> Pyrimidine dimers <br> Chemically altered bases <br> Cross links |
| 4. Removal of alkyl groups without incision (adaptive response) | $O^6$-methyl- and $O^6$-ethylguanine                        |
| 5. Post replication repair                           | Tolerance mechanism. No removal of damage                             |

## 3 Repair of DNA Strand Breaks

DNA strand breaks can be introduced in DNA either directly by ionizing radiation and a number of chemicals or indirectly during the process of excision repair (to be discussed in Sect. 5). The ratio of the number of single strand breaks to double strand DNA breaks varies with the type of ionizing radiation used. For instance in the case of exposure to gamma irradiation the ratio of single to double strand breaks varies between 10 and 100 (van der Schans et al. 1982). However, irradiation with neutrons produces relatively more double strand breaks than single strand breaks (Setlow and Setlow 1972). DNA breaks are relatively fast repaired in mammalian cells in culture, which is illustrated in Fig. 2. These double strand breaks have a half life of only a few minutes.

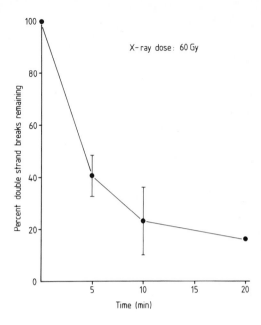

**Fig. 2.** Removal of double strand breaks in DNA from human diploid fibroblasts. Cells were prelabelled in medium containing [3]H-thymidine for 24 h, irradiated with X-rays at 0 °C and incubated in medium at 37 °C for various periods of time. Cells were then scraped off in ice cold PBS and double strand breaks in DNA were determined using the neutral elution method as described by Bradley and Kohn (1979)

# 4 Photoreactivation of Pyrimidine Dimers

A type of DNA lesion which has been investigated very extensively is the pyrimidine dimer, which can be induced by exposure of the DNA to ultraviolet irradiation. The induction of dimers is wavelength dependent. At wavelength below 300 nm cyclobutane type pyrimidine are the major type of DNA lesion. In DNA, three types of pyrimidine dimers can be formed, e.g., thymine dimers, cytosine dimers and thymine-cytosine dimers, the ratio between these being wavelength-dependent (Setlow and Setlow 1972). Once formed, pyrimidine dimers can be monomerized by absorbing another photon of UV-irradiation. However this only occurs at wavelength below 300 nm. At higher wavelength up to 600 nm pyrimidine dimers can be split by an enzymatic process called photoreactivation. The enzyme carrying out the process can only act when exposed to visible light. The wavelength dependence varies with the source of the enzyme. Photoreactivating enzyme has been found in prokaryotic and lower eukaryotic systems. In the case of mammals it has been described for marsupials (Cook and Reagan 1968) and also in human cells (Sutherland 1974). However in the case of human cells the biological consequences of the action of this enzyme are unknown. In higher eukaryotic systems the effect of removal of pyrimidine dimers on cell survival, gene mutations and the induction of chromosomal aberrations and sister chromatid exchanges have been investigated (Natarajan et al. 1980b; van Zeeland et al. 1980). This is illustrated in Fig. 3 which shows the effect of photoreversal of pyrimidine dimers on the survival and mutation induction in *Xenopus* cells exposed to UV-light. A reduction in the frequency of pyrimidine dimers is paralleled by a reduction of UV-light induced biological effects. These results indicate a direct rela-

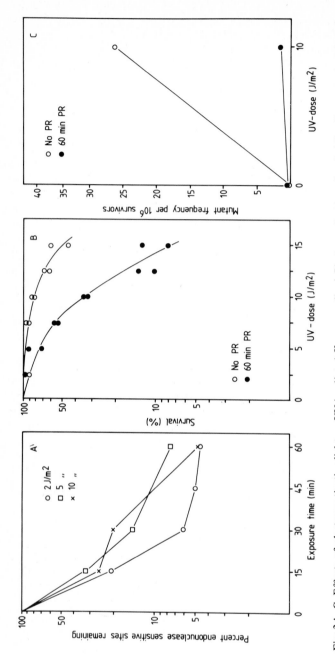

**Fig. 3A–C.** Effects of photoreactivating light on UV-irradiated *Xenopus* cells. **A** Photoreactivation of pyrimidine dimers as a function of the exposure time to photoreactivating light (predominantly 350 nm). Dimers were measured as endonuclease sensitive sites using UV-specific endonuclease V from phage T4. **B** Effect of exposure (60 min) to photoreactivating light on the survival of UV-irradiated *Xenopus* cells. **C** Effect of exposure (60 min) to photoreactivating light on mutation induction in UV-irradiated *Xenopus* cells. Mutation induction was measured as resistance to ouabain ($10^{-6}$ M) after an expression time of 3 days (van Zeeland et al. 1980)

tionship between mutation induction, cell killing and the frequency of pyrimidine dimers in this system (van Zeeland et al. 1980).

## 5 Excision Repair

In contrast with the process of photoreactivation, which is specific for a special type of DNA damage (pyrimidine dimers), the process of excision repair is a mechanism which can operate on a large series of different types of DNA damage. These types range from pyrimidine dimers to DNA damage induced by various chemical mutagens as well as base damage induced by ionizing radiation. The principle of excision repair is shown in Fig. 4. One of the early steps in the process is the incision of the DNA adjacent to a damaged site by an endonuclease. This is followed by removal of a piece of DNA containing the damaged site carried out by an exonuclease. A DNA polymerase is then filling in the DNA repair gap by a process called DNA repair replication. The polymerase is using the opposite strand as template. The remaining nick is then closed by a DNA ligase. For certain types of lesions the DNA is not incised directly, but first an apurinic or apyrimidinic site is generated by a DNA glycosylase which removes the damaged base while leaving the phosphate-deoxyribose backbone intact. Then an AP-endonuclease is carrying out the incision at the AP-site. DNA glycosylases are described which are specific for certain type of altered DNA bases, e.g., 3-alkyl-adenine, hypoxanthine, and uracil (Lindahl 1979). For two types of UV-specific endonucleases, T4 endonuclease V and *Micrococcus luteus* endonuclease, it is known that the incision at the dimer site is preceded by cleavage of the glycosylic bond of one of the pyrimidines which form the dimer. In the case of T4 endonuclease V the same protein carries out the DNA glycosylase step as well as the incision step (Seawell et al. 1980; Haseltine et al. 1980). It is not known whether mammalian UV-specific

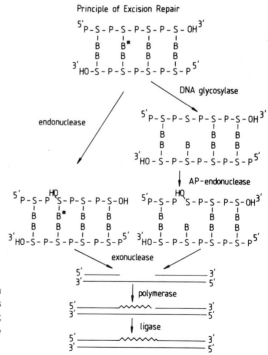

**Fig. 4.** Scheme of the DNA excision repair process. *S* deoxyribose moieties in DNA; *B* the different DNA-bases; *B\** an altered DNA-base which has to be excised

endonucleases also contain a DNA glycosylase activity. It has been recently shown by Rupp et al. (1982) that in *E. coli* the incision of DNA containing pyrimidine dimers does not occur at the site of the dimer, but 7 bases away from the dimer at the 5' site. A second incision is made at a distance of 3 bases at the 3' site of the dimer. A complex of the uvrA, B, and C gene products is necessary for these incisions. There are some indications that the incision process is not the first step in excision repair but that relaxation of the DNA has to occur first, carried out by DNA topoisomerases. Evidence for this comes from the observation that the incision process is inhibited in the presence of inhibitors of DNA topoisomerases such as novobiocin and nalidixic acid (Mattern et al. 1982).

There are generally two types of approaches to monitor excision repair. (i) One can monitor the disappearance of a certain type of DNA-adduct from the DNA. In that case for each type of DNA-adduct a detection method has to be set up. (ii) One can measure one of the steps in the excision repair process, e.g. DNA repair replication. In this approach the methodology is independent of the type of DNA damage induced.

### 5.1 Measurement of DNA Damage and its Removal

UV-induced pyrimidine dimers are the best-known type of DNA damage. The classical technique to detect them is a chromatographical method in which radioactively labelled DNA is hydrolyzed in formic acid and the thymine containing dimers are separated from thymine with two dimensional paper chromatography (Carrier and Setlow 1971) or with a one dimensional chromatographic method using thin layer plates (Cook and Friedberg 1976). The draw back of these methods is that rather high doses of UV-light has to be used in order to get detectable amounts of dimers. The biologically significant dose range for mammalian cells and repair deficient bacterial mutants lies below $10 \, J/m^2$ (254 nm). At this dose only 0.05% of the thymine is present in the dimer region on the chromatogram. A more sensitive method has been developed in which UV-specific endonucleases are used to introduce single-strand scissions in DNA at the sites of pyrimidine dimers. The molecular weight reduction of such DNA, usually determined by velocity sedimentation in alkaline sucrose gradients, may be used to determine dimer frequencies (Paterson et al. 1973; van Zeeland et al. 1981; Ganesan et al. 1981). The sensitivity of the assay is limited by the difficulty of isolating DNA of high molecular weight. However, this difficulty has been circumvented by introducing the endonuclease directly into cells permeabilized by freezing and thawing or by treatment with a detergent and then lysing the cells directly atop the alkaline sucrose gradients (van Zeeland et al. 1981; Ganesan et al. 1981; Ganesan 1973). The data in Fig. 5 show that using T4 endonuclease V, dimer frequencies induced by UV-doses as low as $0.5 \, J/m^2$ are easily measurable. In these experiments the permeabilized cells have been treated with 2 M NaCl which will disrupt the chromatin structure. Under these conditions 90–100% of the pyrimidine dimers are accessible for the endonuclease (van Zeeland et al. 1981). Using this method it can be shown that the removal of pyrimidine dimers from DNA in human fibroblasts is relatively slow compared to the repair of single or double strand breaks

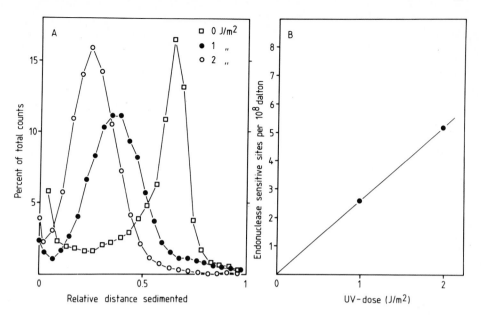

**Fig. 5 A,B.** Measurement of the frequency of pyrimidine dimers in DNA from V79 Chinese hamster cells using T4 endonuclease V as probe. **A** Radioactivity profiles of alkaline sucrose gradients of DNA from V79 cells, prelabelled with $^3$H-thymidine, UV-irradiated as indicated, permeabilized by two cycles of freezing and thawing, exposed for 15 min to 2 M NaCl, incubated for 15 min with T4 endonuclease V and then lysed atop 5–20% alkaline sucrose gradients. **B** The frequency of T4 endonuclease V sensitive sites as a function of the UV-dose. The frequency of endonuclease sensitive sites were obtained from the number average molecular weight calculated from the profiles shown in **A**

(Fig. 6). Twenty four hours after exposure of human cells to 5 J/m$^2$ UV light still 30% of the dimers are left in the DNA.

There are several possibilities to detect lesions in DNA induced by chemical mutagens. In those cases where the lesions introduced in DNA are alkali-labile they can be quantified using alkaline sucrose centrifugation or alkaline elution techniques. However, there is usually more than one type of lesion introduced, which means that alkaline-labile sites might represent a mixture of several types of lesions (Abbondandolo et al. 1982). In addition, apurinic or apyrimidinic sites introduced as a result of excision repair are also alkali-labile. A direct measurement of DNA-adducts introduced by chemical mutagens can be done by using a compound which has been radioactively labelled in the part of the molecule which is transferred to the DNA. The labelled DNA-adducts can then be quantified by hydrolysis of the DNA, chromatographic separation of the different DNA-adducts and quantification by liquid scintillation counting. An example of this method is given in Fig. 7, which shows the frequency of DNA-adducts introduced in DNA in V79 Chinese hamster cells treated with ethylmethanesulphonate, $^3$H-labelled in the ethyl group. Alternatives for this method is the use of immunological techniques using antibodies specific for a certain type of DNA-adduct (Müller and Rajewsky 1980).

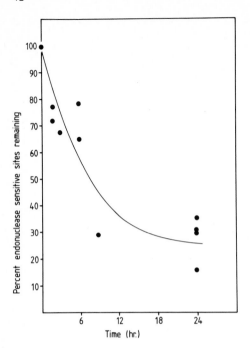

**Fig. 6.** Removal of pyrimidine dimers from DNA in human diploid fibroblasts irradiated with ultraviolet irradiation (5 J/m$^2$). Pyrimidine dimers were measured as endonuclease sensitive sites as described in the legend to Fig. 5

## 5.2 Detection of Specific Steps in Excision Repair

One of the steps most frequently analyzed in excision repair is the process in which the single strand gap is filled up by a polymerase. This process can be followed by monitoring the incorporation of a radioactively labelled precursor for DNA synthesis ($^3$H-thymidine) into DNA. However, this is only possible if one is able to discriminate between the incorporation due to repair replication and incorporation caused by semi-conservative DNA replication carried out by cells in S-phase. A widely used semi-quantitative technique is the autoradiographic method for the detection of excision repair called unscheduled DNA synthesis (UDS). Individual cells are microscopically analyzed for the incorporation of $^3$H-thymidine by counting silver grains present in the photographic emulsion above the cells to be analyzed. The conditions are chosen such that cells in S-phase do have high numbers of grains above their nuclei whereas cells outside the DNA synthetic phase show up with lightly labelled nuclei, caused by DNA repair replication (Cleaver 1981). The classical technique for the quantitative detection of DNA repair replication makes use of the fact that DNA in which thymine is replaced by its analogue bromouracil does have a higher density and can be separated from nonsubstituted DNA using CsCl density gradients (Pettijohn and Hanawalt 1964). Following UV-irradiation cells are grown in medium containing $^3$H-bromodeoxyuridine ($^3$H-BrdUrd) or unlabelled BrdUrd plus a trace of $^3$H-thymidine. Semi-conservative DNA replication is generating hybrid DNA which contains one DNA strand in which thymidine is substituted by BrdUrd. This DNA has a higher density than unsubstituted DNA. However, BrdUrd incorporated into short repair patches by DNA repair replication is not able to increase significantly the density of DNA

**Fig. 7.** The frequency of different ethylation products, expressed as ethylations per nucleotide, in DNA from V79 Chinese hamster cells exposed to $^3$H-labelled ethylmethanesulphonate. Cells were treated in suspension in PBS at 37 °C for 2 h. DNA was purified by phenol extraction and CsCl density centrifugation. Total alkylation level was measured by precipitating with trichloroacetic acid a known amount (±25 μg) of DNA on cellulose nitrate filters and determining the amount of radioactivity after solubilizing the filters with $H_2O_2$ (35%) and perchloric acid (70%). $O^6$-ethylguanine and 7-ethylguanine were determined in acid hydrolysates (0.1 M HCl), analyzed by sephadex G10 chromatography

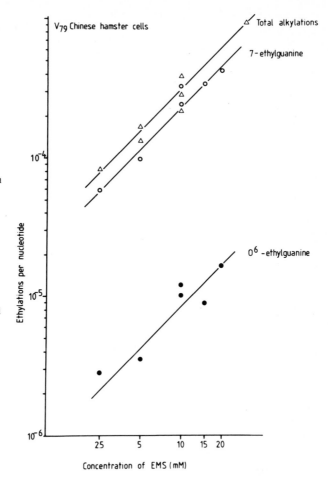

and can therefore be separated from replicating DNA using CsCl density gradients (Fig. 8). The parental DNA can be collected from neutral CsCl gradients and further purified on an alkaline CsCl gradient (Smith et al. 1981).

Figure 9 shows the results of an experiment in which human fibroblasts have been exposed to increasing doses of ultraviolet light and DNA repair replication was measured during the first 4 h following irradiation. The amount of DNA repair replication increases with the UV dose until a saturation occurs at doses between 20 and 30 $J/m^2$. UV-induced repair replication slows down with time after irradiation but is detectable until 24–48 h (Smith 1978). Most of the repair replication observed following exposure to UV-light is due to excision of pyrimidine dimers. This can be visualized in an experiment with *Xenopus* cells where UV-irradiated cells were exposed for 1 h to photoreactivating light or kept for 1 h in the dark followed by the determination of the amount of DNA repair replication. The data (Fig. 10) show that removal of pyrimidine dimers by photoreactivation causes a reduction in the amount of DNA repair replication.

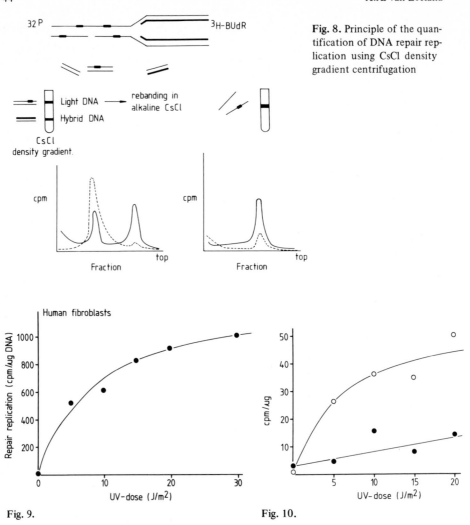

**Fig. 8.** Principle of the quantification of DNA repair replication using CsCl density gradient centrifugation

**Fig. 9.**                                                              **Fig. 10.**

**Fig. 9.** DNA repair replication in human fibroblasts following exposure to different doses of ultraviolet light. DNA repair was measured during the first 4 h after UV-exposure

**Fig. 10.** Repair replication in *Xenopus* laevis cells as a function of the UV-dose, with and without 1 h exposure to photoreactivating light. *Xenopus* cells were prelabelled with $^{32}$P, irradiated with various doses of UV-light and exposed for 1 h to photoreactivating light (●) or kept in the dark (○). During this 1 h period the cells were in MEM supplemented with 15% foetal calf serum buffered with 20 mM *Hepes*. Exposure to PR-light was carried out at room temperature by illuminating the cells through the bottom of the plastic Petri dishes. Following the photoreactivation period the cells were grown for 1 h in medium containing BrdUrd ($10^{-5}$ M) and FdUrd ($10^{-6}$ M). Then 5 μCi/ml $^3$H-thymidine (70 Ci/mmol) was added and cells were lysed 5 h later. Repair replication was quantified using CsCl density gradients and expressed as cpm/μg DNA

One of the questions which has not been answered precisely yet is what is the mechanism of the replication step in excision repair and which DNA polymerase is involved in this process. It is known that polymerase alpha is the enzyme carrying out semiconservative DNA replication and that polymerase gamma replicates mito-chondrial DNA. It was therefore often hypothesized that polymerase $\beta$ might be responsible for the polymerasation step in excision repair. This question has been investigated by using specific inhibitors of one of the DNA polymerases. For instance compounds such as aphidicolin (APC) or cytosine arabinoside are inhibitors of poly-merase $\alpha$ and do not inhibit polymerase $\beta$ or $\gamma$ (Hanoaka et al. 1979; Wist 1979). Initially conflicting results have been reported concerning the effects of APC on DNA repair replication. Ciarrocchi et al. (1979) reported the inhibition by APC of DNA repair replication in human diploid fibroblasts whereas Pedrali-Noy and Spadari (1980) did not find such an inhibiting effect using HeLa cells. We have investigated the effects of APC on UV-induced DNA repair replication in human diploid fibro-blasts as well as in HeLa cells (van Zeeland et al. 1982). The results show that the inhibitory action of APC on DNA repair replication depends on the growth stage of the cells being investigated. In growing human fibroblasts as well as in HeLa cells there was no inhibitory effect of APC on UV-induced repair replication (Table 2). In HeLa cells the amount of repair was even enhanced. However in stationary human fibroblasts APC caused an inhibition of about 85%. In HeLa cells, APC could only inhibit repair replication if also hydroxyurea was present (Hanaoka et al. 1979). Under the conditions where no inhibition of repair replication was found semiconservative DNA replication was blocked completely as shown in Fig. 11. In vitro the effect of APC can be reversed by deoxycytidine triphosphate. We therefore hypothesize that the inhibitory action of APC depends on the size of the deoxy nucleotide triphosphate pools in the cells. It is known that the level of dNTP's in resting cells is relatively low compared with growing cells. Also, HU causes a reduction in pool sizes which might

**Table 2.** Effect of APC on UV-induced repair replication in human diploid fibroblasts (VH-12) and HeLa cells

| Cells | Conditions | APC (5 $\mu$g/ml) | UV-dose (J/m$^2$) | Repair replication ($^3$H/$^{32}$P ratio) |
|-------|-----------|------|---------|---------------------|
| VH-12 | Growing | − | 0 | 0.096 |
| VH-12 | Growing | − | 30 | 2.50 |
| VH-12 | Growing | + | 30 | 2.60 |
| VH-12 | Resting | − | 0 | 0.011 |
| VH-12 | Resting | − | 30 | 4.31 |
| VH-12 | Resting | + | 30 | 0.66 |
| HeLa | Growing | − | 0 | 0.065 |
| HeLa | Growing | − | 30 | 1.40 |
| HeLa | Growing | + | 30 | 1.90 |
| HeLa | Resting | − | 0 | 0.047 |
| HeLa | Resting | − | 30 | 1.85 |
| HeLa | Resting | + | 30 | 0.62 |

Cells were prelabelled by growing them in medium containing 0.3 $\mu$Ci/ml $^{32}$P.
Repair replication was measured over a period of 3 h.
BrdUrd and FUrd were added 1 h before irradiation.
Resting HeLa cells were obtained by inhibiting the growth with 10 mM hydroxyurea, added 30 min before irradiation and present until cell lysis. (van Zeeland et al. 1982)

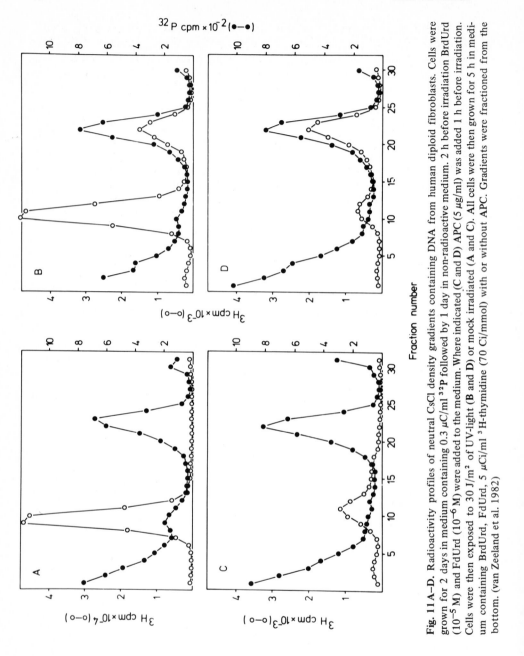

**Fig. 11A–D.** Radioactivity profiles of neutral CsCl density gradients containing DNA from human diploid fibroblasts. Cells were grown for 2 days in medium containing 0.3 μC/ml $^{32}$P followed by 1 day in non-radioactive medium. 2 h before irradiation BrdUrd ($10^{-5}$ M) and FdUrd ($10^{-6}$ M) were added to the medium. Where indicated (**C** and **D**) APC (5 μg/ml) was added 1 h before irradiation. Cells were then exposed to 30 J/m$^2$ of UV-light (**B** and **D**) or mock irradiated (**A** and **C**). All cells were then grown for 5 h in medium containing BrdUrd, FdUrd, 5 μCi/ml $^3$H-thymidine (70 Ci/mmol) with or without APC. Gradients were fractioned from the bottom. (van Zeeland et al. 1982)

explain the effects observed in HeLa cells. In addition, many animal cell mutants resistant to APC have larger dNTP pools (Ayusawa et al. 1979; Sugino and Nakayama 1980; Sabourin et al. 1981). Recently we have also investigated the effect of APC on UV-induced repair replication in human lymphocytes. The experiments were carried out with unstimulated lymphocytes and with lymphocytes stimulated with phytohaemagglutinine. APC strongly inhibited repair replication in unstimulated lymphocytes, whereas the inhibitory effect in stimulated lymphocytes was much smaller (Table 3). These results strengthen our hypothesis concerning the importance of the dNTP pool size for the inhibitory effects of APC. All these results together suggest that polymerase $a$ is involved in DNA repair replication. However, the observation that APC never completely inhibits repair replication, suggests that the $a$-polymerase is not the only enzyme carrying out the polymerisation step and thus polymerase $\beta$ might be involved as well (Smith 1983). In addition it has been shown that even under conditions where no inhibition of repair replication occured the ligation step was not carried out. This was visualized in experiments which showed that the repair label had an enhanced sensitivity to $S_1$-nuclease compared to the bulk DNA (Smith 1983). The involvement of pol $\beta$ in DNA repair has been shown also in experiments in which dideoxythymidine, an inhibitor of this polymerase, has been used. In the presence of dideoxythymidine UV-induced repair breaks stay open for a longer period of time (Mattern et al. 1982).

It has been shown that in the mammalian nucleus, cellular DNA is organized in loops which are supercoiled. These loops are attached to a structure called nuclear matrix or cage. It has been shown that semiconservative DNA replication takes place in close association with the nuclear matrix (Dijkwel 1979). Evidence for this comes from experiments which showed that in isolated nuclei, DNA pulse labelled with $^3$H-thymidine is more resistant to the DNA degrading activity of DNase I than $^{14}$C-labelled parental DNA. Recently it has been shown by Mullenders et al. (1983) that $^3$H-thymidine incorporated into repair patches during excision repair of UV-damage is equally sensitive to DNase I as $^{14}$C-labelled parental DNA. This suggests that excision repair does not take place in close association with the nuclear matrix.

**Table 3.** Effect of APC on UV-induced repair replication in human lymphocytes

| Cells | APC (5 $\mu$g/ml) | UV-dose (J/m$^2$) | Repair replication (cpm/$\mu$g DNA) |
|---|---|---|---|
| Unstimulated | – | 0 | 2.7 |
| Unstimulated | + | 0 | 2.1 |
| Unstimulated | – | 30 | 115.4 |
| Unstimulated | + | 30 | 7.7 |
| Stimulated | – | 0 | 7.5 |
| Stimulated | + | 0 | 11.0 |
| Stimulated | – | 30 | 173.6 |
| Stimulated | + | 30 | 105.2 |

Human lymphocytes were purified using a Ficoll-Histopaque gradient. Unstimulated lymphocytes were incubated in F10 medium in the presence of $10^{-5}$ M BrdUrd and $10^{-6}$ M FdUrd for 1 h, then resuspended in PBS and exposed to UV-light. Repair was measured over a period of 3 h in F10 medium containing BrdUrd, FdUrd, 5 $\mu$C/ml $^3$H-thymidine and APC where indicated. Stimulated lymphocytes were obtained by incubating whole blood in medium containing phytohaemagglutinin for 40 h. The cells were then isolated and treated as described for unstimulated lymphocytes

## 5.3 Effects of Inhibitors of Poly(ADP-Ribose) Synthetase on Excision Repair

Poly(ADP-ribose) synthetase is a nuclear enzyme which couples ADP-ribose units to nuclear proteins (Hayashi and Ueda 1977). The substrate for this enzyme is $NAD^+$. The enzyme has an absolute requirement for DNA containing single strand breaks. The $NAD^+$ level in cells exposed to ionizing radiation drops considerably in the first 2 min following irradiation and recovers then slowly (Halldorsson et al. 1978). Also UV-light causes a drop in $NAD^+$ level in human cells, however this effect is absent in cells from xeroderma pigmentosum patients (Berger et al. 1980). Poly(ADP-ribose) synthetase can be inhibited by several compounds such as 3-aminobenzamide, benz-amide, nicotinamide, and caffeine (Purnell and Whish 1980). These inhibitors prevent the drop of the cellular $NAD^+$ concentration after introduction of DNA-breaks. Also the effect of 3-AB on excision repair has been investigated. There are several reports which show that inhibition of poly(ADP-ribose) synthetase by 3-AB also causes a slow down of the ligation step in excision repair of alkylation damage (James and Lehmann 1982). This is also illustrated in Fig. 12 which shows radioactivity profiles from alkaline sucrose gradients on which cells were lysed which were treated with dimethylsulphate and then allowed to carry out DNA repair in the absence and presence of 3-aminobenzamide. Figure 12 shows that the DNA from cells which have been allowed to carry out repair in the absence of 3-AB has a larger molecular weight than in the presence of 3-AB. We have investigated whether 3-AB has an effect on the polymerization step of excision repair in human fibroblasts and in human lymphocytes (Natarajan et al. 1983). In the case of human lymphocytes irradiated with UV-light,

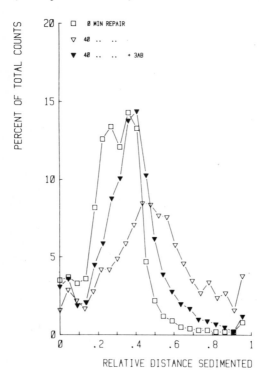

Fig. 12. Radioactivity profiles of alkaline sucrose gradients containing DNA from human diploid fibroblasts. Prelabelled cells were treated with 100 $\mu$M dimethyl-sulphate for 20 min and either lysed directly atop the gradients ($\square$) or incubated in medium for 40 min in the absence ($\triangledown$) or presence ($\blacktriangledown$) of 3 mM 3-aminobenz-amide before lysis on the gradients. Lysis condictions and centrifugation were exactly as described by Durkacz et al. (1980)

3-AB caused an enhancement of unscheduled DNA replication (UDS), measured as $^3$H-TdR incorporation in the presence of hydroxyurea (Fig. 13). However, when repair replication was determined using the CsCl density labelling technique no effect of 3-AB could be visualized (Fig. 14). The discrepancy between the UDS-measurements and the measurements of repair replication can possibly be explained by assuming that 3-AB can block the endogenous synthesis of dTTP in the cell. Small changes in dTTP concentration will have an effect on UDS-measurements because usually very low concentrations of $^3$H-TdR are used. For instance 2 $\mu$C/ml $^3$H-TdR with a specific activity of 80 Ci/mmol results in a $^3$H-TdR concentration of $2.5 \times 10^{-8}$ M. This means that the $^3$H-dTTP concentration is probably even lower. Small changes in the amount of unlabelled dTTP synthesized by the cell will effect the specific activity of the $^3$H-dTTP and therefore the level of UDS observed. In the case of the CsCl density gradient technique, the concentration of the repair label is $10^{-5}$ M, which is considerably higher. Moreover in that case fluorodeoxyuridine was included in the medium which blocks the endogenous synthesis of dTTP.

The size of repair patches synthesized by the excision repair process can also be analyzed on CsCl density gradients (Smith et al. 1981). When DNA containing repair patches labelled with BrdUrd is sheared by sonification, the BrdUrd containing repair

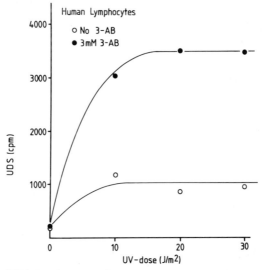

Fig. 13. Effect of 3-aminobenzamide on UV-induced unscheduled DNA synthesis in human lymphocytes. Lymphocytes were isolated from 15 ml human blood using a Ficoll-Histopaque gradient. The cells were resuspended in $F_{10}$ medium containing 20% fetal calf serum and 10 mM hydroxyurea. The cell suspension was subsequently split in 8 equal parts and incubated at 37 °C for 30 min. 3-aminobenzamide (3mM) was added where indicated and incubation was continued for another 30 min. The cells were centrifuged, resuspended in PBS, exposed to various doses of UV-light and resuspended in medium containing 10 mM hydroxyurea and 3 mM 3-aminobenzamide where indicated. The cells were incubated for 30 min at 37 °C after which 5 $\mu$C/ml $^3$H-thymidine (80 Ci/mmol) was added and incubation was continued for 3 h. Cells were centrifuged and washed 3 times with PBS. The amount of radioactivity incorporated was measured as described by Miwa et al. (1981)

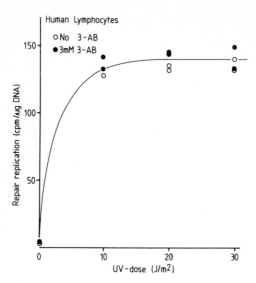

**Fig. 14.** The influence of 3-aminobenz-amide (3 mM) on UV-induced DNA repair replication in human lymphocytes. Repair replication was measured as described for unstimulated lymphocytes in the legend of Table 3. Where indicated 3-aminobenzamide was included in the growth medium 30 min before irradiation and was present during the repair period

patch becomes a significant part of the total piece of DNA. When analyzed on alkaline gradients this DNA will shift towards a higher density compared with unsubstituted DNA. This is illustrated in Fig. 15, which shows the results from an experiment in which human fibroblasts were treated with dimethylsulphate and repair replication was measured in the absence and presence of 3-AB. Sonification of the DNA caused a shift of the $^3$H-profile in the gradient compared with the $^{32}$P-profile. The shift was the same in the absence and presence of 3-AB, indicating that 3-AB did not effect the size of the repair patch induced by dimethylsulphate.

## 6  Removal of Alkyl Groups from the $O^6$-Position of Guanine

Although most chemically induced repairable DNA-adducts are removed by excision repair this is not the case for small alkyl groups at the $O^6$-position of guanine. It has been found in *E. coli* as well as in mammalian systems that methyl and ethyl groups at the $O^6$-position of guanine can be removed by a protein (alkyltransferase) leaving the guanine from which the alkyl group is removed in place. This is a very fast process in which each protein molecules is able to remove only one alkyl group by binding it to a thiol group. In *E. coli* this alkyltransferase can be induced by growing the cells in low concentrations of N-methyl-N'-nitro-N-nitrosoguanidine (Karran et al. 1979). This process is called "adaptive response". There are indications that also in mammalian systems $O^6$-alkyltransferases occur (Yarosh et al. 1983). Also stationary *E. coli* cells do contain some of this alkyltransferase. This can be visualized by treating these cells with increasing concentrations of ethylmethanesulphonate or ethylnitrosourea labelled with $^3$H in the ethyl group. It appears that the ratio of $O^6$-ethylguanine and 7-ethylguanine at low exposure concentrations is much lower than at high exposure concentrations (Fig. 16). At low exposure concentrations a significant part of the

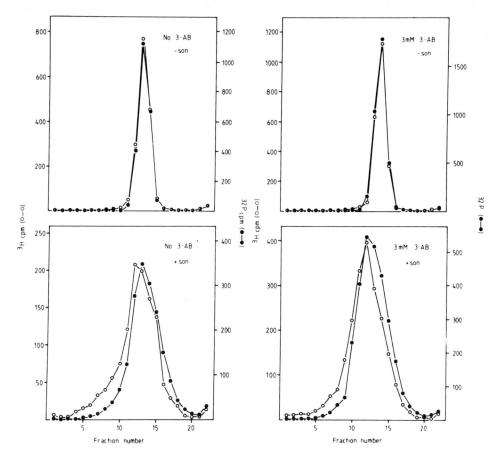

**Fig. 15.** Alkaline CsCl gradients containing DNA from diploid human fibroblasts. $^{32}$P-labelled cells were treated with 400 $\mu$M dimethylsulphate for 20 min in serum-free medium and incubated for 4 h in medium containing $^3$H-BrdUrd (5 $\mu$C/ml, $10^{-5}$ M) and FdUrd ($10^{-6}$ M) with or without 3-aminobenzamide (3 mM). Cells were lysed and unreplicated DNA was purified on two neutral CsCl gradients. Part of the DNA was extensively sonified and subsequently analyzed on alkaline CsCl density gradients as described by Smith et al. (1981)

ethyl groups at the $O^6$-position of guanine is removed. At higher exposure concentrations this process is saturated.

## 7 Post Replication Repair

The term post replication repair is maybe somewhat misleading, because it is not a repair process in the sense that a certain type of DNA damage is removed from the DNA. It is more a recovery process which allows the cell to replicate its DNA despite the fact that this DNA contains several DNA lesions. It has been shown that DNA synthesized in mammalian cells at early times after UV-irradiation is relatively small

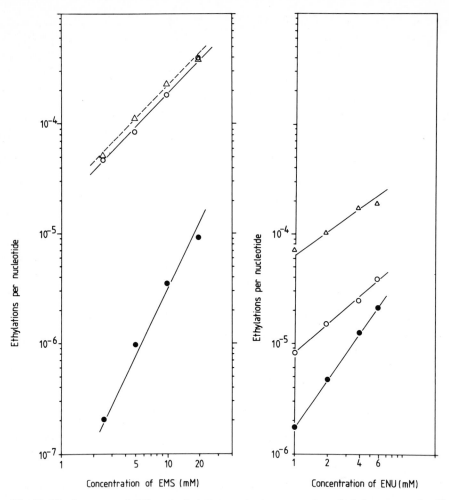

**Fig. 16.** The frequency of different ethylation products, expressed as ethylations per nucleotide, in DNA from *E. coli* exposed to ³H-labelled ethylmethanesulphonate (*left panel*) or ³H-labelled ethylnitrosourea (*right panel*). Stationary *E. coli* cells were treated in PBS at pH = 7.2 for 2 h with EMS or 1 h with ENU. Cells were washed and treated with lysozyme (1 mg/ml) for 15 min at 37 °C and subsequently lysed with sarkosyl. DNA was extracted and analyzed as described in the legend to Fig. 7. (△) total alkylations, (○) 7-ethylguanine, (●) $O^6$-ethylguanine

compared to control cells, whereas at late times cells are able to synthesize high molceular weight DNA even though the parental strand still contains UV-damage (Lehmann 1972; Waters 1979). In microbial systems it has been shown that pyrimidine dimers form a block for the growth of daughter strand DNA which can be overcome by DNA strand exchange (Rupp and Howard-Flanders 1968; Ganesan 1973) or by transdimer synthesis (Caillet-Fauquet et al. 1977). There is no direct evidence that strand exchange plays a significant role in post replication repair in mammalian cells. We have investigated how long pyrimidine dimers do form an absolute block for the growth of nascent DNA strands in UV-irradiated mammalian cells (van Zeeland and Filon 1982). Parental DNA was labelled with $^{14}$C by growing the cells overnight in medium containing $^{14}$C-thymidine. The cells were then UV-irradiated and grown for various periods of time in $^{3}$H-thymidine. The cells were trypsinized, permeabilized by quickly freezing and thawing twice and T4 endonuclease V was subsequently added. Under these conditions T4 endonuclease V will cause single strand breaks in parental DNA at each pyrimidine dimer site. It will also cut stretches of daughter strand DNA present in replicons which were active at the time of UV-irradiation. The length of the $^{3}$H-labelled nascent DNA molecules can then be compared with the interdimer distance found in $^{14}$C-labelled parental DNA, using alkaline sucrose gradients. The results show that in V79 Chinese hamster cells as well as in normal human fibroblasts 15 min after UV-irradiation the length of the newly synthesized DNA is about equal to the interdimer distance found in parental DNA. However, after longer periods of time the newly synthesized DNA becomes rapidly larger. The data also show that in xeroderma pigmentosum cells (complementation group A) the elongation of nascent DNA is somewhat slower than in normal cells and that in the xeroderma pigmentosum-variant cells the elongation is considerably slower, especially during the first 2 h (Fig. 17). These data suggest that pyrimidine dimers do form a block for elongation of nascent DNA molecules during the first 15 min following UV-irradiation but that after this period the cells are able to elongate the nascent DNA past pyrimidine dimers in the parental DNA strand (transdimer synthesis).

## 8 Concluding Remarks

In this chapter a series of DNA repair pathways have been discussed which are available to the cell to cope with the problem of DNA damaged by chemical or physical agents. In the case of microorganisms our knowledge about the precise mechanism of each DNA repair pathway and the regulation of it has been improved considerably when mutants deficient in these repair mechanisms became available. In the case of mammalian cells in culture, until recently there were very little repair deficient mutants available, because in almost all mammalian cells in culture at least the diploid number of chromosomes is present. Therefore the frequency of repair deficient mutants in such populations is very low. Nevertheless because replica plating techniques are improving some mutants from Chinese hamster ovary cells (Thompson et al. 1980) and L5178Y mouse lymphoma cells are now available (Shiomi et al. 1982). In the case of human cells, cultures obtained from patients with certain genetic diseases are available. A number of these cells appear to be sensitive to some chemical or physical mutagens. These include cells from patients suffering from xeroderma pigmentosum, Ataxia telangiectasia, Fanconi's anemia, Cockayne's syndrome (Friedberg et al. 1979). However, only in the case of xeroderma pigmentosum cells, has the sensitivity to ultraviolet light been clearly correlated with a deficiency in excision repair of pyrimidine dimers (Smith and Hanawalt 1978). Furthermore the work with strains obtained from biopsies from man is difficult because these cells generally have low cloning efficiencies and also have only a limited lifespan in vitro. It is therefore very important that more repair deficient mutants will become available from established cell lines from human or animal origin.

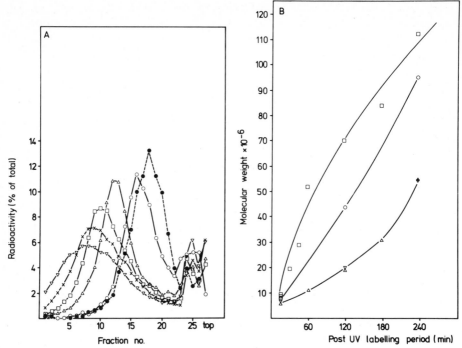

**Fig. 17 A,B. A** Radioactivity profiles of DNA from human diploid fibroblasts on alkaline sucrose gradients. Cells were prelabelled with $^{14}$C-thymidine, irradiated with UV (5 J/m$^2$), grown in medium containing $^3$H-thymidine for different periods of time, permeabilized by freezing and thawing, treated with 2 M NaCl and incubated with T4 endonuclease V before analyzes on a gradient. (●) parental DNA. Nascent DNA was labelled for 15 min (○), 1 h (△), 2 h (□), 3 h (X), or 4 h (▽). **B** The size of nascent DNA as a function of post-UV incubation time in medium containing $^3$H-thymidine. (□) normal cells; (○) XP25RO cells (complementation group A); (△) XP30RO cells (variant); (▽) XP6DU cells (variant). (van Zeeland and Filon 1982

*Acknowledgement.* This work was supported by the Koningin Wilhelmina Fonds, project no 81.89 and 81.91, The J.A. Cohen Interuniversity Institute for Radiopathology and Radiation Protection, the Commission of the European Communities, contract ENV-534-NL and Euratom, contract B10-E-407-81-NL.

# References

Abbondandolo A, Dogliotti E, Lohman PHM, Berends F (1982) Molecular dosimetry of DNA damage caused by alkylation. Single-strand breaks induced by ethylating agents in cultured mammalian cells in relation to survival. Mutat Res 92:361

Ayusawa D, Iwata K, Kozu T, Ikegami S, Seno T (1979) Increase in dATP pool in aphidicolin-resistant mutants of mouse FM3A cells. Biochem Biophys Res Commun 91:946

Berger NA, Sikorski GW, Petzold SJ, Kurohara KK (1980) Defective poly(adenosine diphospho-ribose) synthesis in xeroderma pigmentosum. Biochemistry 19:289

Bradley MO, Kohn KW (1979) X-ray induced DNA double strand break production and repair in mammalian cells as measured by neutral filter elution. Nucleic Acids Res 7:793

Caillet-Fauquet P, Defais M, Radman M (1977) Molecular mechanisms of induced mutagenes. Replication in vivo of bacteriophage ØX174 single stranded, ultraviolet light-irradiated DNA in intact and irradiated host cells. J Mol Biol 117:95

Carrier WL, Setlow RB (1971) The excision of pyrimidine dimers (the detection of dimers in small amounts). In: Grossman L, Moldave K (eds) Methods in enzymology, vol XXI, part D. Academic, New York, p 230

Ciarrocchi G, Jose JG, Linn S (1979) Further characterization of a cell-free system for measuring replicative and repair DNA synthesis with human fibroblasts and evidence for the involvement of DNA polymerase $\alpha$ in DNA repair. Nucleic Acids Res 7:1205

Cleaver JE, Thomas GH (1981) Measurement of unscheduled synthesis by autoradiography. In: Friedberg EC, Hanawalt PC (eds) DNA repair. A laboratory manual of research procedures, vol 1, part B. Dekker, New York, p 277

Cook JS, Regan JD (1968) Photoreactivation and photoreactivating enzyme activity in an order of mammals (Marsupiala). Nature 223:1066

Cook KH, Friedberg EC (1976) Measurement of thymine dimers in DNA by thin-layer chromatography, II. The use of one-dimensional systems. Anal Biochem 73:411

Dijkwel PA, Mullenders LHF, Wanka F (1979) Analysis of the attachment of replicating DNA to a nuclear matrix in mammalian interphase nuclei. Nucleic Acids Res 6:219

Durkacz BW, Omidiji O, Gray D, Shall S (1980) (ADP-ribose)$_n$ participates in DNA excision repair. Nature 283:593

Friedberg EC, Ehmann UK, Williams JI (1979) Human diseases associated with defective DNA repair. Adv Radiat Biol 8:85

Ganesan AK (1973) A method for detecting pyrimidine dimers in the DNA of bacteria irradiated with low dosis of ultraviolet light. Proc Natl Acad Sci USA 70:2753

Ganesan AK, Smith CA, Zeeland AA van (1981) Measurement of the pyrimidine dimer content of DNA in permeabilized bacterial or mammalian cells with endonuclease V of bacteriophage T4. In: Friedberg EC, Hanawalt PC (eds) DNA repair. A laboratory manual of research procedures, vol 1, part A. Dekker, New York, p 89

Halldorsson J, Gray DA, Shall S (1978) Poly(ADP-ribose) polymerase activity in nucleotide permeable cells. FEBS Lett 85:349

Hanaoka F, Kato H, Ikegami S, Ohashi M, Yamada M (1979) Aphidicolin does inhibit repair replication in HeLa cells. Biochem Biophys Res Commun 87:575

Haseltine WA, Gordon LK, Lindan CP, Grafstrom RH, Shaper NL, Grossman L (1980) Cleavage of pyrimidine dimers in specific DNA sequences by a pyrimidine-dimer DNA-glycosylase of M. Luteus. Nature 285:634

Hayaishi O, Ueda K (1977) Poly(ADP-ribose) and ADP-ribolylation of proteins. Annu Rev Biochem 46:95

James MR, Lehmann AR (1982) Role of poly(adenosine diphosphate ribose) in deoxyribonucleic acid repair in human fibroblasts. Biochemistry 21:4007

Karran P, Lindahl T, Griffin P (1979) The adaptive response to alkylating agents involves alteration in situ of $O^6$-methylguanine residues in DNA. Nature 280:76

Kihlman BA (1977) Caffeine and chromosomes. Elsevier, Amsterdam

Lehmann AR (1972) Postreplication repair of DNA in ultraviolet irradiated mammalian cells. J Mol Biol 66:319

Lindahl T (1979) DNA glycosylases, endonucleases for apurinic/apyrimidinic sites and base excision-repair. Prog Nucleic Acid Res Mol Biol 22:135

Mattern MR, Paone RF, Day III RS (1982) Eukaryotic DNA repair is blocked at different steps by inhibitors of DNA topoisomerases and of DNA polymerases $\alpha$ and $\beta$. Biochim Biophys Acta 697:6

Miwa M, Kanai M, Kondo T, Hoshino H, Ishihara K, Sugimura T (1981) Inhibitors of poly(ADP-ribose) polymerase enhance unscheduled DNA synthesis in human peripheral lymphocytes. Biochem Biophys Res Commun 100:463

Mullenders LHF, Zeeland AA van, Natarajan AT (1983) Analysis of the distribution of DNA repair patches in the DNA-nuclear matrix complex from human cells. Biochim Biophys Acta 740:428

Müller R, Rajevsky MF (1980) Immunological quantification by high-affinity antibodies of $O^6$-ethyldeoxyguanosine in DNA exposed to N-ethyl-N-nitrosourea. Cancer Res 40:887

Natarajan AT, Obe G, Zeeland AA van, Palitti F, Meijers M, Verdegaal-Immerzeel PAM (1980a) Molecular mechanisms involved in the production of chromosomal aberrations. II. Utilization of *Neurospora* endonuclease for the study of aberration production by X-rays in G1 and G2 stages of the cell cycle. Mutat Res 69:293

Natarajan AT, Zeeland AA van, Verdegaal-Immerzeel PAM, Filon AR (1980b) Studies on the influence of photoreactivation on the frequencies of UV-induced chromosomal aberrations, sister-chromatid exchanges and pyrimidine dimers in chicken embryonic fibroblasts. Mutat Res 69:307

Natarajan AT, Zeeland AA van, Zwanenburg TSB (1983) Influence of inhibitors of poly(ADP-ribose) polymerase on DNA repair, chromosomal alterations and mutations. In: Miwa M, Hayaishi O, Shall S, Smulson M, Sugimura T (eds) ADP-ribosylation, DNA repair and cancer. Japan Scientific Societies Press, Tokyo, p 227

Paterson MC, Lohman PHM, Sluyter ML (1973) Use of UV endonuclease from *Micrococcus luteus* to monitor the progress of DNA repair in UV-irradiated human cells. Mutat Res 19:245

Pedrali-Noy G, Spadari S (1980) Aphidicolin allows a rapid and simple evaluation of DNA-repair synthesis in damaged human cells. Mutat Res 70:389

Pettijohn DE, Hanawalt PC (1964) Evidence for repair-replication of ultraviolet damaged bacteria. J Mol Biol 9:395

Purnell MR, Whish WJD (1980) Novel inhibitors of poly(ADP-ribose) synthesis. Biochem J 185:775

Rupp WD, Howard-Flanders P (1968) Discontinuities in the DNA synthesized in an excision-defective strain of *Escherichia coli* following ultraviolet irradiation. J Mol Biol 31:291

Rupp WD, Sancar A, Sancar GB (1982) Properties of the uvr ABC endonuclease. Biochimie 64:595

Sabourin CLK, Bates PF, Glatzer L, Chang C, Trosko JE, Boezi JA (1981) Selection of aphidicolin-resistant CHO cells with altered levels of ribonucleotide reductase. Somatic Cell Genet 7:255

Schans GP van der, Centen HB, Lohman PHM (1982) DNA lesions induced by ionizing radiation. In: Natarajan AT, Obe G, Altmann H (eds) DNA repair, chromosome alterations and chromatin structure. Progress In Mutation Research, vol 4. Elsevier Biomedical, Amsterdam, p 285

Seawell PC, Smith CA, Ganesan AK (1980) Den V gene of bacteriophage T4 determines a DNA glycosylase specific for pyrimidine dimers in DNA. J Virol 35:790

Setlow RB, Setlow JK (1972) Effects of radiation on polynucleotides. Annu Rev Biophys Bioeng 1:293

Shiomi T, Hieda-Shiomi N, Sato K (1982) Isolation of UV-sensitive mutants of mouse L5178Y cells by a cell suspension spotting method. Somatic Cell Genet 8:329

Smith CA (1978) Removal of T4 endonuclease V sensitive sites and repair replication in confluent human diploid fibroblasts. In: Hanawalt PC, Friedberg EC, Fox CF (eds) DNA repair mechanisms. Academic, New York, p 311

Smith CA, Hanawalt PC (1978) Phage T$_4$ endonuclease V stimulates DNA repair replication in isolated nuclei from ultraviolet-irradiated human cells, including xeroderma pigmentosum fibroblasts. Proc Natl Acad Sci USA 75:2598

Smith CA (to be published 1983) Analysis of repair synthesis in the presence of inhibitors. In: Collins A, Johnson R (eds) DNA repair and its inhibitors. Dekker, New York

Smith CA, Cooper PK, Hanawalt PC (1981) Measurement of repair replication by equilibrium sedimentation. In: Friedberg EC, Hanawalt PC (eds) DNA repair. A laboratory manual of research procedures, vol 1, part B. Dekker, New York, p 289

Sugino A, Nakayama K (1980) DNA polymerase mutants from a *Drosophila melanogaster* cell line. Proc Natl Acad Sci USA 77:7049

Thompson LH, Rubin JS, Cleaver JE, Whitmore GF, Brookman K (1980) A screening method for isolating DNA repair-deficient mutants of CHO cells. Somatic Cell Genet 6:391

Sutherland BM (1974) Photoreactivating enzyme from human leukocytes. Nature 248:109

Waters R (1979) Repair of DNA in replicated and unreplicated portions of the human genome. J Mol Biol 127:117

Wist E (1979) The role of DNA polymerase $a$, $\beta$, and $\gamma$ in nuclear DNA synthesis. Biochim Biophys Acta 562:62

Yarosh DB, Foote RS, Mitra S, Day III RS (1983) Repair of O$^6$-methylguanine in DNA by de-methylation is lacking in *mer*⁻ human tumor cell strains. Carcinogenesis 4:199

Zeeland AA van, Natarajan AT, Verdegaal-Immerzeel PAM, Filon AR (1980) Photoreactivation of UV induced cell killing, chromosomal aberrations, sister chromatid exchanges, mutations and pyrimidine dimers in Xenopus laevis fibroblasts. Mol Gen Genet 180:495

Zeeland AA van, Smith CA, Hanawalt PC (1981) Sensitive determination of pyrimidine dimers in DNA of UV-irradiated mammalian cells. Introduction of T4 endonuclease V into frozen and thawed cells. Mutat Res 82:173

Zeeland AA van, Bussmann CJM, Degrassi F, Filon AR, Van Kesteren-van Leeuwen AC, Palitti F, Natarajan AT (1982) Effects of aphidicolin on repair replication and induced chromosomal aberrations in mammalian cells. Mutat Res 92:379

Zeeland AA van, Filon AR (1982) Post-replication repair: elongation of daughter strand DNA in UV-irradiated mammalian cells in culture. In: Natarajan AT, Obe G, Altmann H (eds) DNA repair, chromosome alterations and chromatid structure. Progress in Mutation Research, vol 4. Elsevier Biomedical, Amsterdam, p 285

# Structure and Organisation of the Human Genome

H.J. EVANS[1]

## 1 Introduction

It is estimated that at least 10% of ill health in man is a direct consequence of the inheritance of defective or disease predisposing genes (UNSCEAR 1977), and this estimate does not include diseases, such as certain forms of cancer, which may be associated with acquired somatic genetic changes as opposed to inherited defects. Approximately 1% of the live newborn inherit major single gene defects and a proportion of these are a consequence of new mutations. The mutation rate for these changes differs enormously for different genes with at one extreme a gene responsible for a form of X-linked non-specific mental retardation having a mutation rate of around 1 per $10^4$ gametes per generation, whereas other genes may undergo mutation at rates which are at least a thousand fold lower. In addition to the inherited major single gene defects, around 1 in every 150 live newborn babies inherit a chromosomal abnormality which may take the form of an alteration in chromosome number or in chromosome structure, and the majority of these abnormalities are new mutations arising in parental germ lines (Evans 1977a). The incidence of chromosomal mutations in still births and in spontaneous abortuses is very much higher and indeed more than 50% of early abortions are associated with a chromosomal mutation (Boue et al. 1975) and the overall numbers would imply that almost 1 in 10 human gametes carries a chromosomal mutation. Chromosomal and, in some cases, gene mutations can also be readily demonstrated in human somatic cells and chromosomal abnormalities in blood lymphocytes from healthy individuals occur with a frequency of between 1 in $10^2$ to $10^3$ cells. Mutation therefore is by no means a rare event in human somatic and germ cells and the human genome may be a rather less stable entity than is sometimes assumed.

The human genome contains around $3 \times 10^9$ base pairs, equivalent to more than 1 m of DNA, arranged in a haploid complement of 23 chromosomes and each chromosome, on the average, will contain several hundreds of genes or gene clusters. The arrangement and interaction of these genes on specific chromosomes and chromosome regions, their relationships to each other and to non-coding DNA, and to chromosome organelles such as centromeres, secondary constrictions and telomeres, reflect an arrangement arrived at as a result of selection over countless generations. A knowl-

---

1 Medical Research Council, Clinical and Population Cytogenetics Unit, Western General Hospital, Edinburgh, United Kingdom

Mutations in Man, ed. by G. Obe
© Springer-Verlag Berlin Heidelberg 1984

edge of chromosome structure and of the organisation of the genome as a whole is thus a necessary requirement if we are to understand the processes responsible for maintaining genome stability and continuity, and for our appreciation of factors that can influence this stability. Thus, in the present article although certain aspects of genomic structure are considered in the light of the actions of external mutagens on the genome, this discussion will not be confined solely to a consideration of chromosome structure and genome organisation in the context of what we know about mutagenesis.

## 2  General Aspects of Chromatin and Chromosome Architecture

The structural skeleton of chromatin in a human cell nucleus consists of approximately equal parts by weight of DNA, basic histone proteins and acidic non-histone proteins. However, only a proportion of these proteins are essential architectural features of chromosomes and the remainder, together with various RNA species, with and without their associated proteins, cannot be considered as structural components of chromatin. Isolated chromatin can be separated into "condensed" and "dispersed" fractions and their distribution in the interphase nucleus is non-uniform. Condensed fractions contain most of the tightly complexed and relatively transcriptionally inactive heterochromatic regions of the genome, and the dispersed fractions most of the less condensed euchromatic and transcriptionally active regions. The heterochromatic regions may represent condensed regions of single chromosomes, or aggregates from different chromosomes (chromocentres), and may consist of constitutive heterochromatin, characterised by tightly packaged repeated sequence DNA's, or facultative heterochromatin, a "transient" condensed heterochromatic state exemplified by the transcriptionally inactive X chromosome in the female. Some regions of chromatin, including telomeres and heterochromatin, are bound to the nuclear membrane, others are free in the nuclear sap, or are associated by virtue of their gene products, so that the arrangement of chromatin within the nucleus is complex (Agard and Sedat 1983).

The extended DNA in the chromatin of an unreplicated human diploid nucleus would amount to a single duplex some 1.9 m long. This DNA is of course distributed between 46 chromatids and in each chromatid the DNA is present as a single fibre or unineme (reviewed in Evans 1979). At metaphase, human chromosomes range in size from approximately 2 to 10 $\mu$m in length with each chromatid having a diameter of around 0.6 $\mu$m. The smallest chromosomes must therefore contain around 1.4 mm of DNA and the largest around 7.3 cm, so that metaphase chromatids are 10,000 x shorter than the lengths of DNA contained within them. Our knowledge of how this DNA and its associated proteins are packaged within the chromosome is still incomplete, particularly at the higher levels of folding and coiling.

Under the electron microscope, chromatin spread from disrupted nuclei or chromosomes has the appearance of a mesh of single 20 Å diameter fibres connecting regularly spaced blobs of protein to give a "pearls on a string" appearance (Olins and Olins 1974). Each pearl, or nucleosome, consists of a single DNA duplex thread wound around the outside of an octameric sphere of histones containing equal amounts (two flat tetramers) of H2A, H2B, H3, and H4 (Kornberg 1974). Nuclease digestion shows that approximately 200 bp of DNA are associated with each nucleosome, 140 bp being consistently closely associated with the histone octamer ($1\,^3/_4$ times wound around the outside of the complex – Finch and Klug 1977) and the remainder of the DNA, which may vary somewhat in length between different tissues, being associated with histone H1 and serving as a "linker" between adjacent particles (Noll and Kornberg 1977). There have been some indications that nucleosomes may be phased in relation to their associated DNA sequences, but the bulk of evidence (McGhee and Felsenfeld 1983) would suggest a random association between histone beads and DNA sequence.

Nucleosome beads are approximately 100 Å in diameter so that a single nucleo-filament would appear as a 100 Å diameter fibre. This fibre is itself coiled to give a supercoiled solenoid structure (Finch and Klug 1976; Carpenter et al. 1976) which is some 300 Å in diameter and has an average of 6 or 7 nucleosomes for each turn of the coil – and giving a DNA packing ratio of around 40:1. A medium sized human chromatid of 6 $\mu$m length at metaphase contains some 3.3 cm of DNA, and hence a length of 825 $\mu$m of a 300 Å diameter solenoid, so that there must be a further length reduction to pack this solenoid into a metaphase rod approximately 1.7 $\mu$m$^3$ in volume. This further packaging is achieved by further coiling and folding ("looping") of the solenoid, but the precise arrangement of this 300 Å fibre at this level is obscure, although it is clear that this packaging is not uniform along the length of the chromo-some (see Sect. 4). There is, however, much evidence to suggest that this higher level of folding involves the binding of DNA loops to structural acidic proteins which may form a "core" or "scaffold" within the chromosome.

Evidence for the presence of acid-insoluble proteins as structural chromosomal components is long standing, and several models of chromosome structure involving a protein core have been presented (e.g. Stubblefield 1973), usually based on the idea of DNA/protein loops emanating from a core structure and analogous to the type of lateral loop and central axial structures seen in the meiotic lampbrush chromosomes of amphibia. More recent studies have shown that removal of histones from metaphase chromosomes, and hence the loss of primary chromatin structure at the nucleosomal and 300 Å fibre level, does not abolish the gross morphology of metaphase chromo-somes (Paulson and Laemmli 1977; Jeppesen et al. 1978). Moreover, DNA digestion to remove the bulk of the DNA from dehistonised chromosome yields residual struc-tures resembling metaphase chromosomes, but consisting mainly of non-histone protein with a very small amount of protected and tightly associated DNA (Adolph et al. 1977; Jeppesen and Bankier 1979). These studies then strongly suggest the presence of a non-histone protein core associated with the higher level of folding and organisation of chromatin.

The elegant electron micrographs of Paulson and Laemmli were the first to dem-onstrate clear DNA loops in dehistonised chromosomes which appeared to emanate from a central protein core referred to as the scaffold. These loops are probably arranged in a spiral around the axis since a spiral substructure is clearly evident, under certain conditions, in extended chromosomes seen under the light microscope. The loops seen under the electron microscope are approximately 20–50 $\mu$m in length and similar sized loops had been previously inferred from studies on interphase nuclei (Cook and Brazell 1975; Igo-Kemenes and Zachau 1977) and it would appear that interphase DNA loops and mitotic DNA loops are synonymous structures (Warren and Cook 1978).

The conclusion that the sites of anchorage of the DNA loops are to an acidic protein scaffold structure within the chromosome has not been wholly accepted and a number of authors have presented some evidence which they interpret as indicating that at least parts of the scaffold structures seen are artefacts of preparation (e.g. Hadlaczky et al. 1981). Cook and colleagues (McCready et al. 1982) provide evidence for a somewhat different protein-RNA structure which they refer to as the nuclear cage and to which DNA loops of some 75 $\mu$m (220,000 base pairs) are attached during interphase and mitosis. They also emphasise the importance of the cage in controlling transcription and replication and adduce evidence supporting the argument that every gene is in a precisely defined place in the nucleus and attachment to the cage determines whether or not a gene is replicated or transcribed in that these processes occur at the cage attachment sites (Jackson et al. 1981, and see also Robinson et al. 1982; Small et al. 1982).

The conclusion that chromosome scaffolds, or cores, are realities and not artefacts is supported by the demonstration (Howell and Hsu 1979; Satya-Prakash et al. 1980) of a silver staining trypsin-sensitive core protein structure along the length of the chromatids of mammalian somatic chromosomes. The silver staining technique used was essentially that used to stain the lateral elements of the synaptonemal complex in meiotic chromosomes – a firmly established chromosomal axial protein structure (Dresser and Moses 1979; Fletcher 1979).

Recent studies by Laemmli and colleagues report the occurrence of two predominant proteins in isolated scaffolds which account for 40% of the scaffold mass, which itself accounts for 3% to 4% of the total chromosomal protein (Lewis and Laemmli 1982). These proteins are metalloproteins whose chelation results in the dissociation of the scaffold and unfolding of the DNA – a reversible phenomenon (Earnshaw and Laemmli 1983). Gooderham and Jeppesen (1983) have also recently reported the isolation of two rather smaller core proteins, which are not nuclear lamina proteins or tubulins and studies on the binding of these proteins to residual core DNA is in progress.

There is limited evidence in man (see DNA composition and distribution), but abundant evidence in prokaryotes, that genes with related functions are often located close together on the chromosomes, and some evidence in eukaryotes that homologous chromosomes may be located close to each other in the interphase nucleus – so-called somatic pairing. There is no evidence for somatic pairing in mammalian, including human, somatic cell nuclei, but it is pertinent to ask whether there is a fixed, or preferred, arrangement of chromosomes in the interphase nucleus (Comings 1980).

Evidence for a preferred arrangement of chromosomes at mitotic metaphase in plants has been obtained from electron microscope studies of serial sections (Bennett 1982) and this probably reflects a preferred arrangement in interphase; and the incidence with which particular plant chromosomes are involved in X-ray-induced interchange is non-random (see Evans and Bigger 1961). In man there is little information on chromosome distribution at mitosis in cells not exposed to hypotonic solutions and spindle inhibitors, but evidence from UV microirradiation (Cremer et al. 1982; Zorn et al. 1979) and from radiation-induced rearrangements produced during inter- phase (Dutrillaux et al. 1981) suggests that the positions of chromosomes at meta- phase reflect their interphase location. If chromosomes did have a particular prefer- ential disposition in interphase, then it might be predicted that induced exchanges would be non-random reflecting any specific chromosome associations. One "natural" experiment suggests that this is not the case, for in a patient in whom many hundreds of cells were observed to have multiple exchanges giving long chains of chromosomes, the sequences of chromosomes involved in chains was random (Dutrillaux et al. 1978).

## 3 DNA Composition and Distribution

Disrupted human DNA consists of a variety of unique sequence and repeated sequence tracts and their composition, prevalence and distribution within the genome have been studied using a number of approaches. Reassociation experiments on sheared denatured human DNA allowed to reanneal under carefully controlled conditions show at least four kinetically distinct components: (a) largely unique, or so-called single copy, sequences, i.e. DNA that does not contain repeats with a frequency greater than about 30 copies (Britten 1981) and which accounts for somewhere around 70% of the total DNA depending upon the sizes of the fragments and the stringency of the conditions for reassociation; (b) intermediate repetitious DNA containing a mixture of sequences of varying complexities and repetition frequency and comprising some 20% of the total DNA; (c) fast reassociating "satellite" sequences which make up about 5% of the total DNA and; (d) very fast reassociating sequences or "foldback" DNA's which anneal almost immediately following denaturation and consist of inverted repeated sequences (Wilson and Thomas 1974; Dott et al. 1976).

Centrifugation of human DNA to equilibrium in cesium chloride containing $Ag^+$ or $Hg^{2+}$ ions reveals the presence of small satellite DNA peaks equilibrating outside the main band of DNA and Corneo and colleagues originally isolated four human A+T-rich satellite DNA's of different mean densities (Corneo et al. 1972). These satellites consist of families of highly repeated sequences, i.e. the fast reassociating components in reassociation experiments, which are mainly present as large tandemly repeated contiguous tracts of DNA's, which have been shown by in situ hybridisation to be located in the centromerically disposed blocks of heterochromatin (C-bands) in the autosomes and in the long arm of the Y chromosome (Evans et al. 1974; Gosden et al. 1975; Manuelidis 1978).

The use of bacterial restriction endonucleases, the development of cloning techni- ques and of simple DNA sequencing methods, have provided rather different and very powerful tools for analysing human genomic DNA's. Studies on the human satellite DNA's show that they are made up of families of tandemly repeated DNA's which are in fact composed of a mixture of different divergent sequences. For example, although satellite III reassociates as a simple sequence DNA with a Cot $^1/_2$ value of about $10^{-3}$, digestion of this satellite with *Hae* III produces a complex set of fragments many, but by no means all, being simple multiples of a basic 174 nucleotide pair long fragment. One satellite III *Hae* III fragment of 3.4 kb length is specifically located in the Y chromosome (Bostock et al. 1978), but part of this fragment is sufficiently homologous to DNA sequences present on the autosomes to undergo hybridisation with them in situ. By analysing human DNA isolated from human/mouse hybrid cells containing essentially single copies of human chromosomes 1, 7, 11, 15, 22, and X, Beauchamp et al. (1979) were able to show that the distribution of *Hae* III restriction sites within satellite III was different for different human chromosomes. Moreover, similar differences were observed with other restriction enzymes, *Eco* R1 and *Bam* H1, and with the other three satellite DNA's. A study of sequence homologies between the four satellite DNA's shows that each shows varying degrees of homology to one or more of the other three (Mitchell et al. 1979), but that within any given satellite there has evidently been considerable sequence divergence so that different families and subfamilies of a given satellite may be present on different chromosomes. It should be noted that sequences related to these satellites are also to be found in other primates, with satellite III being common to species that diverged some 24 million years ago, but satellites I and II being of more recent origin (Mitchell et al. 1981).

In a study of human genomic DNA principally using reassociation techniques, Schmid and Deininger (1975) showed that repetitive sequence DNA's are distributed throughout 80% or more of the genome. Just over a half of the genome consisted of short, 2 kb long, single copy sequences interspersed with short, around 0.4 kb, repetitive sequences. Other longer single copy DNA's were also found to be interspersed with repetitive sequences and the inverted "foldback" repeated DNA's were randomly positioned throughout the genome. Electron microscope studies (Deininger and Schmid 1976) confirmed that many of the interspersed repeated sequences were about 300 nucleotides in length. Houck et al. (1979), and see also Schmid and Jelinek (1982), then showed by digestion techniques that the human genome contains at least 400,000 copies of a 300 bp repeat sequence identified using the *Alu* restriction enzyme and referred to as *Alu* repeats. This family of short repeats in itself accounts for some 4% of the total genome (see Sect. 12).

The *Alu* family is very prominent among the interspersed repeated sequences in human DNA, but there are many other *non-Alu* repeated sequence families present and the evidence suggests that they could form the majority of the interspersed repeats. One such group of DNA families are the DNA segments released by *Kpn* I digestion which contain four distinct families which appear to be interspersed within the genome (Maio et al. 1981). These *Kpn* I long dispersed repeats show considerable fragment length polymorphism, comprise at least 6% of the total DNA, they are transcribed and their products can be detected in cytoplasmic RNAs (Kole et al. 1983). The functions of many of these repeated sequence DNA's are unknown and although many are transcribed and appear in heterogeneous nuclear RNA (e.g. Pan et al. 1981; Fritsh et al. 1981) their RNA products are often not detectable in cytoplasmic or mRNA's. The possibility exists that some of these interspersed repeats, and particularly the *Alu* and *Kpn* I families, might behave as mobile DNA's and this will be considered elsewhere.

In addition to non-transcribed repeated satellite DNA's (Melli et al. 1975), and the transcribed repeated DNA's of unknown function, there are tandemly repeated sequences of known function, such as the ribosomal genes, which may be separated by spacer sequences and be present in multiple copies at a given chromosomal site (see Sect. 7), and the histone genes, which amount to some 40 copies (Wilson and Melli 1977) located at one site on chromosome 7 (Szabo et al. 1978). Indeed, many of the single copy genes in man may have various numbers, and often large multiples, of processed copies of their coding regions present in other regions of genome (see Sect. 14) and some of the different genes concerned with a given function are sometimes clustered together and have probably evolved following duplication and divergence of a single gene. Examples of such gene clusters at a single site include the genes of the major histocompatibility complex on chromosome 6; the heavy chain immunoglobulin gene clusters on chromosome 14 and the $\kappa$ and $\lambda$ light chain clusters on chromosomes 2 and 20; and the cluster of $\gamma$, $\delta$, and $\beta$ globin genes on chromosome 11, etc.

Following the original demonstrations of the presence of intervening non-coding sequences, or "introns", within and between the coding sequences, or exons, observed first with rabbit (Jeffreys and Flavell 1977) and mouse (Tilghman et al. 1978) $\beta$ globin genes, and then with chick ovalbumin genes (Breathnach et al. 1977), it has become evident that most of the studied genes in the human complement, and in other eukaryotes, are split genes. For example, in the case of the $\beta$-tubulin family of human genes, of which there are some 15 to 20 members, a functional coding gene is split into four coding blocks or exons (Lee et al. 1983), but evidently most of the family are processed pseudogenes that are not transcribed to give functional tubulin. The data of Lee et al. (1983) show that the coding sequences of $\beta$ tubulin amount to a total of 1,332 bp whereas the three intervening exon sequences are larger amounting to 2,459 bp. Of interest is the fact that the first of these intervening sequences contains an *Alu* family sequence and *Alu* sequences have also been found in the intervening sequences of other genes, for example the human *onc* gene, *c-sis*, where there are three *Alu* repeats in the two introns (Dalla Favera et al. 1981). *Alu* repeats are also present at the 5'- or 3'-flanking regions of several human genes as well as in the intergenic regions of gene clusters as in the $\beta$ globin gene family (Duncan et al. 1979).

A number of "single" human genes and gene complexes have now been isolated, mapped and sequenced and some, and in particular those where mutations result in appreciable numbers of individuals with consequent inherited disease, e.g. the globin gene families and the thalassemias (Spritz and Forget 1983), have been much studied. These details will not be referred to here, but these genes and the other coding sequences in the genome whose products are translated into functional proteins must comprise but a minority of the total DNA. Thus, in terms of the over-all organisation and arrangement of DNA in the genome the picture is one of interrupted heterogeneity. Aside from the simple sequence repetitious satellite DNA's, the bulk of which are present as heterochromatic blocks – which does not mean that some of these sequences are not also dispersed in smaller repeat numbers – heterogeneous moderately repeated tandem and inverse repeats, coding and non-coding unique sequences, are all intermingled and distributed throughout the genome. Since functional genes are themselves a mosaic of DNA's, it will be relevant to our discussion to emphasise that base change or damage in non-coding exonic regions of a gene will not be mutational changes in the traditional sense, and that it is probable that only a minority of DNA alterations induced in any given genome as a consequence of mutagen-induced damage will result in a heritable alteration in gene expression.

## 4 Chromosome Bands

The discovery and development in the early 1970's of a variety of methods that revealed the presence of numerous transverse bands in the somatic metaphase chromosomes of man has been reviewed frequently (e.g. Evans, 1977c), as indeed have the various techniques used to produce Q-, G-, and the reverse R-bands (which are essentially similar bands revealed using different techniques) and variants of these bands (e.g. Comings 1978; Sumner 1982). These will not therefore be considered in detail here, except in the context of what they tell us concerning chromosome structure that may have some relevance to the response of human chromosomes to exposure to mutagens.

Some 3–400 specific transverse bands in the metaphase complement are revealed by the Q-, G- or R-banding techniques, but the mechanisms involved in the production of these bands, and the nature of the chromosome organisation underlying them, is still a matter of some debate. A further longitudinal differentiation of the chromosome is high-lighted by techniques that differentially stain the constitutive heterochromatin (C-band techniques, or variants thereof) and by methods that emphasise the centromeric (Cd-band) or nucleolar organiser (N-band) regions of the chromosomes. Centromeres and nucleolar organising regions will, however, be considered elsewhere under separate headings and will not therefore be discussed in this section.

The bulk of the constitutive heterochromatin in man stains intensely with C-band techniques and, in general, it would appear that most of these C-band regions which are located in the centromeric regions of the autosomes, the distal regions of the long arm of the Y chromosome and the satellited regions of the acrocentric autosomes, predominantly contain tightly packaged simple sequence repetitious satellite DNA's (Gosden et al. 1975). C-bands are heritable features of the chromosomes, but are polymorphic in that they may differ in size (and in DNA amount) between different people. In terms of information content and transcriptional activity, these C-band regions are usually considered to be genetical deserts so that induced DNA changes due to the actions of mutagens in these regions might be considered to be of little consequence. This, however, is by no means true in the context of mutagen-induced chromosome aberrations and in terms of the consequences of the transposition of heterochromatin within the genome.

Studies on X-irradiated human cells (Buckton 1976) show that breakages and rearrangements are distributed throughout the genome and occur in C-band regions of the autosomes, but to a far lesser extent in the C-band region of the Y chromosome, which appears to be particularly refractory to X-ray-induced rearrangement. In contrast, studies utilising certain chemical mutagens, and in particular alkylating agents such as Mitomycin C (MMC), show a marked excess of aberrations located in C-band material (cf. Shaw and Cohen 1965). Studies on MMC-induced exchanges in lymphocyte chromosomes of marsupials (Hayman 1980) provide strong support for the notion that the enhanced response of C-band material to mutagen-induced exchange is a consequence of simultaneous replication of large amounts of damaged DNA's having very similar sequences.

The underlying composition and arrangement of DNA's and proteins that give rise to the G-band profiles (and the equivalent Q- and R-bands) are not well understood.

Positive Q- and G-bands generally replicate their DNA late in the S phase (Ganner and Evans 1971; Epplen et al. 1975) – a characteristic of A+T-rich DNA (Bostock and Prescott 1971) – and stain positively with a number of fluorochromes that, at least with DNA in solution, show a preferential binding to A+T-rich DNA. These observations, together with the finding that antiadenosine antibodies bind to positive Q-bands, and anticytosine antibodies to R-bands (Dev et al. 1972; Schreck et al. 1973), are often taken to indicate that positive Q-, and the equivalent G-bands represent A+T-rich regions of the genome, whereas the interbands, or R-bands, reflect DNA's rich in G+C (Jorgenson et al. 1978; Schnedl et al. 1980). However, there are many contradictory findings which include, for example, the demonstration that the antibiotics chromomycin A3 and olivomycin, which bind preferentially to G+C-rich DNA, can give both Q-bands and R-bands (Prantera et al. 1979).

One of the standard techniques for producing G-bands involves the use of proteases (Seabright 1971) and recently there have been a number of reports (Sahasrabuddhe et al. 1978; Mezzanotte et al. 1983; Miller et al. 1983) describing the production of banded profiles in chromosomes exposed to a variety of restriction endonucleases. The data suggest that preferential loss of, and differential access to, different DNA regions in the chromosome are important factors influencing banding by digestion techniques. The notion that G- and R-banding reflects differences in base composition along the chromosome is therefore a very considerable over-simplification of the interpretation of dye-binding studies, which neglects the contribution of a number of factors, not least chromosomal proteins – and in particular the SH containing proteins – and differential condensation within the chromosome (see Sumner 1980, 1983; Burkholder and Duczek 1982). In this regard it should be noted that in the early stages of mitosis more than 1,000 G-banded structures can be resolved in the genome (Francke and Oliver 1978; Yunis et al. 1978; Dutrillaux and Viegas-Pequignot 1981), but positive and negative bands coalesce as prophase proceeds to give around 300 or so bands at metaphase, which must be composite bands, whose patterns of arrangement correspond closely with the patterns of chromomeres on meiotic pachytene chromosomes (Luciani et al. 1975).

Indirect evidence (Ganner and Evans 1971; Yunis et al. 1977) would imply that the bands that are negatively staining with Q- and G-banding techniques, but positive with R-banding methods, are richer in transcribed sequences than the other chromosome regions and, because of the less compacted nature of these regions, it seems likely that they may be more vulnerable to damage by mutagens. Studies on the distribution of breakage or exchange sites in banded chromosomes of cells exposed to X-rays and to various chemical mutagens show that aberration sites can be located in either type of band with a possible preponderance in negative G-bands and especially at the interface between positive and negative bands (Morad et al. 1973, Buckton 1976; Hoo and Parslow 1979; Carrano and Johnston 1977).

## 5 Centromeres (Primary Constrictions)

The centromere, or kinetochore, is an organelle consisting of DNA and proteins which is responsible for the orderly segregation of sister chromatids at mitosis and of homologous chromosomes at meiosis. The single centromere in each human chromosome is a complex structure consisting of DNA and a plate of acidic proteins from which emanate the multiple tubules of the spindle structure (see Ris and Witt 1981). The kinetochore plate is the structure that would appear to be the stained body seen following the use of centromere staining techniques (Evans and Ross 1974), and the body which reacts specifically with antibodies from patients with certain forms of scleroderma (Moroi et al. 1980), but the nature and organisation of the DNA with which this protein structure is associated is unknown. Presumably, each of the 92 chromatids in a dividing human nucleus contains a specific arrangement of DNA sequences that is responsible for organising the kinetochore and it seems probable that the centromeric DNA composition and sequence arrangement is common for all the chromosomes in the complement, but is unique to a single fixed position within each chromosome.

Recently Carbon and colleagues (Clarke and Carbon 1980; Hsaio and Carbon 1981; Fitzgerald-Hayes et al. 1982a) have managed to isolate DNA segments containing the centromeres of several chromosomes of bakers' yeast which were shown to confer mitotic stability and proper meiotic segregation properties when inserted into autonomously replicating plasmids. Two of these centromeric DNA's, on chromosomes 3 and 11, have now been sequenced and both shown to contain an 87–88 bp A–T-rich core segment flanked by two short (11 and 14 bp) sequences. These elements, plus an additional 10 bp region of perfect homology, have an almost identical spatial arrangement within the two centromeres and the findings imply that very small lengths of DNA, of from 600–900 bp long, are necessary for specifying a functional centromere. No equivalent information is available on DNA's in mammalian centromeres, but Fitzgerald-Hayes et al. (1982b) have noted pronounced sequence similarities between yeast centromere DNA's and the Dm 359 satellite DNA of *Drosophila* and have isolated a protein fraction from yeast which binds selectively not only to yeast centromeric DNA's, but also to Dm 359; it is not known whether these fractions bind to mammalian DNA's.

# 6 Telomeres

The concept of a specialised organisation of chromosome termini, or telomeres (Muller 1942; Muller and Herskovitz 1954) followed from the fact that the tips of normal chromosomes, in contrast to the broken ends of ruptured chromosomes, are stable structures that do not normally undergo fusion with other chromosome tips. Although there are a number of examples of permanent healing, or stabilisation, of broken chromosome ends, e.g. in maize sporophytic as opposed to gametophytic tissues (McClintock 1941), in plant holocentric chromosomes (Nordenskiöld 1962; Evans and Pond 1964), in a few instances in *Drosophila* (Roberts 1975) and possibly in mammalian (Hsu 1963), including human (Cooke and Gordon 1965; Shaw 1972), cells, in general, the chromosome ends generated following mutagen-induced chromosome breakage are usually unstable structures which become stabilised only by joining with other broken ends.

In addition to their stability, two other properties of natural chromosome ends, and significantly in contrast to broken ends, is that they are often closely associated with the nuclear membrane and that they promote transient end-to-end associations of non-homologous chromosomes during interphase and meiotic prophase (White 1973). In some rare cases the end-to-end associations are more permanent as indicated by the early studies on the composite chromosome of the nematode *Ascaris* and more recently by the demonstration of the apparent end-to-end fusion of human chromosomes in the lymphocytes of a patient with the Thiberge-Weissenbach syndrome (Dutrillaux et al. 1977).

These properties of telomeres have naturally led to the belief that telomeric chromosome ends were capped by specific DNA sequences and their associated proteins which were common to all natural chromosome ends. A number of speculative models of DNA organisation at telomeres have been proposed, particularly paying regard to problems associated with the replication of chromosome termini (Watson 1972; Cavalier-Smith 1974; Bateman 1975; Dancis and Holmquist 1979), but only recently has any direct evidence emerged concerning telomeric DNA sequence arrangements.

Studies on extrachromosomal ribosomal DNA sequences in a number of primitive eukaryotes have recently provided some evidence concerning the stable terminal ends of linear DNA molecules within the nucleus. In the slime mould *Physarum polycephalum*, 160 to 320 rDNA genes are present as 61 kb molecules which replicate as minichromosomes within the diploid nuclei and segregate without loss to daughter nuclei. Bergold et al. (1983) have now shown that the termini of these mini chromosomes consist of a series of multiple inverted repeats with specific single strand gaps and tightly associated proteins. These palindromic repeated sequences form multiple hairpin structures, but there is not just one consensus terminal sequence at the bends of the hairpins. Tandem repeats and possible hairpin configurations are also present in the extrachromosomal rDNA's of the related *Dictyostelium* (Emery and Weiner 1981) and the ciliate protozoan *Tetrahymena* (Blackburn and Gall 1978). Similar terminal repeats are also found to cap the ends of gene-sized fragments in the macronuclei of a variety of other ciliate protozoans (Boswell et al. 1982) and Szostak and Blackburn (1982) have  shown that these terminal regions can act as telomeres to linear plasmids when these are introduced into yeast. Szostak and Blackburn (1982) have used linear plasmids containing only one capping rDNA sequence derived from *Tetrahymena* to search for and isolate capping sequences in the yeast genome and provide compelling evidence that telomere sequences on the different chromosomes of yeast are largely homologous and are, moreover, very similar in structure to the terminal regions of the ciliate rDNA molecules.

To date, there is no information on telomere organisation in mammalian chromosomes, but a possible telomeric DNA sequence has been isolated from *Drosophila melanogaster* which has been shown by in situ hybridisation to be located at the termini of all chromosomes and to consist of a 3 kb repeat unit homologous to sequences belonging to the *copia* multigene family (Rubin 1978). *Copia* genes are scattered throughout the *Drosophila* genome (see Sect. 12) so that the possible role of copia-like sequences in telomere organisation must await further study.

# 7 Nucleolar Organisers (Secondary Constrictions)

In man, the DNA, and its associated spacer sequences, that codes for the 18 s and 28 s components of ribosomal RNA is present in multiple copies, around 50 or so per genome (Young et al. 1976; Cote et al. 1980) which, as demonstrated by in situ hybridisation (Henderson et al. 1972; Evans et al. 1974), are located in the short arms of each of the five pairs of acrocentric chromosomes. It is at these sites that the nucleoli are organised and these Nucleolar Organising Regions (NOR's) appear as marked constrictions (secondary constrictions) in the short arms. This rDNA is therefore in the stalk or constricted region of the chromosome, may extend into any associated nucleoli, but is not present in the subtended terminal satellite which contains simple sequence repetitious DNA (Gosden et al. 1975). The number of 18 s and 28 s rDNA sequences at any given site may differ between chromosomes (Evans et al. 1974; Miller, Tantravahi et al. 1979), as indeed may the activity at different sites (see below), so that the size of the NOR's may differ between different chromosomes. The DNA's coding for the 5 s rRNA are present as a cluster of 60–90 copies (Coote et al. 1980) on the long arm of chromosome 1 (Steffensen et al. 1974) and their location is not indicated by any associated and obvious morphological feature at that site.

Cytological techniques for staining NOR's (Matsui and Sasaki 1973), and particularly the silver staining technique introduced by Goodpasture and Bloom (1975) and Howell et al. (1975), have been shown to depend upon interaction with acidic proteins located at the NO. Moreover, few or no such proteins are associated with NOR's that have not been involved in transcription and nucleolus formation so that the presence of silver staining typifies the presence of an NOR that was involved in active transcription in the preceding interphase. This correlation was especially clearly demonstrated in the experiments of Miller et al. (1976a,b) on human/murine cell hybrids where NO silver staining only occurred on human NOR's when human rRNA was made and murine chromosomes lost, whereas only mouse chromosomes showed NOR function and silver staining in hybrids which lost human chromosomes. These authors have further shown (Miller et al. 1981) that suppression of rRNA synthesis may, at least in some cases, be a consequence of methylation of the rDNA (see Sect. 15). It should be noted that in any given human cell only a proportion of the total rDNA sequences are involved in transcription. In general, the number of silver staining NOR's in a cell usually ranges between 5 and 10 and the pattern is the same in a given cell type within an individual (Mikelsaar et al. 1977a); may differ between tissues (Reeves et al. 1980); is heritable between cells (Ferraro et al. 1981); and, as indicated by family and twin studies, is heritable between generations (Mikelsaar et al. 1977b; Taylor and Martin-de Leon 1981).

Numerous studies have shown that in man, as in other organisms, fusion takes place between nucleoli during interphase, so that chromosomes bearing active NOR's become associated in space and appear as "satellite associations" at metaphase (see Schwarzacher and Wachtler 1983). This close spatial association of homologous DNA sequences located in non-homologous chromosomes might be expected to predispose to spontaneous exchange between chromosomes at these sites and the fact that such changes do occur is evidenced by the occasional observation of an exchange of polymorphic marked satellites between D and G group chromosomes. There is no evidence to suggest, however, that the rDNA regions are more prone to involvement in exchange, or in any other type of chromosome aberration, following exposure of cells to mutagens.

## 8 Other Secondary Constrictions and Fragile Sites

In addition to the primary constriction of the centromere and the secondary constrictions of the nucleolar organisers, there are various other secondary constriction sites which appear at fixed positions in metaphase chromosomes of the human chromosome complement (Ferguson-Smith and Handmaker 1963). Some of these other secondary constrictions are apparent only in certain individuals, whereas others are more consistent features of the karyotype and some of these are associated with C-band heterochromatin whereas others are not. On the basis of the association between the constrictions and regions of rapidly transcribing rDNA in the NOR regions, it is possible that some of these other secondary constrictions that are not C-band regions may represent other sites of active transcription and the multiple copies of sequences coding for the tRNA's seem likely possibilities, however, no evidence concerning this suggestion has been forthcoming.

In recent years there have been an increasing number of reports of secondary constrictions that are mechanically weak and prone to breakage, either in the living cell or as a consequence of culture conditions in vitro or of methods used to produce chromosome preparations, and these have been referred to as "fragile sites". Since the original report by Dekaban (1965) there have been numerous publications reporting the presence of fragile sites in at least ten different chromosomes in the complement. These sites are fixed, permanent and heritable properties of the chromosome phenotype and provide useful markers in family studies. Sutherland (1977, 1979) originally showed that the expression of some of these fragile sites in lymphocyte chromosomes was dependent upon the composition of the tissue culture medium used and indeed there is little evidence that these fragile sites are present in cells in vivo as opposed to cells in culture.

Interest in fragile sites has been especially stimulated by the association of one form of fragile site, on Xq28, with a particular form of X-linked mental retardation (Lubs 1969; Harvey et al. 1977; Jacobs et al. 1980). The expression of this, and indeed many other, fragile site is very dependent upon culture conditions and these sites are best expressed in cells grown in tissue culture media deficient in folic acid and thymidine. Beek et al. (1983) have shown that the incidence of lymphocytes with micronuclei increases when such cells are cultured under conditions of folic acid deprivation, implying that fragile sites undergo breakage in such cells cultured in vitro. There is, however, no information on the nature of the DNA's and chromosomal proteins associated with these folic acid-dependent fragile sites (see Chap. Obe and Beek, p. 177ff.).

# 9 Replicating Chromatin

A great deal of evidence on mutation and chromosome aberration induction following exposure of cells to chemical mutagens, has shown not only that DNA lesions may be converted into mutations and chromosome structural rearrangements as a consequence of the DNA replication process, but also that cells are most sensitive to the induction of mutational damage (Tong et al. 1980) and chromosomal structural change (Evans 1977b), if exposed during the early part of the replicating, or S, phase of the cell cycle.

At the DNA level, replication appears to be controlled at two levels. First, the positively staining G- and C-band chromosome regions (see Sect. 4), together with the inactive X chromosome in female cells, undergo replication late in the S phase of the cycle – and this is a general property of heterochromatin. Second, replication is initiated, and proceeds bidirectionally, at a large number of sites in each chromosome so that there are several thousand replicating units, or replicons, in each chromatid (see Evans 1977c, for review). The organisation of replicons is a matter of some speculation and they may be equivalent to the supercoiled DNA loops associated with the nuclear cage and referred to earlier (McCready et al. 1980). Recently, Heintz et al. (1983) have made use of a Chinese hamster cell line containing 1,000 copies of a 135 kb early replicating sequence which includes a DNA replication initiation site within each of the amplified sequences. These authors have cloned, and restriction mapped, the initiation site for replication for each amplicon and show that all derive from a single locus within each repeat, but whether other replication initiation sites for other regions in the genome have a similar sequence organisation, or indeed whether a wide variety of different sequence organisations can serve as replication initiation sites, is unknown.

DNA replication is closely coupled with histone synthesis (Detke et al. 1979) and new histone beads become associated with nascent DNA at the replication fork (Worcel et al. 1978), so that there are no long strands of protein-free DNA (McKnight and Miller 1977). Each histone octamer is a conserved unit that does not consist of a mixture of old and new histones and is stable over a number of cell cycles (Leffak et al. 1977). The distribution of these histone beads as between the two daughter semi-conservatively segregating DNA molecules is non-random, but each strand receivers a mixture of old and new histone octamers (Tsanev 1982).

Little is known concerning the duplication and segregation of structural non-histone proteins, but they presumably retain their role as core proteins during replication and may well be also responsible for disposing daughter DNA templates to the outside of the replicating complex. This regular disposition is evident in harlequin-stained diplochromosomes and parallels the mesosome directed segregation of duplexes in bacteria (see Evans 1979).

The changes in the organisation of DNA and its associated proteins during replication will of themselves alter the probability of DNA damage ensuing following mutagen exposure. Replicating chromatin is indeed characterised by its increased sensitivity to digestion with micrococcal nuclease and DNAase I (Cremisi 1979), presumably because of an increased access to internucleosomal linker DNA in relaxed solenoids and the exposed DNA's at the sites of replication forks. However, an enhanced mutagen sensitivity during the S phase of the cycle does not solely follow from the altered structure of replicating chromatin.

Various studies have shown that a gross imbalance in nucleotide pools may result in an increased "spontaneous" mutation frequency (Kaufmann and Davidson 1979) and the presence of excess thymidine, or its analogues, enhances the response of cells to mutation induction by alkylating agents (Peterson and Peterson 1979). Of interest in this connection are three recent reports demonstrating that mutagen-induced damage to nucleotides in precursor pools may occur far more readily than in the protected bases of the packaged DNA and that such bases may be incorporated into replicating DNA. Topal and Baker (1982) and Topal et al. (1982) first showed that methylation at the N-1 position of adenosine following exposure to N-methyl-N-nitrosourea occurs some 13,000 times more frequently than methylation at the same site in duplex DNA and their data show that DNA precursor nucleotides have a very considerably greater availability for alkylation than incorporated bases. These reports were followed by Dodson et al. (1982) who went on to show, using T7 bacteriophage and $O^6$-methyldeoxyguanosine triphosphate, that $O^6$-methylguanine is incorporated into newly synthesised DNA and results in mutation frequencies some 10 to 70 times higher than in controls. These experiments emphasise that S phase sensitivity increases to mutagens in human cells may not necessarily reflect only changes in chromosome structure, but may also involve the incorporation of mutagen damaged bases into replicating DNA.

## 10 Transcribing Chromatin

There is much evidence to indicate that one of the most sensitive phases in the life of a cell to the induction of mutations and chromosome structural change is the DNA replication phase. The possibility is also entertained that chromatin exposed to mutagens whilst being actively transcribed is also at greater risk to external mutagens than genetically quiescent chromosome regions, but there is no direct evidence in support of this contention. Less than 10%–20% of the DNA in the human genome is ever transcribed, and at any one time only a minority of this DNA is in a form that may even predispose it to transcription, let alone be actively involved in the process. Nevertheless, by analogy with what is known about the structure and organisation of transcribing chromatin in for example the lampbrush chromosomes of amphibia (Miller and Hamkalo 1972), the rDNA regions in chromosomes of the fruit fly (McKnight and Miller 1976), or the puffs in dipteran polytene chromosomes (Daneholt et al. 1976), then it is to be expected that considerable structural changes may be associated with those regions of the human genome that are involved in transcription in the interphase nucleus.

Nuclease digestion experiments with pancreatic DNase I on a variety of eukaryote species show that actively transcribed genes are sensitive to cutting whereas inactive genes are not (Weintraub and Groudine 1976; Mathis et al. 1980; Chung et al. 1983). Many of the DNase I hypersensitive sites are tissue and developmental stage specific and must reflect an aspect of chromosome structure associated with function in the sense of development or differentiation, e.g. as in the case of the heat shock genes in *Drosophila* (Wu et al. 1979) or in the immunoglobulin kappa light chain genes in the mouse (Chung et al. 1983). However, DNase sensitivity is not a prerogative of active genes (Miller, Turner et al. 1978). Studies with DNase II, which primarily cuts between nucleosomes, shows that the conformation of active nucleosomes differs from the bulk of nucleosomes and that this may be due, at least in part, to the binding of the HMG proteins 14 and 17 (Weisbrod et al. 1980). These two proteins are clearly involved in the transcription process and fluorescent labelled antibodies to them reveal that they are specifically associated with transcriptionally active puffs in salivary gland polytene chromosomes (Mayfield et al. 1978). Quite a lot of evidence implies that part of the transcriptional process may involve the creation of a nucleosome-free region proximal to the polymerase complex (e.g. Scheer 1978; Lamb and Daneholt 1979) so that there is a change in both milieu and structure of the chromosome in these regions which may result in a diminished level of protection against the actions of mutagens at that site. Of particular relevance in the context of mutagenesis is that studies on the chick globin gene have revealed the presence of single strand nicks at the 5' end of the transcribing gene that are sensitive to cutting with single strand specific $S_1$ nuclease; whether such nicks render the DNA at that site more susceptible to induced scission or mutagenesis following exposure to mutagens is not known.

## 11  Z-DNA

Although fibre diffraction studies on DNA show the presence of a variety of different molecular forms, the structure in solution has usually been considered to be a regular B form. Studies in recent years, however, have seen quite profound changes in our view of DNA structure and it is evident that the DNA duplex can exist in a variety of physical states in addition to the simple right-handed B form of the classical Watson and Crick model. Of special interest has been the structure that is referred to as Z-DNA (Wang et al. 1979).

On the basis of studies on the crystal structure of hexanucleotides of d(CpG), Wang et al. (1979) showed that this kind of deoxypolymer, with the simple alternating purine-pyrimidine sequence, crystallised as a left-handed double helical molecule with Watson-Crick base pairs and an anti-parallel organisation of the sugar phosphate chain. The left-handed stacking was then shown by Arnott et al. (1980) to be apparent in DNA fibres containing poly(dG-dT)·poly(dA-dC) and it has become evident that Z-DNA is a reversible structural form which is only entered into by tracts of DNA containing regularly alternating purines and pyrimidines. The z-conformation is stabilised in vitro under conditions of high salt, or under conditions of low salt if the DNA is methylated at the C5 position as was demonstrated in tracts of poly(dC-dG) (Behe and Felsenfeld 1981). Singleton et al. (1982) further showed that Z-DNA formation is induced by supercoiling in the plasmid pRW751 under physiological conditions of ionic strength and super helical density. Thus although Z-DNA is normally unstable in physiological salt conditions in vitro, it clearly can be stabilised by methylation, by supercoiling and by the presence of cationic polyamines, such as spermidine, that are normally present in nuclei.

Z-DNA is readily demonstrated with synthesised oligonucleotides in vitro, but recent evidence indicates that this form of DNA may also be present in vivo. Rich and his colleagues (Nordheim et al. 1981) purified Z-DNA antibodies from sera from immunised rabbits and, by indirect immunofluorescence, showed that these antibodies bound specifically to the interband regions of *Drosophila* salivary gland chromosomes. Binding occurred even in the presence of poly (dG-dC)·poly(dG-dC) in the B conformation, but no binding was evident if the chromosome preparations were preincubated with Z-forms of poly (dG-dC) DNA. These results thus imply that Z-DNA is present and that it is more abundant, or more accessible, in the interband regions of *Drosophila* chromosomes and it is these regions that are known to contain the structural genes in this species. Nordheim et al. (1982) have also recently shown that there exist in *Drosophila* classes of proteins that have the ability to recognise and bind specifically to Z-DNA. At least four major protein species are involved and it is suggested that they are likely to play a role in stabilising Z-DNA within the genome. If Z-DNA may be involved in the regulation of gene expression (see below) then such proteins may well have an important function in regulation.

Of particular interest in terms of chromosome structure is the recent report by Walmsley et al. (1983) utilising a d(TpG) hybridisation probe in studies on the yeast genome. These authors describe the presence of at least 30 regions of potential Z-forming tracts, in the form of $d(TpG)_n$, dispersed throughout the genome with one subset of these sequences located at the telomeric ends of the yeast chromosomes. Whether potential Z-DNA sequences are characteristic of telomeric regions of eukaryotic chromosomes in general is, however, yet to be established.

Are Z-DNA sequences relevant to man? Hamada and Kakunaga (1982) have recently shown that a cloned human cardiac muscle actin gene has an intron with a sequence of 50 alternating dT and dG residues and using this, and a probe specific for poly(dT-dG), they have shown that the human genome contains a very large number (up to $10^5$) of highly dispersed potential Z-DNA forming sequences. That such sequences may be present in a Z form is implied by the recent report by Viegas-Pequignot et al. (1982) who used anti Z-DNA antisera, similar to those utilised on *Drosophila* chromosomes, on the metaphase chromosomes of the gerbil. In this work it was shown that the antibodies bound relatively specifically to certain heterochromatic regions in the complement and preliminary studies on human chromosomes indicated an affinity for the large heterochromatic region on chromosome 9. However, the authors note that various euchromatic parts of the chromosomes also bound the antibody and that there was no correlation between the richness of G–C bases and the fixation of the anti-Z-DNA antibody.

It is evident that demethylation of Z-DNA renders it unstable and makes it revert to its B form. If Z-DNA exists throughout the genome, then its demethylation would allow unwinding of B-DNA at some distance removed from the Z region, or it could allow for the binding of DNA binding proteins, and in this way could act as a trigger for transcription. There is evidence that the junctions between these Z and B forms of DNA are sensitive to S1 nuclease and that there may be a single stranded break at those sites (Singleton et al. 1982): whether this would predispose to spontaneous or mutagen-induced chromosome structural rearrangements being preferentially involved at such sites is unknown, but Z-DNA could be very important in terms of being responsible for genetic position effects in chromosome translocations.

In the context of mutagenesis it is also important to note that a significant difference between B-DNA and Z-DNA is that in the latter form there is a vastly increased accessibility of the guanine N7 and C8 atoms. There are many mutagens that are relatively specific for reacting with the O6, N7, or C8 sites in guanine so that Z-DNA structure should increase the reactivity of guanine considerably. It should also be noted that the GC base pair forms part of the outer surface of the Z-DNA, so that the C5 position of cytosine is far more accessible to any enzyme than it is in B-DNA.

## 12 Mobile DNA's

The classical studies of McClintock on transposable controlling elements in the corn plant *Zea mays* (McClintock 1952) were the first to indicate the presence of chromosomally located mobile and transmissible elements in the genome which induced chromosome breaks, duplications, a variety of other structural rearrangements, as well as changes in gene expression at many loci. These observations were made before the first demonstrations of mobile DNA's in bacteria, in the form of plasmids and bacterial viruses (bacteriophage), which could insert themselves into the genome, but also detach themselves completely and replicate independently in the cell (see Hayes 1968). The realisation that processes similar to those in corn plants also occurred in bacteria, was then stimulated by the original demonstrations of Jordan et al. (1968) and Shapiro (1969) that certain highly pleiotropic mutations in the *gal* operon of *E. coli* were a consequence of the insertion of discrete DNA segments (Insertion Sequences, or IS's) that were mobile within and between genomes and were thus able to jump from one chromosomal site to another. It later became evident that closely spaced pairs of IS's could move as units (Transposons or Tn's) carrying along the DNA sequences that lay between them and thus acting as effective gene transferers.

The possibility of a more widespread occurrence of mobile DNA segments in eukaryotes followed directly from studies on the integration, and its consequences, of viral sequences into mammalian genomes, and from the genetic studies, originally by Green, on highly mutable loci in *Drosophila* and, in particular, on those elements operating at the white locus (Green 1973, 1976). Detailed studies on *Drosophila* led to the identification of, initially, two repeated gene families, *copia* and *412*, that were present as 100–200 copies of 5 kb stretches of DNA in single genomes and whose number and chromosomal locations varied between strains and individuals, thus pointing to their mobility within and between genomes (Finnegan et al. 1978; Rubin et al. 1981). A number of *copia*-like transposable elements have now been described (Spradling and Rubin 1981, 1982); they account for more than 5% of the total genome of *Drosophila melanogaster* and they share several important structural features in common with the integrated proviral form of retroviruses (see below). *Copia*-like elements have direct repeat sequences a few hundred base pairs long at each end and these long terminal repeat sequences are themselves terminated in short inverted repeats of around ten-base pairs long (Rubin et al. 1981). Moreover, they can exist in the cytoplasm as circular DNA's (Flavell and Ish-Horowicz 1981) and are transcribed into long polyadenylated RNA's. Although no infectious retroviruses have been found in *Drosophila*, Shiba and Saigo (1983) have recently described the presence of virus-like particles containing *Copia* RNA in *Drosophila* cells maintained in tissue culture. These particles are in many ways similar to the intracisternal A-type particle of mouse cells (see below) in that they are non-infectious, transcribed, dispersed, repeated DNA sequences, that have a reverse transcriptase activity, are transposable and cause mutations by inserting into genes within their near vicinity.

Dispersed repeated DNA sequences similar to the *Drosophila Copia*-like elements were originally reported by Cameron et al. (1979) to be also present in yeast. Further studies on yeast, and particularly in the *his*4 locus, have shown that transposable elements can result in gene inactivation due to insertion (Fink et al. 1981), or can transfer the *his*4C gene into an essential gene located on a different chromosome where it then acts as a recessive lethal (Greer et al. 1981). Indeed, the studies on the *his*4 locus in these mutant yeast strains bearing *Tyl* transposable elements shows that these elements cause a high frequency and wide variety of chromosome aberrations including translocations, deletions, duplications, and transpositions (Roeder and Fink 1980).

A major group of DNA sequences that come under the heading of mobile DNA's in mammalian cells are the proviral forms of the retroviruses. These are RNA viruses that have a reverse transcriptase activity and replicate through DNA intermediates to give double stranded DNA reverse transcripts, termed proviruses, which integrate into host cell DNA (Temin 1976). Analyses of both integrated and free proviral DNA's show that they have a short inverted repeat sequence at each terminus of the linear molecule which extends into a longer directly repeated terminal sequence (Long Terminal Repeat, or LTR) which may be from 300–1,300 bp in length. The LTR sequences contain a promoting TATA box region, a polyadenylation signal and a poly(A) acceptor sequence. Proviruses integrate as intact elements at a variety of sites in the host genome and, as in the case of bacterial IS and *Drosophila* and yeast transposable elements, integration is accompanied by duplication of a small number of host sequences on either side of the inserted provirus. Temin (1974) originally proposed that retroviruses evolved from movable genetic elements that are part of the normal cellular genome, so that retroviruses could be considered as transposons that had evolved such that they could escape from one cellular genome and enter another one. The striking structural and functional resemblances between the retroviruses and transposable elements from bacteria, yeast and *Drosophila* certainly point to common requirements and indeed to a common origin (Temin 1980). One possibly significant difference, however, is the recent demonstration by Collins and Rubin (1983) that members of the fold-back transposable element families in *Drosophila* contrast with prokaryotic transposable elements in showing a high frequency of *precise* excisions – a feature which is of rare occurrence in prokaryotic species.

Endogenous retroviral sequences are found in multiple DNA copies in the chromosomes of probably all vertebrates where they may make up a considerable part of the genome. For example, in the mouse it is estimated that something like 0.3% of the total genome may have C-type retrovirus related sequences and up to 100 proviral copies have been demonstrated to exist in the mouse Y chromosome (Phillips et al. 1982). They are transmitted vertically in the same way as Mendelian genes, in contrast with the horizontal transmission of exogenous viruses, and in most cases it would appear that new proviral copies appear to be a consequence of exogenous infection for there is but little evidence for transposition of retroviral sequences within the host genome in the absence of RNA intermediates and an infectious cycle. Infection, however, can result in integration in the DNA of germ line cells with very high frequencies (Jaenisch et al. 1981).

The fact that proviruses are capable of insertion into a variety of different regions in the host genome means that on occasion their integration may be a mutagenic process because of the simple physical disruption of functional host DNA sequences. Retroviruses are also capable of affecting expression of cellular genes by enhancing transcription from a proximal position under the influence of the LTR. Hayward et al. (1981) and Blair et al. (1981) showed that insertion of an LTR could activate the expression of adjacent cellular *onc* genes and lead to a malignant transformation and Varmus et al. (1981) showed that integration of Moloney leukaemia virus into the *src* gene of an RSV-transformed cell line influenced *src* transcription and resulted in reversion of the transformed phenotype. A similar activation of a cellular oncogene, in this case the *c-mos* gene in a mouse myeloma, was recently demonstrated following insertion of an IS-like element immediately 5' to the junction with *mos* (Rechavi et al. 1982). The influence of endogenous retroviral sequences on genes not involved in malignant transformation has also been observed. Jenkins et al. (1981) have shown that a spontaneous coat-colour mutation in mouse is associated with insertion of a retroviral genome, and a recessive-lethal mutation leading to early embryonic death has been induced experimentally by insertion of Moloney leukaemia virus into the germ line of mice (Jaenisch et al. 1983).

Initial interest in the interactions between retroviruses and their host genomes centred on their involvement in the initiation of malignant tumours in chickens and mice. More recently there has been very considerable interest following from the demonstration that the oncogenic transforming sequences in these acutely transforming retroviruses, referred to as *v-onc* genes, are identical or closely similar to normal genes in vertebrate, including human, hosts which in a normal state may be involved in differentiation, but under certain circumstances are involved in malignant transformation and are referred to as *c-onc* genes (Cooper 1982; Santos et al. 1982). As we have referred to elsewhere (see Sect. 14), retroviruses have the ability to acquire and transduce foreign genes as a consequence of incorporation of mRNA into virus particles, reverse transcription and recombination. In consequence, the cellular gene loses its introns and it would appear that by this mechanism non-transforming retroviruses have acquired oncogenic potential by recombining with cellular genes that are involved in differentiation. To date at least 12 different *c-onc* genes have been identified in the human genome and their positions assigned to, or located within, 10 different human chromosomes (Evans 1983).

Many of the retroviral sequences have been inserted over evolutionary time periods and are stable, others such as some of the C-type viruses in mice show a diversity of integration sites indicating the repeated occurrence of germ line infections and Jaenisch et al. (1981) have shown that exposure of mouse embryos to exogenous Moloney leukaemia virus results in the incorporation of inherited proviral sequences in high frequency. Inserted sequences which are apparently stable could well perhaps become less stable in cells subjected, for example, to the stresses of mutagen exposure and one group of stable inserted elements in the mouse, the intracisternal A-particle genes (IAP's) may be of interest in this connection. The IAP's are retroviral-like elements, but appear to have no infectious or recognised extracellular phase, are very similar to the virus-like *copia* RNA particle in *Drosophila*, and they are present in about 1,000 copies per genome. Recently, Kuff et al. (1983) have shown that these particles are responsible, by insertion, for mutations involving the $\kappa$ immunoglobulin light chain gene in mouse hybridoma cells, and emphasise that the family of IAP-related genes must be considered as potential movable elements and a source of insertional mutations in the mouse.

The human genome contains at least 400,000 copies of a 300 bp repeat sequence that acts as a substrate for the *Alu* restriction enzyme and are referred to as *Alu* repeats (Schmid and Jelinek 1982). Since the genome contains $3 \times 10^9$ bp then these repeats must be one of the most predominant groups of repetitive sequences in the human genome and account for approximately 4% of the total DNA. *Alu* repeats are present in all mammalian genomes and a 40 bp sequence appears to have been conserved throughout mammalian evolution. A 14 bp segment in the repeat is homologous to a sequence near to the replication origin of several DNA viruses and the repeat is present in transcripts generated by both RNA polymerase II and III. Thus, a role for the *Alu* repeat sequences in the initiation of DNA replication and in nuclear RNA processing has been postulated (Jelinek et al. 1980; Calabretta et al. 1981). Of interest in the context of mobile DNA, however, is that most *Alu* elements are flanked by direct repeats of sequences some 7–20 nucleotides long (Haynes et al. 1981; Grimaldi and Singer 1982) and since they may be transcribed by RNA polymerase III, the promotor sequences are themselves transcribed: thus the DNA's derived from these transcripts can themselves be transcribed by RNA polymerase III. The *Alu* repeats are therefore prime candidates for mobile elements within the genome and evidence in support of their mobility has come from studies on African green monkey aDNA satellite sequences (Grimaldi and Singer 1982) and on spacer sequences in mouse ribosomal genes (Kominami et al. 1983). Of particular interest is the possibility that pairs of *Alu* sequences might be able to act as pairs of insertion sequences in being able to transfer intervening DNA segments, *inter-Alu* DNA, around the genome, in other words acting in exactly the same way as bacterial transposons. Some evidence for the support of this possibility has recently been obtained.

Small polydispersed circular DNA's which are not of mitochondrial origin have been observed in a variety of vertebrates and mammals including monkey (Krolewski et al. 1982) and man (Smith and Vinograd 1972). In a recent study of various human cell types and tissues, Calabretta et al. (1982) have shown that both chromosomal and extra-chromosomal forms of *inter-Alu* DNA in normal and neoplastic human tissues show considerable inter-tissue and inter-individual polymorphisms. In an extension of this work Schmookler Reis et al. (1983) have further shown that a DNA sequence situated between *Alu*-repeat clusters is progressively amplified in extra-chromosomal DNA circles with the ageing of human cells cultured in vitro and with cell ageing in vivo. As indicated, an *inter-Alu* sequence is structurally analogous to prokaryote transposons, but although the case for transposability of *inter-Alu* sequences is strong, such transposition has yet to be definitively demonstrated. Indeed, in considering the number, type and distribution of *Alu*-like elements in man and mouse Sharp (1983) has pointed out that the duplication and insertion of *Alu* elements has been, and is, continuously occurring, but the rate of change is slow and these events may not constitute a significant mutation liability.

The receptiveness of the genome for integration of extra-genomic DNA is of course exploited in the numerous studies on gene transfer in mammalian cells. Genes have been introduced into cells following co-precipitation with calcium phosphate and "passive" uptake (Wigler et al. 1979), by microinjection (Capecchi 1980), following cell fusion (Ruddle 1981), by packaging in liposomes (Schaefer-Ridder et al. 1982) or in viral vectors (Mulligan et al. 1979) or in empty capsids of polyoma virus (Slilaty and Aposhian 1983). In the case of DNA-mediated gene transfer, a major limiting factor has been shown to be cellular uptake and it is clear that within the cell the inter-molecular recombination and ligation of introduced DNA's occurs with an extraordinarily high efficiency (Folger et al. 1982; Miller and Temin 1983). This is by no means surprising, since it is becoming evident that an ever increasing number of species make considerable use of programmed genomic rearrangements as a normal part of development, differentiation, and adaptation. These include the DNA rearrangements involved in: (1) antigenic variation of flagella antigens of *Salmonella* (Silverman and Simon 1980); (2) surface antigens of the protein coat in trypanosomes (Williams et al. 1979; Bernards et al. 1983); (3) the transposable elements in yeast which are blocks of structural genes, or "casettes", whose position in the genotype determines whether or not they are expressed and which are the controlling elements in determining mating type (Hicks et al. 1977); and (4) the well-known DNA rearrangements in the immunoglobulin gene clusters of B lymphocytes in mammals which result in antibody diversity (Leder et al. 1980; Rabbitts et al. 1980; Tonegawa et al. 1980).

Although much emphasis has been placed upon the RNA viruses and retrovirus-like elements in terms of their mobility within and between genomes, we must not forget that infectious DNA as well as RNA viruses have also long been known to play an important role in inducing transpositions within the genomes of infected cells. Following the original demonstration that the DNA virus *Herpes simplex* induced damage in the chromosomes of infected Chinese hamster cells (Hampar and Ellison 1961), infections with a wide spectrum of DNA as well as RNA viruses were reported to be associated with the development of chromosomal aberrations in infected cells (Nichols 1970). These included among the DNA virus groups various adenoviruses, herpesviruses, papavoviruses and poxvirus and these, together with a whole range of RNA viruses – including those responsible for some common childhood infections, e.g. mumps and measles – have all been recorded as being responsible for the induction of a variety of chromosomal aberrations (Harnden 1974). In many instances it would appear that the chromosome aberrations develop in lytically infected cells which are destined for an early death, but aberrations are also present in cells that are non-permissive for the virus in question and which may not acquire malignant properties, as well as in those cells that undergo viral transformation and eventually acquire a malignant status.

In a number of instances it would appear that persistence of the virus genome within the host cell results in a continuing genomic instability that is usually consider- ed to be a consequence of viral insertion and recombination with host DNA. Molecular evidence for such insertions and associated host DNA rearrangements have been well documented for a number of viruses. For example, in the case of the DNA papavovirus SV40, transformed rodent cells have been shown to contain variable numbers of copies and sizes of SV40 genomes that are combined with host sequences and integrat- ed at various sites in the genome, with the integration involving extensive amplifica- tion and rearrangement of the viral DNA as well as alterations in associated host DNA sequences (Botchan et al. 1980; Clayton and Rigby 1981; Sager et al. 1981). Integra- tion may be random in the genome and although no single DNA sequence in either the host or virus genomes serves as an obligatory site of integrative recombination, Stringer (1982) has shown that homologous sequences of no more than 6 bp length are necessary as between the host and viral genomes and that the integration process generates chromosomal deletions in the host genome of at least 3 kb in length. It may be of interest to note that in contrast with the RNA retroviruses, and DNA viruses in the adenovirus (Steenbergh et al. 1977) and herpesvirus (Sheldrick and Berthelot 1977) families, SV40 DNA is not bordered or flanked by inverted terminal repeat sequences. It is not known whether these terminal repeats are important for the integration of the large viruses such as the Epstein Barr virus of the herpes group, but recent evidence on EB transformed human lymphoblastoid cells show that probably complete viral genomes are integrated into host DNA, and that integration does not involve host sequences homologous to defined internal repeats within the viral genome (Henderson et al. 1983).

The facts that many mutations in prokaryotes are consequences of the integration of transposable DNA segments and that infectious and non-infectious segments similar in structure and behaviour to bacterial transposons exist in eukaryotes, raise two questions of practical importance to those interested in spontaneous and induced mutagenesis in man. First, to what extent do "spontaneous" mutations in man owe their origin to transposable DNA segments? And, in particular, are they involved in any of the inherited chromosomal instability syndromes (Ray and German 1981), or in those conditions which show particularly high mutation rates (e.g. neurofibro- matosis or X-linked non-specific mental retardation, where mutation rates are of the order of $1 \times 10^{-4}$ per gamete). Second, are any of the mutations that may result following exposure to a known physical or chemical mutagen a consequence of activation, relocation and reinsertion of potentially mobile DNA segments within the genome?

Direct information on the first of the above questions is completely lacking, but the inherited chromosomal instabilities, where the instability is not associated with any known infectious virus, would seem to be prime candidates for study. With regard to the second question it has frequently been noted that X-ray induced leuka- emias in certain mouse strains are associated with X-ray activation of provirus present in the irradiated cells (e.g. Wald et al. 1964) and activation of infectious EB virus can be readily induced in human lymphoblastoid cells exposed to the base analogues 5-bromo deoxyuridine (Gerber 1972) or 5-azacytidine (Ben-Sasson and Klein 1981), or even to the tumour promoting phorbol ester TPA (Zur Hausen et al. 1978). It would seem then not unlikely that such treatments with base analogues or other mutagens might possibly result in mutations not only because of their direct inter-action with coding DNA sequences at the mutational sites, but as a consequence of their possible effect in initiating the redistribution of endogenous, and potentially mobile, DNA segments within the genome. Evidence for such an effect involving a transposable element at the white locus in *Drosophila* has been obtained from studies on larvae exposed to X-rays or to ethyl methane sulfonate (Rasmuson et al. 1978), but whether transposable elements may play a role in any enhanced sensitivity of mammalian cells and/or genes to mutagens has yet to be established.

## 13 Gene Amplification

Sequence amplification is not a prerogative of functionally inert (or apparently inert) simple sequence repetitious DNA, but is evident at all levels of DNA organisation. Indeed, the phrase "unique sequence DNA" is in many cases something of a misnomer, particularly in view of the fact that many genes are present as clusters, or families, of closely identical sequences and that processed copies of many apparently single genes appear to be dispersed throughout the genome (see "pseudogenes"). However, of special interest in terms of chromosome structure are those functional DNA sequences that are transcriptionally active and which become amplified in response to a variety of intrinsic or extrinsic stimulii.

Classical examples of amplified sequences in the human genome that are transcribed and functional are of course those that code for the 18 s + 28 s and the 5 s ribosomal RNA's. The former are coded for by sequences located at the nucleolar organizer regions of the 5 pairs of acrocentric autosomes (Evans et al. 1974) each containing an estimated 3 to 4 copies of the 18 s + 28 s genes (Cote et al. 1980), and the latter are confined to a distal region on chromosome 1q (Steffensen et al. 1974) and are present as a cluster of around 60-90 copies (Cote et al. 1980). In these cases the amplification is a consequence of tandem duplication of sequences, the amplification normally having occurred over a long time span in evolutionary terms and in response to the requirement for a large amount of gene product necessary for the efficient production of cellular proteins. Amplification can, however, occur over very much shorter time periods and in response to demands upon the cell for the increased production of a specific gene product.

In 1968 Brown and Dawid demonstrated a specific amplification of ribosomal genes in *Xenopus* oocytes – a developmentally programmed process resulting in the accumulation of a large number of ribosomes to enable protein synthesis to occur at a rapid rate following fertilisation. This was followed by the demonstration by Meneghini et al. (1971) that administration of the hormone ecdysone to *Rhyncosciara* larvae resulted in selective amplification of regions of salivary gland polytene chro-mosomes in the form of DNA puffs. The first evidence for specific amplification of DNA sequences in mammalian cells followed from the discovery of homogeneously stained regions (HSR) that developed in chromosomes of human or Chinese hamster cells surviving a selection regime involving stepwise increases in the antifolate drug methotrexate that was associated with a concomitant overproduction of the target enzyme dihydrofolate reductase (Biedler and Spengler 1976). This elevation of enzyme concentration was then clearly shown to reflect amplification of the sequences coding for the enzyme in question (Alt et al. 1978). The amplified genes can be either stable or unstable when cells are subsequently grown in the absence of methotrexate. Stable cells are usually characterised by the presence of HSR, but in unstable cells the amplified genes are present in extrachromosomal self replicating elements which lack centromeres and are referred to as double minutes (DM's) (Balaban-Malenbaum and Gilbert 1977; Kaufman et al. 1979). DM's had previously been recorded in mouse (Hellström et al. 1962) and human (Cox et al. 1965) malignant cells and their instability was first clearly demonstrated by Donner and Bubenik (1968) and by Levan et al. (1977) in a mouse tumour cell line which when grown in an ascitic form contained large numbers of DM's which were lost when the cells were cultured in vitro. DM's and HSR's have since been shown to be present in a wide variety of malignant human cells (Hartley and Toolis 1980), but their association with specific amplified genes and gene products in these cells has yet to be documented. In addition to DM's, Levan et al. (1978) have also reported the presence of additional "C-bandless chromosomes" (CM's) which have centromeres and are more stable than, but in other respects are similar to, DM's in the mouse tumour cell lines.

The implications are that HSR, DM's, and CM's may all be cytological manifestations of gene amplification in response to extrinsic selection factors in the case of cell culture experiments, or of inherent environmental conditions in proliferating malignant cells in vivo. Various mechanisms may be involved in such gene amplification including; differential replication resulting in tandem duplications within the chromosome; differential replication of the "onion skin" type resulting in extrachromosomal copies of the amplified sequences; excision and extrachromosomal replication; or structural transpositions followed by, in the case of DM's, unequal segregation at mitosis. Each of these mechanisms may be involved under different circumstances and examples of tandem duplications include the ribosomal gene compensation seen in the bobbed mutants of *Drosophila* (Endow 1980), which contrasts with the extrachromosomal amplification of ribosomal genes in *Xenopus* oocytes that we have already referred to. It may also be readily imagined that if the incidence of spontaneous chromosome breakage and rearrangement in cells struggling to survive under strenuous conditions is not negligible, then rearrangements which involve duplications of sequences favouring cell survival will be selected. Such rearrangements do not need to be precise and for example in the case of methotrexate selection the unit of amplification has been shown to differ between different cell clones and to contain DNA additional to those sequences that specifically code for dihydrofolate reductase (Bostock and Tyler-Smith 1981).

The demonstration of the amplification of the dihydrofolate reductase gene under conditions of intense selection, prompted the question of whether this phenomena was a regular cellular response to unusual demands for a much increased level of gene product on the part of other genes in the genome? This may well be the case for Wahl et al. (1979) have reported that selection for resistance to an inhibitor of aspartyl transcarbaminase in Syrian hamster cell lines is the result of an up to 200-fold amplification of the DNA coding for this enzyme. A similar amplification has also been reported for the metallothionein I gene in cadmium resistant cultured mouse cells (Beach and Palmiter 1981) and for the HPRT gene in a thioguanine resistant mouse neuroblastoma cell line (Melton et al. 1981). Of more general relevance, however, are the ever increasing number of demonstrations of gene amplification in response to specific requirements associated with development and differentiation and examples include amplification of a group of genes specifying chorion (egg shell) proteins in *Drosophila* follicle cells (Spradling and Mahowald 1980) and of actin genes during myogenesis in the chick (Zimmer and Schwartz 1982).

In the context of chromosome structure, the phenomenon of gene amplification once again underlines the propensity for plasticity of the genome architecture and although there are, as yet, only a few clearly documented cases where additional gene copies appear, and indeed disappear, rapidly in response to a transient requirement during development, it would seem not unlikely that gene amplification may not be an exceptional phenomenon. Indeed, regulation of gene expression as a consequence of gene dosage, as opposed to modulation of transcription by other mechanisms, is evident in species ranging from bacteria to man and "dosage compensation" is well-known to *Drosophila* and mammalian geneticists. It would therefore not seem improbable that gene amplification could well play a part not only as a transient process during development, but as a more permanent entity in differentiated cells programmed to produce large amounts of a specific protein or proteins.

## 14 Pseudogenes

Many of the genes in man, and other eukaryotes, can be grouped into families where each family has member genes with homologous, or nearly homologous, DNA sequences. Within each family, certain member genes may be genetically inactive and these are referred to as pseudogenes. If the multigene family is spatially arranged as a cluster, as in the β-globin cluster of man and other mammals (Lacy and Maniatis 1980), or the 5 s RNA gene clusters in *Xenopus* (Miller et al. 1978), then the pseudogenes may well have arisen by a process of gene duplication, or unequal crossing over, followed by genetic drift. Other pseudogenes, however, may be spatially dispersed from the functional coding homologue in the family.

One group of pseudogenes has specific structural features suggesting that they were generated from a functional gene sequence via an RNA intermediate and these have been referred to as "processed genes" (Hollis et al. 1982). These processed genes have precisely lost the intervening intron sequences found in the corresponding functional genes and they retain homology to the 3' end of the functional gene through the polyadenylation signal to the site of poly(A) addition. Processed genes are therefore homologous to the spliced mRNA of structural genes. Since the original description of a processed $\alpha$-globin gene in the mouse (Nishioka et al. 1980) a large number of processed genes have been identified and characterised. These include, for example: two human β-tubulin pseudogenes that have lost their intervening sequences and are flanked by direct repeats 11-16 base pairs long and include a 3 poly(A) tract downstream from a polyadenylation signal (Wilde et al. 1982a,b); processed genes bearing

the human Cµ coding region (Battey et al. 1982); a human metallothionein gene
(Karin and Richards 1982); and processed genes for rat α-tubulin (Lemischka and
Sharp 1982), etc. In many instances it has been shown that the processed gene is not
found in the chromosome that encodes for its normal functional counterpart. For
instance, the α-globin processed gene in the mouse was shown to be located on chromosome 15, whereas the active mouse α-globin locus is on chromosome 11 (Nishioka
1980; Leder et al. 1981). Similarly, a processed human immunoglobulin E gene has
been reported to be localised to chromosome 9 rather than chromosome 14, which is
the normal site of the active immunoglobulin locus (Battey et al. 1982).

   The general features of these processed genes indicate that they arise from processed
RNA species that are originally transcribed from functional genes and are then
conveyed and inserted into new chromosome locations where they form a significant
element of the total genomic structure. The mechanisms whereby these processed
genes arise, however, are somewhat speculative, but a number of features of several
of these genes are similar to integrated transposons. For example, they are often
flanked by short direct repeat sequences and they are dispersed in the sense that they
do not occur in large tandem arrays. These facts together with the finding that some,
but not all, of the pseudogenes terminate in poly(A) sequences have led to the suggestion that they represent a class of transposable element that passes through an
RNA intermediate. Of course, if integration of a processed gene does not require a
specific integration element and is a fundamental property of any RNA, or its cDNA
copy, then processed genes could arise from any transcribed sequence.

   In 1981 Van Ardsell et al. described three small nuclear RNA (snRNA) pseudogenes in the human genome which were flanked by direct repeats. More recent studies
have revealed a great abundance of pseudogenes relative to the normal transcribed
genes of these snRNA species where, for example, more than 1,000 U1 pseudogenes
appear to derive directly from 100 bona fide U1 genes (Denison et al. 1981; Manser
and Gesteland 1982) and there are pseudogenes for U1, U2, U3, and U4 sRNAs that
appear to have been created by the insertion of cDNA copies of the sRNA at new
chromosomal sites (Bernstein et al. 1983). Most important, Bernstein et al. have
further shown that the U3 snRNA can function as a self-priming template for AMV
reverse transcriptase. The existence of endogenous retroviral sequences in avian and
murine genomes (Bishop 1978) and in primate genomes (Todaro et al. 1978) suggest
that a reverse transcriptase activity might be found in normal human cells. An RNA-
dependent DNA polymerase activity can be detected in most normal human placental
tissue and retroviral-related sequences have been identified in human DNA's (Martin
et al. 1981). Another possibility is that the reverse transcription of cellular RNA's
might occur during retroviral infection of human germ line cells and retroviral particles
could possibly function as transducing vectors for horizontal transmission of somatic
RNA species such as immunoglobulin transcripts into germ line cells.

   In summary then, although it is not at all clear how processed genes are integrated
into the genome, the sequence of events presumably must involve (1) the initiation
of an aberrant transcript that begins at some distance 5' from the genes normal initiation site; (2) the processing of the transcript so that coding sequences are spliced and
poly(A) added; (3) the conversion by reverse transcriptase of this RNA into DNA;
and (4) the integration of this cDNA into chromosomal DNA. It is possible that the

integration of this "stowaway DNA" into the chromosome requires its linkage to specific insertion sequences, such as for example a retrovirus sequence or an Alu or snRNA element which would provide the appropriate structure for insertion into the genomic DNA. What is clear is that although the existence of processed genes has only been recognised in the last few years, they must account for a significant proportion of the "non-repeat" DNA sequences in the genome and it seems probable that a substantial proportion of functional coding sequences have sleeping partners located elsewhere in the genome.

## 15 DNA Methylation

5-methyl cytosine (5 MeC) is a modified DNA base that is found in all mammalian genomes where it may be present at a concentration of one Mole percent (Vanyushin et al. 1970). Although the bulk of 5-MeC is in CpG dinucleotides that are concentrated in simple sequence repetitive DNA's (Harbers et al. 1975; Miller, Schnedl et al. 1974), methylated cytosine is present throughout all sequences. The pattern of 5-methyl cytosine residues in DNA is associated with gene expression and changes in methylation pattern of cytosine residues are evident in relation to the expression of specific genes in differentiated tissues. Methylation patterns are certainly tissue-specific, and structural genes and related DNA sequences are hypomethylated in tissues in which they are actively expressed, e.g., globin (Waalwijk and Flavell 1978; McGhee and Ginder 1979), ovalbumin and conalbumin (Mandel and Chambon 1979), immunoglobulin (Yagi and Koshland 1981) and delta-crystallin genes (Jones et al. 1981). However, these same DNA sequences are hypermethylated in tissues not expressing these genes (Razin and Riggs 1980). A similar correlation is evident both for transcriptionally active amplified rRNA genes in amphibia, which are hypomethylated whereas their inactive counterparts in somatic cells are hypermethylated (Bird and Southern 1978), and for transcriptionally active and inactive rDNA sequences in human and murine cells (Miller, Tantravahi et al. 1981). This correlation between hypomethylation and active gene expression is also apparent for viral sequences in the host DNA, or present as episomes, in infected cells (Desrosiers et al. 1979; Sutter and Doerfler 1980). However, this correlation is not universal, for Gerber-Huber et al. (1983) have recently reported that specific genes in the toad *Xenopus* can be actively transcribed even when they are fully methylated so that changes in methylation pattern are not a prerequisite for gene activation.

The importance of methylation is emphasised in experiments in which cells are exposed to 5-azacytidine which is incorporated into DNA and cannot be methylated since a nitrogen atom is present in place of the normal carbon atom at the 5 position on the pyrimidine ring. In such experiments (Jones and Taylor 1980; Venolia et al. 1982) altered gene expression is clearly evident and indications from these experiments are that methylation serves to inhibit transcription. The pattern of DNA methylation in a nucleus is conserved throughout cell division so that patterns are copied over many divisions during development (Bird 1978; Wigler et al. 1981; Harland 1982). The mechanism whereby a given pattern of methylation is established are not yet clear, but it is proposed that cells contain "maintenance" methyl transferases that are capable of efficient modification of hemi-methylated sites formed during DNA replication (Riggs 1975).

The abundant evidence indicating that hypermethylation results in gene inactivation naturally led to the suggestion that the inactive X chromosome in the human complement was "turned off" because of methylation. However, attempts to demonstrate hypermethylation in inactive X chromosomes using fluorescence tagged antibodies that specifically bound to methylated groups were unsuccessful in revealing any differences between the inactive and the normal X chromosome (Miller et al. 1982). An alternative approach to this question has been to use the inactive X chromosomal DNA to see if it functions in HPRT transformation assays. Liskay and Evans (1980) used DNA from hamster or mouse cells which had an HPRT$^-$ allele on the active X chromosome and an HPRT$^+$ allele on the inactive X chromosome and demonstrated that the inactive X chromosome did not function in HPRT transformation. These experiments have been confirmed and extended by Chapman et al. (1982) and Venolia and Gartler (1983) who have clearly shown that DNA from the inactive X chromosome functions very poorly in HPRT transformation, all the evidence being consistent with the idea that DNA modification plays an important role in the maintenance of X chromosome inactivation: a conclusion supported by studies on the reactivation of genes on inactive X chromosomes exposed to 5-azacytidine (Shapiro and Mohandas 1983).

The evidence that specific methylation patterns control gene expression in normal cells led naturally to the suggestion that alterations in these patterns might result in aberrant gene expression in neoplastic cells. Recently, Wilson and Jones (1983) have shown that the formation of alkali-labile sites following exposure of DNA to chemical carcinogens lessens its ability to accept methyl groups and inhibits the action of methyl transferase. The degree of methyl transferase inhibition was found to be far greater than that expected simply from the number of lesions induced by carcinogens in the DNA and it was shown that carcinogenic agents resulted in heritable changes in 5-methyl cytosine patterns through a variety of mechanisms, including not only adduct formation and formation of apurinic sites and single strand breaks, but also by direct inactivation of DNA methyl transferase. At the level of the neoplastic cell, there have been several studies on DNA methylation in cultured cells derived from human or animal malignancies which have shown either no change, an increase, or a decrease in the degree of methylation of total DNA. Feinberg and Vogelstein (1983), however, have now reported on DNA methylation levels in specific genes located on three different chromosomes in primary human tumours and adjacent normal tissues, and show that substantial hypomethylation is evident in certain genes in cancer cells as compared with their normal counterparts. Whether such changes in methylation patterns are responsible for the abnormal gene expression that may characterise a neoplastic transformation is, however, yet to be established.

## 16 Concluding Comments

In this review I have attempted to sketch out our knowledge of some of the main structural features of the human genome and to emphasise areas where rapid advances are being made in our understanding of the functional organisation of the chromosome. I have deliberately avoided considering genome organisation in meiotic and specialised germ cells, where there are specific changes associated with the pairing and exchange of chromosomes at meiosis and the packaging of chromosomes into sperm heads – since these are but variations on the overall pattern of the functioning genome in somatic cells.

The chromosome as an organelle is essentially a similar structure in plants, animals and man and this similarity extends not only to structure, but, and to varying degrees, to gene content and function. With the development of chromosome banding and in vitro cell genetics techniques, many workers have related chromosome banding profiles, and also linkage groups for a wide range and number of genes, as between different primate and other mammalian species. The fact that we can identify homologous chromosome structures and gene arrangements in at least 20% of the genome of man and the domestic cat (Nash and O'Brien 1982) and that there are many linkage groups in man that are also common to mouse (Pearson et al. 1982), points strongly to the conservation of genomic structure over substantial periods of evolutionary time. This conservation is a reflection of efficient functioning, for, as we have seen, the genome is not as stable a structure as has sometimes been imagined. Despite the complexity of the genome there is clearly order both within and between its defined chromosomes and a better understanding of this order will surely enable us to obtain a better insight into the processes and consequences of mutation in man.

## References

Adolph KW, Cheng SM, Paulson JR, Laemmli UK (1977) Isolation of a protein scaffold from mitotic HeLa cell chromosomes. Proc Natl Acad Sci USA 74:4937–4941

Agard DA, Sedat JW (1983) Three-dimensional architecture of a polytene nucleus. Nature 302: 676–681

Alt FW, Kellems RE, Bertino JR, Schimke RT (1978) Selective multiplication of dihydrofolate reductase genes in methotrexate-resistant variants of cultured murine cells. J Biol Chem 253:1357–1370

Arnott S, Chandrasekaran R, Birdsall DL, Leslie AGW, Ratliff RL (1980) Left-handed DNA helices. Nature 283:743–745

Balaban-Malenbaum G, Gilbert F (1977) Double minute chromosomes and homogeneously staining regions in chromosomes of a human neuroblastoma cell line. Science 198:739–742

Bateman AJ (1975) Simplification of palindromic telomere theory. Nature 253:379–380

Battey J, Max EE, McBride WO, Swan D, Leder P (1982) A processed human immunoglobulin ε gene has moved to chromosome 9. Proc Natl Acad Sci USA 79:5956–5960

Beach LR, Palmiter RD (1981) Amplification of the metallothionein-I gene in cadmium-resistant mouse cells. Proc Natl Acad Sci USA 78:2110–2114

Beauchamp RS, Mitchell AR, Buckland RA, Bostock CJ (1979) Specific arrangements of human satellite III DNA sequences in human chromosomes. Chromosoma 71:153–166

Beek B, Jacky PB, Sutherland GR (1983) Heritable fragile sites and micronucleus formation. Ann Genet 26:5–9

Behe M, Felsenfeld G (1981) Effects of methylation on a synthetic polynucleotide: the B-Z transition in poly(dG-m⁵dC)·poly(dG-m⁵dC). Proc Natl Acad Sci USA 78:1619–1623

Ben-Sasson SA, Klein G (1981) Activation of the Epstein Barr virus genome by 5-aza-cytidine in latently infected human lymphoid lines. Int J Cancer 28:131–135

Bennett MD (1982) Nucleotypic basis of the spatial ordering of chromosomes in eukaryotes and the implications of the order for genome evolution and phenotypic variation. In: Dover GA, Flavell RB (eds) Genome evolution. Academic, London, p 239

Bergold PJ, Campbell GR, Littau VC, Johnson EM (1983) Sequence and hairpin structure of an inverted repeat series at termini of the Physarum extra-chromosomal rDNA molecule. Cell 32:1287–1299

Bernards A, Michels PAM, Lincke CR, Borst P (1983) Growth of chromosome ends in multiplying trypanosomes. Nature 303:592–597

Bernstein LB, Mount SM, Weiner AM (1983) Pseudogenes for human small nuclear RNA U3 appear to arise by integration of self-primed reverse transcripts of the RNA into new chromosomal sites. Cell 32:461–472

Biedler JL, Spengler BA (1976) Metaphase chromosome anormaly: association with drug resistance and cell-specific products. Science 191:185–187

Bird AP (1978) Use of restriction enzymes to study eukaryotic DNA methylation: II. The symmetry of methylated sites supports semi-conservative copying of the methylation patterns. J Mol Biol 118:49–60

Bird AP, Southern EM (1978) Use of restriction enzymes to study eukaryotic DNA methylation: I. The methylation pattern in ribosomal DNA from Xenopus laevis. J Mol Biol 118:27–47

Bishop JM (1978) Retroviruses. Annu Rev Biochem 47:35–88

Blackburn EH, Gall JG (1978) A tandemly repeated sequence at the termini of the extrachromosomal ribosomal RNA genes in Tetrahymena. J Mol Biol 120:33–53

Blair DG, Oskarsson M, Wood TG, McClements WL, Fischinger PJ, Vande Woude GG (1981) Activation of the transforming potential of a normal cell sequence: a molecular model for oncogenesis. Science 212:941–943

Bostock CJ, Gosden JR, Mitchell AR (1978) Localisation of a male-specific DNA fragment to a sub-region of the human Y-chromosome. Nature 272:324–328

Bostock CJ, Prescott DM (1971) Shift in buoyant density of DNA during the synthetic period and its relation to euchromatin and heterochromatin in mammalian cells. J Mol Biol 60:151–162

Bostock CJ, Tyler-Smith C (1981) Gene amplification in methotrexate resistant mouse cells. II. Rearrangement and amplification of non-dihydrofolate reductase gene sequences accompany chromosomal changes. J Mol Biol 153:219–236

Boswell RE, Klobutcher LA, Prescott DM (1982) Inverted terminal repeats are added to genes during macronuclear development in Oxytricha nova. Proc Natl Acad Sci USA 79:3255–3259

Botchan M, Stringer J, Mitchison T, Sambrook J (1980) Integration and excision of SV40 DNA from the chromosome of a transformed cell. Cell 20:143–152

Boue J, Boue A, Lazer P (1975) Retrospective and prospective epidemiological studies of 1,500 karyotyped spontaneous human abortions. Teratology 17:11–26

Breathnach R, Mandel JL, Chambon P (1977) Ovalbumin gene is split in chicken DNA. Nature 270:314–319

Britten R (1981) DNA sequence organization and repeat sequences. In: Bennett MD, Bobrow M, Hewitt G (eds) Chromosomes today, vol 7. Allen and Unwin, London, p 9

Brown DD, Dawid IB (1968) Specific gene amplification in oocytes. Science 160:272–280

Buckton KE (1976) Identification with G and R banding of the position of breakage points induced in human chromosomes by in vitro X-irradiation. Int J Radiat Biol 29:475–488

Burkholder GD, Duczek LL (1982) The effect of chromosome banding techniques on the proteins of isolated chromosomes. Chromosoma 87:425–435

Calabretta B, Robberson DL, Barrera-Saldena HA, Lambrou TP, Saunders GF (1982) Genome instability in a region of human DNA enriched in Alu repeat sequences. Nature 296:219–225

Calabretta B, Robberson DL, Maizel AL, Saunders GF (1981) mRNA in human cells contains sequences complementary to the Alu-family of repeated DNA. Proc Natl Acad Sci USA 78: 6003–6007

Cameron JR, Loh EY, Davis RW (1979) Evidence for transposition of dispersed repetitive DNA families in yeast. Cell 16:739–751

Capecchi MR (1980) High efficiency transformation by direct microinjection of DNA into cultured mammalian cells. Cell 22:479–488

Carpenter BG, Baldwin JP, Bradbury EM, Ibel K (1976) Organisation of sub units in chromatin. Nucleic Acids Res 3:1739–1746

Carrano AV, Johnston GR (1977) The distribution of Mitomycin C-induced sister chromatid exchanges in the euchromatin and heterochromatin of the Indian Muntjac. Chromosoma 64:97–107

Cavalier-Smith T (1974) Palindromic base sequences and replication of eukaryotic chromosome ends. Nature 250:467–470

Chapman VM, Kratzer PG, Siracusa LD, Quarantillo BA, Evans R, Liskay RM (1982) Evidence for DNA modification in the maintenance of X-chromosome inactivation of adult mouse tissues. Proc Natl Acad Sci USA 79:5357–5361

Chung S-Y, Folsom V, Wooley J (1983) DNase 1-hypersensitive sites in the chromatin of immunoglobulin κ light chain genes. Proc Natl Acad Sci USA 80:2427–2431

Clarke L, Carbon J (1980) Isolation of a yeast centromere and construction of functional small circular chromosomes. Nature 287:504–509

Clayton CE, Rigby PWJ (1981) Cloning and characterization of the integrated viral DNA from three lines of SV40 transformed mouse cells. Cell 25:547–559

Collins M, Rubin GM (1983) High frequency precise excision of the Drosophila foldback transposable element. Nature 303:259–260

Comings DE (1978) Methods and mechanisms of chromosome banding. Methods Cell Biol 17: 115–132

Comings DE (1980) Arrangement of chromatin in the nucleus. Hum Genet 53:131–143

Cook PR, Brazell IA (1975) Super coils in human DNA. J Cell Sci 19:261–279

Cooke P, Gordon RR (1965) Cytological studies on a human ring chromosome. Ann Hum Genet 29:147–150

Cooper GM (1982) Cellular transforming genes. Science 218:801–806

Corneo G, Zardi L, Polli E (1972) Elution of human satellite DNAs on a methylated albumin kiesselguhr chromatographic column: isolation of satellite DNA 4. Biochim Biophys Acta 269:201–204

Cote BD, Uhlnebeck OC, Steffensen DM (1980) Quantitation of in situ hybridization of ribosomal ribonucleic acids to human diploid cells. Chromosoma 80:349–367

Cox D, Yuncken C, Spriggs A (1965) Minute chromatin bodies in malignant tumours of childhood. Lancet II:55–58

Cremer T, Cremer C, Schneider T, Baumannh H, Kirsch-Volders M (1982) Analysis of chromosome positions in the interphase nucleus of Chinese hamster cells by laser-UV irradiation experiments. Hum Genet 62:241

Cremisi C (1979) Chromatin replication revealed by studies of animal cells and papovaviruses (Simian virus 40 and Polyoma virus). Microbiol Rev 43:297–319

Dalla Favera R, Gelmann EP, Gallo RC, Wong-Staal F (1981) A human onc gene homologous to the transforming gene (v-sis) of simian sarcoma virus. Nature 292:31–35

Dancis BM, Holmquist GP (1979) Telomere replication and fusion in eukaryotes. J Theor Biol 78:211–224

Daneholt B, Case ST, Hyde J, Nelson L, Weislander L (1976) Production and fate of Balbiani ring products. Prog Nucleic Acid Res Mol Biol 19:319–334

Deininger PL, Schmid CW (1976) An electron microscope study of the DNA sequence organization of the human genome. J Mol Biol 106:773–790

Dekaban A (1965) Persisting clone of cells with an abnormal chromosome in a woman previously irradiated. J Nucl Med 6:740–746

Denison RA, Van Arsdell SW, Bernstein LB, Weiner AM (1981) Abundant pseudogenes for small nuclear RNAs are dispersed in the human genome. Proc Nat Acad Sci USA 78:810–814

Desrosiers RC, Mulder C, Fleckenstein B (1979) Methylation of Herpes virus saimiri DNA in lymphoid tumor cell lines. Proc Natl Acad Sci USA 76:3839–3843

Detke S, Lichtler A, Phillips I, Stein J, Stein G (1979) Reassessment of histone gene expression during cell cycle in human cells by using homologous H4 histone cDNA. Proc Natl Acad Sci USA 76:4995–4999

Dev VG, Warburton D, Miller OJ, Miller DA, Erlanger BF, Beiser SM (1972) Consistent pattern of binding of anti-adenosine antibodies to human metaphase chromosomes. Exp Cell Res 74: 288–293

Dodson LA, Foote RS, Mitra S, Masker WE (1982) Mutagenesis of bacteriophage T7 *in vitro* by incorporation of $O^6$-methylguanine during DNA synthesis. Proc Natl Acad Sci USA 79: 7440–7444

Donner L, Bubenik J (1968) Minute chromatin bodies in two mouse tumours induced in vivo by Rous sarcoma virus. Folia Biol 14:86–88

Dott PJ, Chuang CR, Saunders GF (1976) Inverted repeated sequences in the human genome. Biochemistry 15:4120–4125

Dresser ME, Moses MJ (1979) Silver staining of synaptonemal complexes for light and electron microscopy. Exp Cell Res 121:416–419

Duncan C, Biro PA, Choudary PV et al. (1979) RNA polymerase III transcriptional units are interspersed among human non-α-globin genes. Proc Natl Acad Sci USA 76:5095–5099

Dutrillaux B, Aurias A, Couturier J, Croquette MF, Viegas-Pequignot E (1977) Multiple telomeric fusions and chain configurations in human somatic chromosomes. In: de la Chapelle A, Sorsa M (eds) Chromosomes today, vol 6. Elsevier/North-Holland, Amsterdam, p 37

Dutrillaux B, Croquette MF, Viegas-Pequignot E, Aurias A, Coget J, Couturier J, Lejeune J (1978) Human somatic chromosome chains and rings. Cytogenet Cell Genet 20:70–77

Dutrillaux B, Viegas-Pequignot E (1981) High resolution R- and G-banding on the same preparation. Hum Genet 57:93–95

Dutrillaux B, Viegas-Pequignot E, Aurias A, Mouthuy M, Prieur M (1981) Non random position of metaphasic chromosomes: a study of radiation induced and constitutional chromosome rearrangements. Hum Genet 59:208–210

Earnshaw WC, Laemmli UK (1983) Architecture of metaphase chromosomes and chromosome scaffolds. J Cell Biol 96:84–93

Emery HS, Weiner AM (1981) An irregular satellite sequence is found at the termini of the linear extrachromosomal rDNA in *Dictyoistelium discordeum*. Cell 26:411–419

Endow SA (1980) On ribosomal gene compensation in *Drosophila*. Cell 22:149–155

Epplen JT, Siebers J-W, Vogel W (1975) DNA replication patterns of human chromosomes from fibroblasts and amniotic fluid cells revealed by a Giemsa staining technique. Cytogenet Cell Genet 15:177–185

Evans HJ (1977a) Chromosome anomalies among live births. J Med Genet 14:309–312

Evans HJ (1977b) Molecular mechanisms in the induction of chromosome aberrations. In: Scott D, Bridges BA, Sobels FH (eds) Progress in genetic toxicology. Elsevier/North-Holland, Amsterdam, p 57–74

Evans HJ (1977c) Some facts and fancies relating to chromosome structure in man. In: Harris H, Hirschhorn K (eds) Advances in human genetics, vol 8. Plenum, New York, p 347–438

Evans HJ (1979) Chromosome structure in man and related organisms. In: Dion AS (ed) Concepts of the structure and function of DNA, chromatin and chromosomes. Symposia Specialists, Miami, p 81

Evans HJ (1983) Chromosomes and cancer: from molecules to man – an overview. In: Rowley J, Ultmann JE (ed) Chromosomes and cancer: from molecules to man. Academic, New York (Bristol-Myers Cancer Symposia, vol. 5), p. 333–352

Evans HJ, Bigger TRL (1961) Chromatid aberrations induced by gamma irradiation. II. Non randomness in the distribution of chromatid aberrations in relation to chromosome length in *Vicia faba* root-tip cells. Genetics 46:277–289

Evans HJ, Buckland RA, Pardue ML (1974) Location of genes coding for 18 s and 28 s ribosomal RNA in the human genome. Chromosoma 48:405–426

Evans HJ, Gosden JR, Mitchell AR, Buckland RA (1974) Location of human satellite DNAs on the Y chromosome. Nature 251:346–347

Evans HJ, Pond V (1964) The influence of the centromere on chromosome fragment frequency under chronic irradiation. Portug Acta Biologica, serie A 8, pp 125–146

Evans HJ, Ross A (1974) Spotted centromeres in human chromosomes. Nature 249:861–862

Feinberg AP, Vogelstein B (1983) Hypomethylation distinguishes genes of some human cancers from their normal counterparts. Nature 301:89–92

Ferguson-Smith MA, Handmaker SD (1963) The association of satellited chromosomes with specific chromosomal regions in cultured human somatic cells. Ann Hum Genet 27:143–156

Ferraro M, Lavia P, Pellicia F, de Capoa A (1981) Clonal inheritance of rRNA gene activity: cytological evidence in human cells. Chromosoma 84:345–351

Finch JT, Klug A (1976) Solenoidal model for superstructure in chromatin. Proc Natl Acad Sci USA 73:1897–1901

Finch JT, Klug A (1977) X-ray and electron microscope analyses of crystals of nucleosome cores. Cold Spring Harbor Symp Quant Biol 42:1–9

Fink G, Farabaugh P, Roeder G, Chaleff D (1981) Transposable elements (Ty) in yeast. Cold Spring Harbor Symp Quant Biol 45:575–580

Finnegan DJ, Rubin GM, Young MW, Hogness DS (1976) Repeated gene families in *Drosophila melanogaster*. Cold Spring Harbor Symp Quant Biol 42:1053–1063

Fitzgerald-Hayes M, Buhler J-M, Cooper T, Carbon J (1982a) Isolation and subcloning analysis of functional centromere DNA (*CEN11*) from yeast chromosome XI. Mol Cell Biol 2:82–87

Fitzgerald-Hayes M, Clarke L, Carbon J (1982b) Nucleotide sequence comparisons and functional analysis of yeast centromere DNAs. Cell 29:235–244

Flavell AJ, Ish-Horowicz D (1981) Extrachromosomal circular copies of the eukaryotic transposable element *copia* in cultured *Drosophila* cells. Nature 292:591–595

Fletcher JM (1979) Light microscope analysis of meiotic prophase chromosomes by silver staining. Chromosoma 72:241–248

Folger KR, Wong EA, Wahl G, Capecchi MR (1982) Patterns of integration of DNA microinjected into cultured mammalian cells: evidence for homologous recombination between injected plasmid DNA molecules. Mol Cell Biol 2:1372–1387

Francke U, Oliver N (1978) Quantitative analysis of high resolution trypsin-Giemsa bands on human prometaphase chromosomes. Hum Genet 45:137–165

Fritsh EF, Shen CKJ, Lawn RM, Maniatis T (1981) The organization of repetitive sequences in mammalian globin gene clusters. Cold Spring Harbor Symp Quant Biol 45, part 2, pp 761–775

Ganner E, Evans HJ (1971) The relationship between patterns of DNA replication and of quinacrine fluorescence in the human chromosome complement. Chromosoma 35:326–341

Gerber P (1972) Activation of Epstein Barr virus by 5-bromodeoxyuridine in "virus-free" human cells. Proc Natl Acad Sci USA 69:83–85

Gerber-Huber S, May FEB, Westley BR, Felber BK, Hosbach HA, Andres A-C, Ryffel GU (1983) In contrast to other xenopus genes the estrogen-inducible vitellogenin genes are expressed when totally methylated. Cell 33:43–51

Gooderham K, Jeppesen P (1983) Chinese hamster metaphase chromosomes isolated under physiological conditions. Exp Cell Res 144:1–14

Goodpasture C, Bloom SE (1975) Visualization of nucleolar organizer regions in mammalian chromosomes using silver staining. Chromosoma 53:37–50

Gosden JR, Mitchell AR, Buckland RA, Clayton RP, Evans HJ (1975) The location of four human satellite DNAs on human chromosomes. Exp Cell Res 92:148–158

Green MM (1973) Some observations and comments on mutable and mutator genes in Drosophila. Genetics [Suppl] 73:187–194

Green MM (1976) Mutable and mutator loci. In: Ashburner M, Novitski E (eds) The genetics and biology of Drosophila, vol 1b. Academic, New York, p 929

Greer H, Igo M, de Bruijn F (1981) Transposable elements involving the *his 4* region of yeast. Cold Spring Harbor Symp Quant Biol 45:567–574

Grimaldi G, Singer MF (1982) A monkey *Alu* sequence is flanked by 13-base- pair direct repeats of an interrupted a-satellite DNA sequence. Proc Natl Acad Sci USA 79:1497–1500

Hadlaczky G, Sumner AT, Ross A (1981) Protein-depleted chromosomes. II. Experiments concerning the reality of chromosome scaffolds. Chromosoma 81:557–567

Hamada H, Kakunaga T (1982) Potential Z-DNA forming sequences are highly dispersed in the human genome. Nature 298:396–398

Hampar B, Ellison SA (1961) Chromosomal aberrations induced by animal virus. Nature 192: 145–147

Harbers K, Harbers B, Spencer JH (1975) Nucleotide clusters in deoxyribonucleic acids. XII. The distribution of 5-methylcystosine in pyrimidine oligonucleotides of mouse L-cell satellite DNA and main band DNA. Biochem Biophys Res Commun 66:738–746

Harland RM (1982) Inheritance of DNA methylation in microinjected eggs of *Xenopus laevis*. Proc Natl Acad Sci USA 79:2323–2327

Harnden DG (1974) Viruses, chromosomes and tumors; the interaction between viruses and chromosomes. In: German J (ed) Chromosomes and cancer. Wiley, New York, p 152

Hartley SE, Toolis F (1980) Double minute chromosomes in a case of acute myeloblastic leukemia. Cancer Genet Cytogenet 2:275–280

Harvey J, Judge C, Wiener S (1977) Familial X-linked mental retardation with an X chromosome abnormality. J Med Genet 14:46–50

Hayes W (1968) The genetics of bacteria and their viruses: studies in basic genetics and molecular biology, 2nd edn. Wiley, New York

Hayman D (1980) Factors affecting the frequency and pattern of mitomycin C-induced chromatid exchanges. Chromosoma 78:341–352

Haynes SR, Toomey TP, Leinwand L, Jelinek WR (1981) The Chinese hamster *Alu*-equivalent sequence: a conserved, highly repetitious, interspersed deoxyribonucleic acid sequence in mammals has a structure suggestive of a transposable element. Mol Cell Biol 1:573–583

Hayward WS, Neel BG, Astrin SM (1981) Activation of a cellular *onc* gene by promoter insertion in ALV-induced lymphoid leukosis. Nature 290:475–480

Heintz NH, Milbrandt JD, Greisen KS, Hamlin JL (1983) Cloning of the initiation region of a mammalian chromosomal replicaon. Nature 302:439–441

Hellstrom I, Hellstrom KE, Sjogren HO (1962) Further studies on superinfections of polyoma-induced mouse tumors with polyoma virus in vitro. Virology 16:282–300

Henderson A, Ripley S, Heller M, Kieff E (1983) Chromosome site for Epstein-Barr virus DNA in a Burkitt tumor cell line and in lymphocytes growth-transformed in vitro. Proc Natl Acad Sci USA 80:1987–1991

Henderson AS, Warburton D, Atwood KC (1972) Localization of ribosomal DNA in the human chromosome complement. Proc Natl Acad Sci USA 69:3394–3398

Hicks JB, Strathern JN, Herskowitz I (1977) The cassette model of mating type interconversion. In: Bukhari AI, Shapiro JA, Adhya SL (eds) DNA insertion elements, plasmids, and episomes. Cold Spring Harbor Laboratory, New York, p 457

Hollis GF, Hieter PA, McBride OW, Swan D, Leder P (1982) Processed genes: a dispersed human immunoglobulin gene bearing evidence of RNA-type processing. Nature 296:321–325

Holmquist GP, Dancis B (1979) Telomere replication, kinetochore organizers, and satellite DNA evolution. Proc Natl Acad Sci USA 76:4566–4570

Hoo JJ, Parslow MI (1979) Relation between the SCW points and the DNA replication bands. Chromosoma 73:67–74

Houck CM, Rinehart FP, Schmid CW (1979) A ubiquitous family of repeated DNA sequences in the human genome. J Mol Biol 132:289–306

Howell WM, Denton TE, Diamond JR (1975) Differential staining of the satellite regions of human acrocentric chromosomes. Experientia 31:260–262

Howell WM, Hsu TC (1979) Chromosome core structure revealed by silver staining. Chromosoma 73:61–66

Hsaio C-L, Carbon J (1981) A direct selection procedure for the isolation of functional centromeric DNA. Proc Natl Acad Sci USA 78:3760–3764

Hsu TC (1963) Longitudinal differentiation of chromosomes and the possibility of interstitial telomeres. Exp Cell Res [Suppl] 9:73–85

Igo-Kemenes T, Zachau HG (1977) Domains in chromatin structure. Cold Spring Harbor Symp Quant Biol 42:109–118

Jackson DA, McCready SJ, Cook PR (1981) RNA is synthesised at the nuclear cage. Nature 292: 552–555

Jacobs PA, Glover TW, Mayer M, Fox P, Gerrard JW, Dunn HG, Herbst DS (1980) X-linked mental retardation: a study of 7 families. Am J Med Genet 7:471–489

Jaenisch R, Harbers K, Schnieke A, Lohlers J, Chumakoo I, Jahner D, Grotkopp D, Hoffmann E (1983) Germline integration of Moloney murine leukemia virus at the *Mov 13* locus leads to recessive lethal mutation and early embryonic death. Cell 32:209–216

Jaenisch R, Jahner D, Nobis P, Simon I, Lohler J, Harbers K, Grotkopp D (1981) Chromosomal position and activation of retroviral genomes inserted into the germ line of mice. Cell 24: 519–529

Jeffreys AJ, Flavell RA (1977) The rabbit β-globin gene contains a large insert in the coding sequence. Cell 12:1097–1108

Jelinek WR, Toomey TP, Leinwand L, Duncan CH, Biro PA, Choudary PV, Weissman SM, Ribin CM, Houck CM, Deininger PL, Schmid CW (1980) Ubiquitous, interspersed repeated sequences in mammalian genomes. Proc Natl Acad Sci USA 77:1398–1402

Jenkins NA, Copeland NG, Taylor BA, Lee BK (1981) Dilute (*d*) coat colour mutation of DBA/2J mice is associated with the site of integration of an ecotropic MuLV genome. Nature 293: 370–374

Jeppesen PGN, Bankier AT (1979) A partial characterization of DNA fragments protected from nuclease degradation in histone depleted metaphase chromosomes of the Chinese hamster. Nucleic Acids Res 7:49–67

Jeppesen PGN, Bankier AT, Sanders L (1978) Non-histone proteins and the structure of metaphase chromosomes. Exp Cell Res 115:293–302

Jones FE, DeFeo D, Piatigorsky J (1981) Transcription and site-specific hypomethylation of the δ-crystallin in embryonic chicken lens. J Biol Chem 256:8172–8176

Jones PA, Taylor SM (1980) Cellular differentiation, cytidine analogues and DNA methylation. Cell 20:85–93

Jordan E, Saedler H, Starlinger P (1968) O$^c$ and strong polar mutations in the *gal* operon are insertions. Mol Gen Genet 102:353

Jorgenson KF, Van de Sande JH, Lin CC (1978) The use of base pair specific DNA binding agents as affinity labels for the study of mammalian chromosomes. Chromosoma 68:287–302

Karin M, Richards RI (1982) Human metallothionein genes – primary structure of the metallothionein-II gene and a related processed gene. Nature 299:797–802

Kaufman RJ, Brown PC, Schimke RT (1979) Amplified dihydrofolate reductase genes in unstably methotrexate-resistant cells are associated with double minute chromosomes. Proc Natl Acad Sci 76:5669–5673

Kaufmann ER, Davidson RL (1979) Bromodeoxyuridine mutagenesis in mammalian cells is stimulated by purine deoxyribonucleosides. Somatic Cell Genet 5:653–663

Kole LB, Haynes SR, Jelinek WR (1983) Discrete and heterogeneous high molecular weight RNAs complementary to a long dispersed repeat family (a possible transposon) of human DNA. J Mol Biol 165:257–286

Kominami R, Muramatsu M, Moriwaki K (1983) A mouse type 2 *Alu* sequence (M2) is mobile in the genome. Nature 301:87–89

Kornberg RD (1974) Chromatin structure: a repeating unit of histones in DNA. Science 184: 868–871

Krolewski JJ, Bertelsen AH, Humayun MZ, Rush MG (1982) Members of the *Alu* family of interspersed, repetitive DNA sequences are in the small circular DNA population of monkey cells grown in culture. J Mol Biol 154:399–415

Kuff EL, Feenstra A, Lueders K, Smith L, Hawley R, Hozumi N, Shulman M (1983) Intracisternal A-particle genes as movable elements in the mouse genome. Proc Natl Acad Sci USA 80:1992 to 1996

Lacy E, Maniatis T (1980) The nucleotide sequence of a rabbit β-globin pseudogene. Cell 21: 545–553

Lamb MM, Daneholt B (1979) Characterization of active transcription units in Balbiani rings of Chironomus tentans. Cell 17:835–848

Leder A, Swan D, Ruddle F, D'Eustachio P, Leder P (1981) Dispersion of α-like globin genes of the mouse to three different chromosomes. Nature 293:196–200

Leder P, Max EE, Seidman JG, Kwan S-P, Scharff M, Nau M, Norman B (1980) Recombination events that activate, diversify, and delete immunoglobulin genes. Cold Spring Harbor Symp Quant Biol 45:859–865

Lee MG-S, Lewis SA, Wilde CD, Cowan NJ (1983) Evolutionary history of a multigene family: an expressed human β-tubulin gene and three processed pseudogenes. Cell 33:477–487

Leffak IM, Grainger R, Weintraub H (1977) Conservative assembly and segregation of nucleosomal histones. Cell 12:837–845

Lemischka I, Sharp PA (1982) The sequences of an expressed rat α-tubulin gene and a pseudogene with an inserted repetitive element. Nature 300:330–335

Levan A, Levan G, Mandahl N (1978) A new chromosome type replacing the double minutes in a mouse tumor. Cytogenet Cell Genet 20:12–23

Levan G, Mandahl N, Bengtsson BO, Levan A (1977) Experimental elimination and recovery of double minute chromosomes in malignant cell populations. Hereditas 86:75–90

Lewis CD, Laemmli UK (1982) Higher-order metaphase chromosome structure: evidence for metalloprotein interactions. Cell 29:171–181

Liskay RM (1980) Inactive X chromosome DNA does not function in DNA-mediated cell transformation for the hypoxanthine phosphoribosyl-transferase gene. Proc Natl Acad Sci 77: 4895–4898

Lubs HA (1969) A marker X chromosome. Am J Hum Genet 21:231–244

Luciani JM, Morazzani M-R, Stahl A (1975) Identification of pachytene bivalents in human male meiosis using G-banding technique. Chromosoma 52:275–282

Maio JJ, Brown FL, McKenna WG, Musich PR (1981) Toward a molecular paleontology of primate genomes. II. The Kpn 1 families of alphoid DNAs. Chromosoma 83:127–144

Mandel JL, Chambon P (1979) DNA methylation: specific variations in the methylation pattern within and around ovalbumin and other chicken genes. Nucleic Acids Res 7:2081–2103

Manser T, Gesteland RF (1982) Human U1 loci: genes for human U1 RNA have dramatically similar genomic environments. Cell 29:257–264

Manuelidis L (1978) Chromosomal localization of complex and simple repeated human DNAs. Chromosoma 66:23–32

Martin MA, Bryan T, Rasheed S, Khan AS (1981) Identification and cloning of endogenous retroviral sequences present in human DNA. Proc Natl Acad Sci USA 78:4892–4896

Mathis D, Oudet P, Chambon P (1980) Structure of transcribing chromatin. Prog Nucleic Acid Res Mol Biol 24:1–55

Matsui S, Sasaki M (1973) Differential staining of nucleolus organizers in mammalian chromosomes. Nature 246:148–150

Mayfield JE, Serunian LA, Silver LM, Elgin SCR (1978) A protein released by DNAase I digestion of Drosophila nuclei is preferentially associated with puffs. Cell 14:539–544

McClintock B (1941) The stability of broken ends of chromosomes in Zea mays. Genetics 26: 234–282

McClintock B (1952) Chromosome organization and genic expression. Cold Spring Harbor Symp Quant Biol 16:13–47

McCready SJ, Goodwin J, Mason DW, Brazell IA, Cook PR (1980) DNA is replicated at the nuclear cage. J Cell Sci 46:365–386

McCready SJ, Jackson DA, Cook PR (1982) Attachment of intact superhelical DNA to the nuclear cage during replication and transcription. In: Natarajan AT, Obe G, Altmann H (eds) DNA repair, chromosome alterations and chromatin structure. Elsevier, Amsterdam, p 113–130 (Progress in Mutation Research, vol 4)

McGhee JD, Felsenfeld G (1983) Another potential artefact in the study of nucleosome phasing by chromatin digestion with micrococcal nuclease. Cell 32:1205–1215

McGhee JD, Ginder GD (1979) Specific DNA methylation sites in the vicinity of the chicken-globin genes. Nature 280:419–420

McKnight SL, Miller OL (1976) Ultrastructural patterns of RNA synthesis during early embryogenesis of Drosophila melanogaster. Cell 8:305–319

McKnight SL, Miller OL (1977) Electron microscopic analysis of chromatin replication in the cellular blastoderm *Drosophila melanogaster* embryo. Cell 12:795–804

Melli M, Ginelli E, Corneo G, diLernia R (1975) Clustering of the DNA sequences complementary to repetitive nuclear RNA of HeLa cells. J Mol Biol 93:23–38

Melton DW, Konecki DS, Ledbetter DH, Hejtmancik JF, Caskey CT (1981) In vitro translation of hypoxanthine/guanine phosphoribosyltransferase mRNA: characterization of a mouse neuroblastoma cell line that has elevated levels of hypoxanthine/guanine phosphoribosyltransferase protein. Proc Natl Acad Sci USA 78:6977–6980

Meneghini R, Armelin HA, Balsamo J, Lara FJS (1971) Indication of gene amplification in *Rhynchosciara* by RNA-DNA hybridization. J Cell Biol 49:913–916

Mezzanotte R, Ferrucci L, Vanni R, Bianchi U (1983) Selective digestion of human metaphase chromosomes by Alu 1 restriction endonuclease. J Histochem Cytochem 31:553–556

Mikelsaar AV, Schmid M, Krone W, Schwarzacher HG, Schnedl W (1977a) Frequency of Ag-stained nucleolus organizer regions in the acrocentric chromosomes of man. Hum Genet 37:73–77

Mikelsaar AV, Schwarzacher HG, Schnedl W, Wagenbichler P (1977b) Inheritance of Ag-stainability of nucleolus organizer regions. Hum Genet 38:183–188

Miller CK, Temin HM (1983) High-efficiency ligation and recombination of DNA fragments by vertebrate cells. Science 220:605–609

Miller DA, Choi Y-C, Miller OJ (1983) Chromosome localization of highly repetitive human DNA's and amplified ribosomal DNA with restriction enzymes. Science 219:395–397

Miller DA, Dev VG, Tantravahi R, Miller OJ (1976a) Suppression of human nucleolus organizer activity in mouse-human somatic hybrid cells. Exp Cell Res 101:235–243

Miller DA, Okamoto E, Erlanger BF, Miller OJ (1982) Is DNA methylation responsible for mammalian X chromosome inactivation? Cytogenet Cell Genet 33:345–349

Miller DM, Turner P, Nienhuis AW, Axelrod DE, Gopalakrishnan TV (1978) Active conformation of the globin genes in uninduced and induced mouse erythroleukemia cells. Cell 14:511–521

Miller JR, Cartwright EM, Brownlee GG, Fedoroff NV, Brown DD (1978) The nucleotide sequence of oocyte 5 s DNA in Xenopus laevis. II. The GC-rich region. Cell 13:717–725

Miller OJ, Miller DA, Dev VG, Tantravahi R, Croce CM (1976b) Expression of human and suppression of mouse nucleolus activity in mouse-human somatic cell hybrids. Proc Natl Acad Sci USA 73:4531–4535

Miller OJ, Schnedl W, Allen J, Erlanger BF (1974) 5-methylcytosine localised in mammalian constitutive heterochromatin. Nature 251:636–637

Miller OJ, Tantravahi R, Miller DA, Szabo YLC, Prensky P (1979) Marked increase in ribosomal RNA gene multiplicity in a rat hepatoma cell line. Chromosoma 71:183–195

Miller OJ, Tantravahi U, Katz R, Erlanger BF, Guntaka RV (1981) Amplification of mammalian ribosomal RNA genes and their regulation by methylation. In: Arrighi FE, Rao PN, Stubblefield E (eds) Genes, chromosomes, and neoplasia. Raven, New York, p 253

Miller OL, Hamkalo BA (1972) Visualization of RNA synthesis on chromosomes. Int Rev Cytol 33:1–25

Mitchell AR, Beauchamp RS, Bostock CJ (1979) A study of sequence homologies in four satellite DNAs of man. J Mol Biol 135:127–149

Mitchell AR, Gosden JR, Ryder OA (1981) Satellite DNA relationships in man and the primates. Nucleic Acids Res 9:3235–3249

Morad M, Jonasson J, Lindsten J (1973) Distribution of Mitomycin C induced breaks on human chromosomes. Hereditas 74:273–282

Moroi Y, Peebles C, Fritzler MJ, Steigerwald J, Tan EM (1980) Autoantibody to centromere (kinetochore) in scleroderma sera. Proc Natl Acad Sci USA 77:1627–1631

Muller HJ (1942) Induced mutations in *Drosophila*. Cold Spring Harbor Symp Quant Biol 9:151–167

Muller HJ, Herskowitz IH (1954) Concerning the healing of chromosome ends produced by breakage in *Drosophila melanogaster*. Amer Nat 88:177–208

Mulligan RC, Howard BH, Berg P (1979) Synthesis of rabbit β-globin in cultured monkey kidney cells following infection with a SV40 β-globin recombinant genome. Nature 277:108–114

Nash WG, O'Brien SJ (1982) Conserved regions of homologous G-banded chromosomes between orders in mammalian evolution: carnivores and primates. Proc Natl Acad Sci USA 79:6631 to 6635

Nichols W (1970) Virus-induced chromosomal abnormalities. Annu Rev Microbiol 24:479–500

Nishioka Y, Leder A, Leder P (1980) Unusual α-globin-like gene that has cleanly lost both globin intervening sequences. Proc Natl Acad Sci USA 77:2806–2809

Noll M, Kornberg RD (1977) Action of micrococcal nuclease on chromatin and the location of histone H1. J Mol Biol 109:393–404

Nordenskiold H (1962) Studies of meiosis in *Luzula purpurea*. Hereditas 48:503–519

Nordheim A, Pardu ML, Lafer EM, Moller A, Stollar BD, Rich A (1981) Antibodies to left-handed Z-DNA bind to interband regions of *Drosophila* polytene chromosomes. Nature 294:417–422

Nordheim A, Tesser P, Azorin F, Kwon YH, Moller A, Rich A (1982) Isolation of *Drosophila* proteins that bind selectively to left-handed Z-DNA. Proc Natl Acad Sci USA 79:7729–7733

Olins AL, Olins DE (1974) Spheroid chromatid units (ν bodies). Science 183:330–332

Pan J, Elder JT, Duncan CH, Weissman SM (1981) Structural analysis of interspersed repetitive polymerase III transcription units in human DNA. Nucleic Acids Res 9:1151–1170

Paulson JR, Laemmli UK (1977) The structure of histone-depleted chromosomes. Cell 12:817 to 828

Pearson PL, Roderick TH, Davisson MT, Lalley PA, O'Brien SJ (1982) Report of the committee on comparative mapping. Cytogenet Cell Genet 32:208–220

Peterson AR, Peterson H (1979) Facilitation by pyrimidine deoxyribonucleosides and hypoxanthine of mutagenic and cytotoxic effects of monofunctional alkylating agents in Chinese hamster cells. Mutation Res 61:319–331

Phillips SJ, Birkenmeier EH, Callahan R, Eicher EM (1982) Male and female mouse DNAs can be discriminated using retroviral probes: Nature 297:241–243

Prantera G, Bonaccorsi S, Pimpinelli S (1979) Simultaneous production of Q and R bands after staining with chromomycin A$_3$ or Olivomycin. Science 204:79–80

Rabbits TH, Bentley DL, Dunnick W, Forster A, Matthyssens GEAR, Milstein C (1980) Immunoglobulin genes undergo multiple sequence rearrangements during differentiation. Cold Spring Harbor Symp Quant Biol 45:867–878

Rasmuson B, Svahlin H, Rasmuson A, Montell I, Olofsson H (1978) The use of a mutationally unstable X-chromosome in *Drosophila melanogaster* for mutagenicity testing. Mutation Res 54:33–38

Ray JH, German J (1981) The chromosome changes in Bloom's syndrome, ataxia-telangiectasia, and Fanconi's anemia. In: Arrighi FE, Rao PN, Stubblefield E (eds) Genes, chromosomes, and Neoplasia. Raven, New York, p 351

Razin A, Riggs AD (1980) DNA methylation and gene function. Science 210:604–610

Rechavi G, Givol D, Canaani E (1982) Activation of a cellular oncogene by DNA rearrangement: possible involvement of an IS-like element. Nature 300:607–611

Reeves BR, Casey G, Harris H (1980) Variations in the activity of nucleolar organizers in different tissues. 7th Int Chromosome Conference, Oxford (unpublished abstracts, p 48)

Riggs AD (1975) X-inactivation, differentiation and DNA methylation. Cytogenet Cell Genet 14:9–25

Ris H, Witt PL (1981) Structure of the mammalian kinetochore. Chromosoma 82:153–170

Roberts PA (1975) In support of the telomere concept. Genetics 80:135–142

Robinson SI, Nelkin BD, Vogelstein B (1982) The ovalbumin gene is associated with the nuclear matrix of chicken oviduct cells. Cell 28:99–106

Roeder GS, Fink GR (1980) DNA rearrangements associated with a transposable element in yeast. Cell 21:239–249

Rubin GM (1978) Isolation of a telomeric DNA sequence from *Drosophila melanogaster*. Cold Spring Harbor Symp Quant Biol 42:1041–1046

Rubin GM, Brorein WJ, Dunsmuir P, Flavell AJ, Levis R, Strobel E, Toole JJ, Young E (1981) *copia*-like transposable elements in the *Drosophila* genome. Cold Spring Harbor Symp Quant Biol 45:619–628

Ruddle FH (1981) Keynote address: somatic cell genetics – past, present, and future. In: Arrighi FE, Rao PN, Stubblefield E (eds) Genes, chromosomes, and neoplasia. Raven, New York, p7

Sager R, Anisowicz A, Howell N (1981) Genomic rearrangements in a mouse cell line containing integrated SV40 DNA. Cell 23:41–50

Sahasrabuddhe CG, Pathak S, Hsu TC (1978) Responses of mammalian metaphase chromosomes to endonuclease digestion. Chromosoma 69:331–338

Santos E, Tronick SR, Aaronson SA, Pulciani S, Barbacid M (1982) T24 human bladder carcinoma oncogene is an activated form of the normal human homologue of BALB- and Harvey-MSV transforming genes. Nature 298:343–347

Satya-Prakash KL, Hsu TC, Pathak S (1980) Behaviour of the chromosome core in mitosis and meiosis. Chromosoma 81:1–8

Schaefer-Ridder M, Wang Y, Hofschneider PH (1982) Liposomes as gene carriers: efficient transformation of mouse L cells by thymidine kinase gene. Science 215:166–168

Scheer U (1978) Changes of nucleosome frequency in nucleolar and nonnucleolar chromatin as a function of transcription: an electron microscope study. Cell 13:535–549

Schmid CW, Deininger PL (1975) Sequence organisation of the human genome. Cell 6:345–358

Schmid CW, Jelinek WR (1982) The Alu family of dispersed repetitive sequences. Science 216:1065–1070

Schmookler Reis RJ, Lumpkin CK, McGill JR, Riabowol KT, Goldstein S (1983) Extrachromosomal circular copies of an *"inter-Alu"* unstable sequence in human DNA are amplified during in vitro and in vivo ageing. Nature 301:394–398

Schnedl W, Dann O, Schweizer D (1980) Effects of counterstaining with DNA binding drugs on fluorescent banding patterns of human and mammalian chromosomes. Eur J Cell Biol 20:290–296

Schreck RR, Warburton D, Miller OJ, Beiser SM, Erlanger BF (1973) Chromosome structure as revealed by a combined chemical and immunochemical procedure. Proc Natl Acad Sci USA 70:804–807

Schwarzacher HG, Wachtler F (1983) Nucleolus organizer regions and nucleoli. Hum Genet 63:89–99

Seabright M (1971) A rapid banding technique for human chromosomes. Lancet 2:971–972

Shapiro JA (1969) Mutations caused by the insertion of genetic material into the galactose operon of *Escherichia coli*. J Mol Biol 40:93

Shapiro LJ, Mohandas T (1983) DNA methylation and the control of gene expression on the human X chromosome. Cold Spring Harbor Symp Quant Biol, part 2: Structure of DNA, pp 631–637

Sharp PA (1983) Conversion of RNA to DNA in mammals: *Alu*-like elements and pseudogenes. Nature 301:471–472

Shaw MW (1972) Human chromosome abnormalities revisited. Amer J Human Genet 24:227–228

Shaw MW, Cohen MM (1965) Chromosome exchanges in human leukocytes induced by Mitomycin C. Genetics 51:181–190

Sheldrick P, Berthelot N (1977) Inverted repetitions in the chromosome of herpes simplex virus. Cold Spring Harbor Symp Quant Biol 39:667–678

Shiba T, Saigo K (1983) Retrovirus-like particles containing RNA homologous to the transposable element *copia* in *Drosophila melanogaster*. Nature 302:119–124

Silverman M, Simon M (1980) Phase variation: genetic analysis of switching mutants. Cell 19:845–854

Singleton CK, Klysik J, Stirdivant SM, Wells RD (1982) Left-handed Z-DNA is induced by supercoiling in physiological ionic conditions. Nature 299:311–316

Slilaty SN, Aposhian HV (1983) Gene transfer by polyoma-like particles assembled in a cell-free system. Science 220:725–727

Small D, Nelkin BD, Vogelstein B (1982) Non-random distribution of repeated DNA sequences with respect to supercoiled loops and the nuclear matrix. Proc Natl Acad Sci USA 79:5911 to 5915

Smith CA, Vinograd J (1972) Small polydisperse circular DNA of HeLa cells. J Mol Biol 69:163 to 178

Spradling AC, Mahowald AP (1980) Amplification of genes for chorion proteins during oogenesis in *Drosophila melanogaster*. Proc Natl Acad Sci 77:1096—1100

Spradling AC, Rubin GM (1981) Drosophila genome organization: conserved and dynamic aspects. Ann Rev Genet 15:219—264

Spradling AC, Rubin GM (1982) Transposition of cloned P elements into *Drosophila* germ line chromosomes. Science 218:341—347

Spritz RA, Forget BG (1983) The thalassemias: molecular mechanisms of human genetic disease. Am J Hum Genet 35:333—361

Steenbergh PH, Maat J, Ormondt HV, Sussenbach J (1977) The nucleotide sequence at the termini of adenovirus type 5 DNA. Nucleic Acids Res 4:4371—4389

Steffensen DM, Duffey P, Prensky W (1974) Localisation of 5 S ribosomal RNA genes on human chromosome 1. Nature 252:741—743

Stringer JR (1982) DNA sequence homology and chromosomal deletion at a site of SV40 DNA integration. Nature 296:363—366

Stubblefield E (1973) The structure of mammalian chromosomes. Int Rev Cytol 35:1—60

Sumner AT (1980) Dye binding mechanisms in G-banding of chromosomes. J Microsc 119:397 to 406

Sumner AT (1982) The nature and mechanisms of chromosome banding. Cancer Genet Cytogenet 6:59—87

Sumner AT (1983) The role of protein sulphydryls and disulphides in chromosome structure and condensation. In: Kew Chromosome Conference II. Allen and Unwin, London, p 1—9

Sutherland GR (1977) Fragaile sites on human chromosomes: demonstration of their dependence on the type of tissue culture medium. Science 197:265—266

Sutherland GR (1979) Heritable fragile sites on human chromosomes. I. Factors affecting expression in lymphocyte culture. Am J Hum Genet 31:125—135

Sutter D, Doerfler W (1980) Methylation of integrated adenovirus type 12 DNA sequences in transformed cells is inversely correlated with viral gene expression. Proc Natl Acad Sci USA 77:253—256

Szabo P, Yu LC, Borun T, Varricchio F, Siniscalco M, Prensky W (1978) Localisation of the histone genes in man. Cytogenet Cell Genet 22:359—363

Szostak NW, Blackburn E (1982) Cloning yeast telomeres on linear plasmid vectors. Cell 29:245—255

Taylor EF, Martin-de Leon PA (1981) Familial silver staining patterns of human nucleolus organizer regions (NOR). Am J Hum Genet 33:67—76

Termin HM (1974) On the origin of RNA tumor viruses. Annu Rev Genet 8:155—177

Temin HM (1976) The DNA provirus hypothesis. Science 192:1075—1080

Temin HM (1980) Origin of retroviruses from cellular moveable genetic elements. Cell 21:599 to 600

Tilghman SM, Tiemeier DC, Seidman JG, Peterlin BM, Sullivan M, Maizel JV, Leder P (1978) Intervening sequence of DNA identified in the structural portion of a mouse β-globin gene. Proc Natl Acad Sci USA 75:725—729

Todaro GJ, Benveniste RE, Sherwin SA, Sherr CJ (1978) MAC-1, a new genetically transmitted type C virus of primates: "low-frequency" activation from stumptail monkey cell cultures. Cell 13:775—782

Tonegawa S, Sakano H, Maki R, Traunecker A, Heinrich G, Roeder W, Kurosawa Y (1980) Somatic reorganization of immunoglobulin genes during lymphocyte differentiation. Cold Spring Harbor Symp Quant Biol 45:839—858

Tong C, Fazio M, Williams GM (1980) Cell cycle-specific mutagenesis at the hypoxanthine phosphoribosyltransferase locus in adult rat liver epithelial cells. Proc Natl Acad Sci USA 77:7377—7379

Topal MD, Baker MS (1982) DNA precursor pool: a significant target for $N$-methyl-$N$-nitrosourea in C3H/10T$^1$/$_2$ clone 8 cells. Proc Natl Acad Sci USA 79:2211–2215

Topal MD, Hutchinson CA, Baker MS (1982) DNA precursors in chemical mutagenesis: a novel application of DNA sequencing. Nature 298:863–865

Tsanev R (1982) Replication of chromatin. In: Natarajan AT, Obe G, Altmann H (eds) DNA repair, chromosome alterations and chromatin structure. Elsevier, Amsterdam, p 131–145 (Progress in Mutation Research, vol 4)

UNSCEAR (1977) Sources and effects of ionizing radiation. United Nations, New York

Van Arsdell SW, Denison RA, Bernstein LB, Weiner AM (1981) Direct repeats flank three small nuclear RNA pseudogenes in the human genome. Cell 26:11–17

Vanyushin BF, Tkacheva SG, Belozersky AN (1970) Rare bases in animal DNA. Nature 225: 948–949

Varmus HE, Quintrell N, Ortiz S (1981) Retroviruses as mutagens: insertion and excision of a nontransforming provirus alter expression of a resident transforming provirus. Cell 25:23–36

Venolia L, Gartler SM (1983) Comparison of transformation efficiency of human active and inactive X-chromosomal DNA. Nature 302:82–83

Venolia L, Gartler SM, Wassman ER, Yen P, Mohandas T, Shapiro LJ (1982) Transformation with DNA from 5-azacytidine-reactivated X chromosomes. Proc Natl Acad Sci USA 79:2352 to 2354

Viegas-pequignot E, Derbin C, Lemeunier F, Taillandier E (1982) Identification of left-handed Z-DNA by indirect immunomethods in metaphasic chromosomes of a mammal. *Gerbillus nigeriae* (Gerbillidae, Rodentia). Ann Genet 25:218–222

Waalwijk C, Flavell RA (1978) DNA methylation at CCGG sequence in the large intron of the rabbit-globin gene: tissue specific variations. Nucleic Acids Res 5:4631–4641

Wahl GM, Padgett RA, Stark GR (1979) Gene amplification causes over-production of the first three enzymes of UMP in $N$-(phosphonacetyl)-L-aspartase-resistant hamster cells. J Biol Chem 254:8679–8689

Wald N, Upton AC, Jenkins VK, Borges WH (1964) Radiation-induced mouse leukaemia: consistent occurrence of an extra and a marker chromosome. Science 143:810–831

Walmsley RM, Szostak JW, Petes TD (1983) Is there left-handed DNA at the ends of yeast chromosomes? Nature 302:84–86

Wang AH-J, Quigley GJ, Kolpak FJ, Crawford JL, Van Boom JH, Van Der Marel G, Rich A (1979) Molecular structure of a left-handed double helical DNA fragment at atomic resolution. Nature 282:680–686

Warren AC, Cook PR (1978) Supercoiling of DNA and nuclear conformation during the cell cycle. J Cell Sci 30:211–226

Watson JD (1972) Origin of concatameric T7 DNA. Nature New Biol 239:197–201

Weintraub H, Groudine M (1976) Chromosomal subunits in active genes have an altered conformation. Science 193:848–856

Weisbrod S, Groudine M, Weintraub H (1980) Interaction of HMG 14 and 17 with actively transcribed genes. Cell 19:289–301

White MJD (1973) Animal cytology and evolution. University Press, Cambridge

Wigler M, Levy D, Perucho M (1981) The somatic replication of DNA methylation. Cell 24:33–40

Wigler M, Sweet R, Sim GK, Wold B, Pellicer A, Lacy E, Maniatis T, Silverstein S, Axel R (1979) Transformation of mammalian cells with genes from procaryotes and eucaryotes. Cell 16: 777–785

Wilde CD, Crowther CE, Cowan NJ (1982b) Diverse mechanisms in the generation of human β-tubulin pseudogenes. Science 217:549–552

Wilde CD, Crowther CE, Cripe TP, Lee MG-S, Cowan NJ (1982a) Evidence that a human β-tubulin pseudogene is derived from its corresponding mRNA. Nature 297:83–84

Williams RO, Young JR, Majiwa PAO (1979) Genomic rearrangements correlated with antigenic variation in *Trypanosoma brucei*. Nature 282:847–849

Wilson DA, Thomas CA (1974) Palindromes in chromosomes. J Mol Biol 84:115–139

Wilson MC, Melli M (1977) Determination of the number of histone genes in human DNA. J Mol Biol 110:511–535

Wilson VL, Jones PA (1983) Inhibition of DNA methylation by chemical carcinogens in vitro. Cell 32:239–246

Worcel A, Han S, Wong ML (1978) Assembly of newly replicated chromatin. Cell 15:969–977

Wu C, Bingham PM, Livak KJ, Holmgren R, Elgin SCR (1979) The chromatin structure of specific genes: I. Evidence for higher order domains of defined DNA sequence. Cell 16:797–806

Yagi M, Koshland ME (1981) Expression of the J chain gene during B cell differentiation is inversely correlated with DNA methylation. Proc Natl Acad Sci USA 78:4907–4911

Young BD, Hell A, Birnie GD (1976) A new estimate of human ribosomal gene number. Biochim Biphys Acta 454:539–548

Yunis JJ, Kuo MT, Saunders GF (1977) Localization of sequences specifying messenger RNA to light-staining G-bands of human chromosomes. Chromosoma 61:335–344

Yunis JJ, Sawyer JR, Ball DW (1978) The characterisation of high resolution G banded chromosomes of man. Chromosoma 67:293–307

Zimmer WE, Schwartz RJ (1982) Amplification of chicken actin genes during myogenesis. In: Schimke RT (ed) Gene amplification. Cold Spring Harbor Laboratory, New York, p 137

Zorn C, Cremer C, Cremer T, Zimmer J (1979) Unscheduled DNA synthesis after partial UV irradiation of the cell nucleus. Distribution in interphase and metaphase. Exp Cell Res 124:111–119

Zur Hausen H, O'Neill F, Freeze U, Hecker E (1978) Persisting oncogenic herpes virus induced by the tumour promotor TPA. Nature 272:373–375

# Gene or Point Mutations

F. VOGEL[1]

## 1 Introduction

Subdivision òf mutations into chromosome and genome mutations, on the one hand, and gene or point mutations, on the other, goes back to a time when the microscope was the only technical means for visualizing genetic changes directly and the genetic code was still unknown. Mutations identified not directly by examining the chromosomes themselves, but only indirectly, from their effects on the phenotype, were named "gene or point mutations". Meanwhile, this distinction has been blurred somewhat since we have learned that point mutations comprise a variety of different events at the molecular level, for example single base substitutions ("point mutations in the strict sense"), small deletions sometimes leading to frame shifts; chain elongations; or molecular recombinations. Still, the notion of "point mutation" remains useful for many practical purposes, provided that one keeps in mind that it describes reality at a somewhat preliminary level.

In the following, we shall depict first and very briefly the most common phenotypic effects of point mutations (Sect. 2). Due to lack of space, problems of gene action on genotype-phenotype relations will only be touched upon very briefly. Assessment and interpretation of mutation rates will be discussed; this will lead us to the problem of how mutations might be utilized for monitoring human populations for a possible increase of mutation rate due to mutagenic agents (ionizing radiation or chemical mutagens).

This entire chapter will be limited to human mutations throughout; data from other objects will be mentioned only occasionally.

## 2 Phenotypes Caused by Mutations

The number of nucleotide pairs in the haploid human genome is estimated at about $3.5 \times 10^9$. The length of a transcribed DNA sequence – a classical "gene" – can be calculated from protein sequence data, and also from DNA sequencing. Assuming an average gene length of $\approx 1,000$ nucleotides, this would accommodate $\approx 3.5 \times 10^6$

1 Institut für Anthropologie und Humangenetik, University of Heidelberg, Im Neuenheimer Feld 328, 6900 Heidelberg, FRG

Mutations in Man, ed. by G. Obe
© Springer-Verlag Berlin Heidelberg 1984

such genes in the human genome. This estimate, however, is much too high, since it neglects introns, pseudogenes, control regions, highly repetitive sequences such as satellite DNA, and other interspersed and non-transcribed DNA-sequences. The true number of classical genes in humans – and in all other mammals – is still unknown; informed guesses assume an order of magnitude of $\approx 50,000$ to $100,000$.

All these genes are subject to occasional mutations. They might occur at various sites; the primary molecular event might be quite different; and so are the phenotypic consequences. An amazing variety has been uncovered by thorough analysis of a few examples, especially the hemoglobin genes (Bunn et al. 1977; Fairbanks 1980). Only a relatively small fraction of these mutations can be identified at the level of phenotypes, the most conspicuous being the hereditary diseases with simple modes of inheritance.

## 2.1 Hereditary Diseases with Simple Modes of Inheritance

Table 1 (McKusick 1982) gives the estimated number of human traits with simple modes of inheritance, mostly hereditary diseases. It shows a conspicuous increase within the last about 20 years. In part, this increase is due to discovery of entirely new diseases; mostly, however, it is caused by splitting up of conditions hitherto regarded as homogenous into subtypes with similar or even identical phenotypes but different genetic and biochemical causes. This discovery of "genetic heterogeneity" is a welcome and very useful side effect of successful probing into the causes and mechanisms of diseases. It is easy to predict that the number of discernible hereditary conditons will continue to increase in future; in addition to refinement in methods, the introduction of medical genetic services in new countries and populations might contribute to such a development, since in some populations other, especially recessive, genes might be more common than in Europe or North America. China, for example, with its population of one billion is almost entirely terra incognita for medical genetics.

Table 1. Number of loci identified by mendelizing phenotypes. (McKusick 1982)

| Phenotype | Verschuer (1958) | Mendelian inheritance in man | | | | | |
|---|---|---|---|---|---|---|---|
| | | 1966 | 1968 | 1971 | 1975 | 1978 | 1981 |
| Autosomal dominant | 285 | 269 (+568) | 544 (+449) | 415 (+528) | 583 (+635) | 736 (+753) | 918 (+883) |
| Autosomal recessive | 89 | 237 (+294) | 280 (+349) | 365 (+418) | 466 (+481) | 521 (+596) | 601 (+692) |
| X-linked | 38 | 68 (+51) | 68 (+55) | 86 (+64) | 93 (+78) | 107 (+98) | 116 (+123) |
| Total | 412 | 574 (+913) | 692 (+853) | 866 (+1,010) | 1,142 (+1,194) | 1,364 (+1,447) | 1,635 (+1,698) |
| Grand total | 412 | 1,487 | 1,545 | 1,876 | 2,336 | 2,811 | 3,333 |

## 2.1.1 Autosomal Recessive Diseases

Table 1 shows 601 confirmed and 692 suspected conditions with autosomal recessive mode of inheritance. With very few exceptions, homozygotes of these conditions are rare or very rare. However, since frequencies of genotypes in human populations follow the Hardy-Weinberg law, heterozygotes are much more common. The frequency of a gene A might be denoted p, frequency of its allele a, q; p + q = 1. Then, the three genotypes occur in a randomly mating population in frequencies AA = $p^2$; Aa = 2 pq; aa = $q^2$.

Therefore, most homozygous children occur in matings between two heterozygotes. For the health condition of a population deviations of fitness in heterozygotes, for example small increases of liabilities for common diseases or behavioral abnormalities, such as described increasingly in various conditions (see Vogel 1983), might be even more important than the more conspicuous, and often severe, diseases known in homozygotes.

In some hundred recessive conditions, the basic biochemical defects have been uncovered. Usually, an enzyme defect was found. As a rule, enzyme activity in heterozygotes is reduced by roughly 1/2 in comparison with normal homozygotes, whereas in homozygotes of the mutant allele, only some very little rest activity is usually left. This means that for these enzymes, half the normal activity is usually sufficient to maintain near-normal function. The genetic blocks in the metabolism of the amino acids phenylalanine and tyrosine provide classical examples for enzyme defects and their various phenotypic consequences. Figure 1 gives an (oversimplified) overview.

Block A is classical phenylketonuria (PKU). The defects found in the enzyme phenylalanine hydroxylase differ slightly from one case to the other (see Bartholomé 1979; Scriver 1980). Moreover, there are other hyperphenylalaninemias of different, and sometimes still unknown, origin. Almost since this condition has been discovered in the 1930's, slight anomalies have been suspected in the heterozygotes, as well (see Vogel and Motulsky 1979). More recently, well-controlled studies point to a slight difference in test intelligence between heterozygotes and appropriate controls: their average verbal I.Q. appears to be slightly lower than the control value (Thalhammer et al. 1977).

Block B, a defect in the complex tyrosinase system, leads to a lack of the body pigment melanin normally produced from tyrosine by this enzyme. Albinism phenotypes caused by such mutations are widespread not only in humans, but in other animals as well. An albinism mutation is one of the seven "specific loci" mutations included in the most widely used test system for point mutations in the mouse (Chap. Ehling, p. 292ff.). Block C leads to tyrosinosis, another (group of) hereditary diseases, and block D to alcaptonuria, the abnormality that induced Garrod (1902) not only to describe for the first time a simple mode of inheritance for a human trait, but also to develop his concept of "inborn errors of metabolism". E, F, and G are defects in the synthesis of the thyroid hormone, all leading to hypothyreoidism with goitre (see Stanbury 1974).

These blocks in amino acid metabolism offer a good example for a variety of mechanisms and phenotypic effects of enzyme blocks. The clinical signs of PKU are

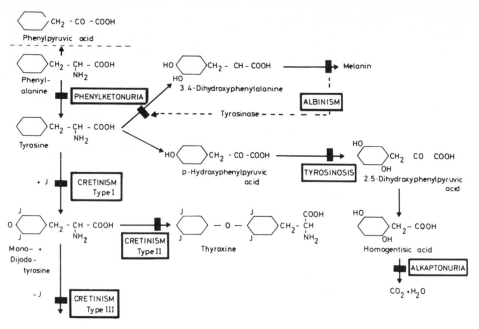

**Fig. 1.** Genetic blocks in metabolic pathways of some aromatic amino acids. The diagram is somewhat oversimplified. Genetic blocks leading to phenylketonuria, albinism, alkaptonuria, tyrosinosis, and three types of hereditary cretinism are included. (Vogel and Motulsky 1979)

caused by an overflow of a compound (phenylalanine) that can not be metabolized properly. Albinism and the various forms of hypothyreoidism, on the other hand, are due to a deficiency of metabolites. In distinction to the other defects, alkaptonuria leads to relatively minor ill-effects, obviously due to the fact that the enzyme controls a peripheral last step of metabolism.

Some other enzyme defects are more complex, affecting common units of two (or more) different enzymes, cofactors or additional components (for details see Stanbury 1978; Vogel and Motulsky 1979). Sometimes, analysis of genetic blocks has contributed appreciably to the elucidation of metabolic pathways. The lysosomal enzyme necessary for stepwise degradation of glycosaminoglycans provide an example; the mucopolysaccharidoses played an essential part in their analysis (Neufeld 1974, Cantz and Gehler 1976).

An especially interesting group of autosomal recessive diseases are the (proven and suspected) defects in DNA repair. In Xeroderma pigmentosum, an early stage of excision repair, the ability to excise thymine dimers, is deficient and this leads to reduction of the insertion of new bases (repair replication; unscheduled DNA synthesis; for details see Chap. van Zeeland, p. 35ff.). These conditions has become an interesting model system in human genetics for several reasons: It has helped in disentangling the various steps of DNA repair; and it has shown the usefulness of cell fusion techniques for demonstrating genetic heterogeneity (defects at different gene loci complement each other in fused cells, restoring the deficient function; mutations at the same loci do not complement).

Other suspected "reparatoses" are the chromosome instability syndromes. An unusually high rate of "spontaneous" chromosome breaks was first described in Franconi's anemia, a panmyelopathy leading, among other symptoms, to progressive anemia with fatal outcome (Schroeder et al. 1964; Schroeder-Kurth 1982). Chromosome instability is also found in Bloom's syndrome (German 1965, 1974) and ataxia-teleangiectasia (Hecht et al. 1966; see also Hecht 1977). All three conditions enhance the risks for various types of cancer, obviously due to somatic mutations. In Ataxia-teleangiectasia (and possibly also in Xeroderma) heterozygotes might also have an increased cancer risk (Swift 1976, 1979). Such an increased risk was also suspected in Fanconi's anemia but so far evidence is inconclusive.

Bloom's syndrome has provided us with the first "mutator" mutation in humans: A mutation that enhances the mutation rate at other gene loci (see Mohn and Würgler 1977; for details see Sect. 3.5).

Another common feature of these syndromes is the usually high effect of clastogenic agents on chromosome breakage (see also Chap. Natarajan, p. 156ff.).

## 2.1.2 Autosomal Dominant Diseases

The latest version of McKusick's catalogue enumerates 1801 autosomal dominant conditions, 918 confirmed, and 883 suspected. The often-cited rule of thumb that structural defects are mostly dominant, whereas recessives used to be enzyme defects (Lenz 1961), oversimplifies matters since there are too many exceptions. But it is certainly true that molecular mechanisms underlying major phenotypic manifestations in heterozygotes are more diverse and much less well-understood than the enzyme defects leading to many recessive diseases (Table 2). Anomalies in formation, aggregation, or degradation of structural and functional proteins, defects of membranes and receptors, or diminished feedback inhibition of enzymes might manifest themselves in heterozygotes, leading to dominant disease. For most of the characteristic conditions with pleiotropic manifestation of a gene primarily affecting a basic process in embryogenesis, the basic defects are still unknown. Some examples for dominant diseases will be described briefly in Sect. 3.2 (classical mutation rates). It is an unlucky coincidence that for those dominant phenotypes that offer themselves most

Table 2. Some mechanisms of dominant disease. (Vogel and Motulsky 1979)

| Mechanism | Example |
| --- | --- |
| Abnormal aggregation of protein subunits | Abnormal fibrinogens |
| Disturbance of multimeric protein function by abnormal subunits | Unstable hemoglobins |
| Diminished feedback inhibition by end product due to enzyme deficiency | Uroporphyrinogen synthetase deficiency in acute intermittent porphyria |
| Cell receptor defects | LDL cholesterol receptor defects in familial hypercholesterolemia |
| Cell membrane defects | Hereditary spherocytosis |
| Deposition of abnormal fibrillar protein | Hereditary amyloidosis (Portuguese type) |

readily for epidemiological studies and especially for mutation rate estimates the molecular mechanisms are still unknown.

Of potential interest for the problem of (somatic) mutation are the numerous examples of dominant inheritance for an increased susceptibility for common cancer – either on the basis of a precancerous condition (such as the various types of intestinal polyposis) – or without such harbinger (see Mulvihill 1977). Unfortuantely, the molecular mechanisms involved are still largely unknown.

### 2.1.3 X-Linked Diseases

McKusick (1983) knows 239 X-linked mutations, among them 116 sufficiently confirmed. Within this group, the most common anomalies are the disturbances in red-green color vision [deficiencies of the "red" component, protanopia and protanomaly; and those of the "green" component, deuteranopia and deuteranomaly: see Francois 1972]. In Caucasoid populations, they usually have a gene frequency and hence a frequency among male hemizygotes, of more than 5% altogether. The highest incidence of a severe hereditary disease in the X-linked group is Duchenne muscular dystrophy (see also Sect. 3 for the mutation rate and other aspects of mutation). All over the world, this disease, which has remained virtually untreatable so far, presents some of the most intriguing problems to the medical geneticist. The heterozygotes, for example, can sometimes be recognized by slight clinical anomalies or biochemical deviations from the norm. Some heterozygotes, however, even if confirmed by genetic evidence, for example as mothers of more than one affected son, usually have a completely normal phenotype. This can be explained by the Lyon hypothesis (Lyon 1961; see for discussion Lyon 1968; Vogel and Motulsky 1979); – i.e. random inactivation of one X-chromosome in early embryonic development. Analysis of an X-linked enzyme defect, hypoxanthine-guanine phosphoribosyl transferase-(HPRT)-deficiency, at the cellular level has contributed much to our knowledge not only on X-chromosomal activation but on spontaneous and induced mutations as well (see Sect. 3.5).

According to recent results, inactivation does not extend over the entire X-chromosome; the end of the short arm (bands Xp 22.13; 22.3; homologous to a part of the Y chromosome, and comprising, for example, the genes for the Xg blood types and the enzyme STS) remains active; see Schempp and Meer (1983).

In some instances, deviations from the norm in heterozygotes are so strong, and so important for health, that the condition must be regarded as X-linked dominant[2]. Even here, however, manifestation is, as a rule, milder in female heterozygotes than in male hemizygotes. In extreme cases, impairment of embryonic development in male hemizygotes is so severe that affected embryos die even before birth, and are aborted. This leads to pedigrees in which the condition is transmitted from affected mothers to 50% of their daugters only. In addition, the sex ratio in these sibships is shifted strongly to the female side and there are many miscarriages. The best-known examples are a skin disease, Incontinentia pigmenti, and a complex malformation

---

2  Usage of the terms "dominance" and "recessivity" in human genetics deviates somewhat from experimental genetics, following practical requirements. For details, see Vogel and Motulsky (1979)

syndrome, the oro-facio-digital (OFD) syndrome (for other examples see Wettke-Schäfer and Kantner 1983). For Incontinentia pigmenti, a few male cases have been observed, who show the same mosaic pattern as the affected females; and are, without exceptions, sporadic. They are now explained as half-chromatid mutations (Gartler and Francke 1975; Lenz 1975; Fig. 2).

### 2.1.4 Frequency of Diseases with Simple Modes of Inheritance in Human Populations

Despite a few attempts at estimating overall population frequencies of diseases with simple modes of inheritance (Stevenson 1959; Trimble et al. 1974), no really reliable figures are available. The following figures might at least hit the right order of magnitude: Autosomal-dominant and X-linked diseases together: $\approx 1\%$; autosomal recessives: $\approx 1\%_0$. In addition, one or a few percent might be ill-defined dominant conditions with very incomplete penetrance, variable expressivity and non-characteristic phenotype. Rough frequency estimates for the most common dominant and X-linked diseases are given in Table 3.

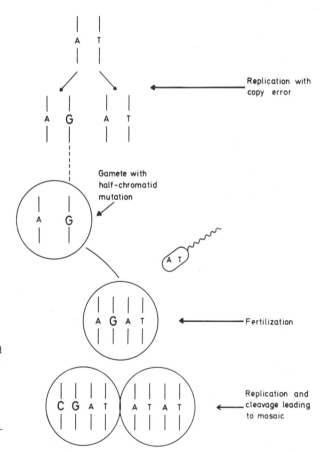

**Fig. 2.** Principle of half-chromatid mutations. The base replacement in one nucleotide half-strand occurred during the last DNA replication before meiosis. In the first cleavage division, the mutant cell-strand gives use to a cell clone having a gene mutation. (Vogel and Motulsky 1979)

**Table 3.** Estimated incidences at birth per 1,000 newborns for some of the most common dominant and X-lined diseases. (After Carter 1977)

| Incidence/1,000 live births (order of magnitude) | Diseases |
| --- | --- |
| 2.0 | Monogenic familial hypercholesterolemia |
| 1.0 | Dominant otosclerosis |
| 0.8 | Adult polycystic kidney disease |
| 0.5 | Multiple exostoses |
| 0.4 | Huntington's chorea |
| 0.2 | Hereditary spherocytosis, neurofibromatosis, Duchenne-type muscular dystrophy |
| 0.1 | Hereditary polyposis, dominant form of blindness, dominant form of early childhood onset deafness, dentinogenesis imperfecta, hemophilia A, dominant ichtyosis, nonspecific X-linked mental retardation |
| 0.04 | Osteogenesis imperfecta, Marfan's syndrome |
| 0.03 | Basilar impression, hereditary retinoblastoma, hemophilia B |
| 0.02 | Achondroplasia, acute intermittent porphyria, X-linked deafness, ocular albinism, nystagmus |
| 0.01 | Tuberous sclerosis, Ehlers-Danlos syndrome, osteopetrosis tarda, variegate porphyria, cleft lip and/or palate with mucous pits of lip, X-linked imperforate anus, X-linked aqueduct stenosis, hypogammaglobulinemia, hypophosphatemic rickets, anhidrotic ectodermal dysplasia, amelogenesis imperfecta |

In diseases impairing reproductive fitness of their bearers very much, such as Duchenne muscular dystrophy, achondroplasia or the Marfan syndrome, the population frequencies are maintained largely by new mutations (see the mutation rates in Sect. 3.2). For the more common diseases with late onset, such as monogenic hypercholesteralemia, mutation rates, and hence details of the interplay between mutation and selection, are completely unknown. In Huntington's chorea, new mutations appear to be very rare; practically all cases can be linked to large pedigrees (Wendt and Drohm 1972).

## 2.2 Genetic Polymorphisms

Mutations might lead not to phenotypes that are more or less "deficient", causing a hereditary disease in the dominant or recessive state, but are variants "in the normal range". There is no sharp limit between "normal" and "abnormal" mutations. On the one hand, a mutation usually leading to an abnormal phenotype might occasionally cause a very mild phenotypic deviation (variable expressivity), or manifestation might be wanting altogether (incomplete penetrance). On the other hand, however, there is increasing evidence for the influence of "normal" variants of cell surface antigens (blood groups; HLA types) and serum proteins or enzymes on "multifactorial" disease liabilities, susceptibilities to behavioral anomalies, such as addiction, or drug metabolism (see Motulsky et al. 1978; Vogel and Motulsky 1979).

Conventionally, a polymorphism is assumed if a monogenic trait exists in a population in at least two phenotypes (and presumably at least three genotypes), neither of which occurs with a frequency of less than 1%–2%. Often, there are more than two alleles and more than two phenotypes for a single locus.

In addition to alleles occurring in polymorphic proportion, rare alleles leading to "normal" phenotypes are observed frequently. Such alleles have been found especially for enzymes or other proteins that can be identified by electrophoresis; for their utilization for mutation screening see Sect. 3.6.

## 2.2.1 Blood Groups

Blood groups are special structures of the surface of red blood cells that can be identified by cell agglutination through specific antibodies (for reference see Race and Sanger 1975; Prokop and Göhler 1976). The ABO system has been known for more than 80 years, since individuals usually carry the antibodies against antigens not present on their own blood cells, very probably the result of early exposure to ubiquitous, antigen-carrying microorganisms. The ABO system has a widespread influence on susceptibility to many common diseases, including, for example, many cancers, thromboembolic diseases, peptic ulcers and very probably a number of common infections (see Vogel and Helmbold 1972; Mourant et al. 1978).

Blood groups systems are involved in serological mother-child incompatibility. Here, the Rh system is known best but other systems, including the ABO groups, might lead to the same complications. In addition to the well-known polymorphic antigens, many "private" blood groups, i.e. antigens identified only in single families or individuals are known, as well as their counterparts, "public" antigens present in almost all individuals of a population, and identified through an antibody formed by an occasional, antigen-negative individual.

There are hardly any sufficiently confirmed instances of new mutations for blood group alleles, ABO or otherwise, indicating very low mutation rates for these mutations. This is understandable since the biochemical differences produced by different alleles of one blood group system are very specific (for details see Watkins 1980).

## 2.2.2 Serum Groups and Isoenzymes

The second group of human polymorphisms is formed by serum protein groups (see Prokop and Göhler 1976; Putnam 1975) and enzymes (see Harris and Hopkinson 1976) found in serum and blood cells, especially erythrocytes. With introduction of many new electrophoretic techniques, the spectrum of known proteins has kept increasing. In part, this is certainly due to the easy availability of human blood for examination. Our knowledge of protein polymorphisms in other tissues is much less complete. Table 4 (from Vogel and Motulsky 1979) gives an overview over the most important blood group, serum protein and enzyme polymorphisms. According to Harris and Hopkinson (1972), 71 different gene loci for enzymes have been tested for the presence of polymorphic variants; in 20 (28.2%), such variants were actually found. Comparably high proportions of polymorphic loci were described in other

**Table 4.** Most important human polymorphisms

| Name | Main alleles | Remarks |
|------|--------------|---------|
| Erythrocyte surface antigens (blood groups) | | |
| ABO | $A_1$, $A_2$, B, O | |
| ABH secretor | Se, se | Interaction with the Lewis system |
| MNSs | MS, Ms, NS, Ns | There are some other, closely linked antigens; Hunter and Henshaw especially in blacks |
| P | $P_1$, $P_2$, p | Allele p is very rare |
| Rhesus | Gene complexes CDe, cde, cDE, $C^W$De, Cde, cdE, CDE, and others in varying combinations | Causes maternal-fetal incompatibility |
| Lewis | Le, le | Interaction with ABH secretor locus |
| Duffy | $Fy^a$, $Fy^b$, Fy | Amorphic allele Fy common in negroids |
| Kidd | $Jk^a$, $Jk^b$ | Very few individuals with Jk(a-b-) |
| Diego | $Di^a$, $Di^b$ | Allele $Di^a$ present only in Amerindians and Mongoloid populations |
| Lutheran | $Lu^a$, $Lu^b$ | |
| Kell | K, k | Other, closely linked loci, for example Sutter ($Js^a$) |
| Xg | $Xg^a$, Xg | X-linked |
| Serum protein groups | | |
| Haptoglobins | $Hp^{1S}$, $Hp^{1F}$, $Hp^2$ | Many rare variants are known; genetic and nongenetic ahaptoglobinemia occurs |
| Transferrins | $Tf^C$, $Tf^B$, $Tf^D$ | Different D and B variants have been described, all of them are rare. D variants occur mainly in blacks |
| Group-specific components (Gc groups) | $Gc^1$, $Gc^2$ | Special variants described, for example $Gc^{Chip}$ in Chippewa Indians, $Gc^{Ab}$ in Australian aborigines; subdivision of common alleles possible by isoelectric focusing |
| Protease inhibitor (Pi) | $Pi^M$, $Pi^F$, $Pi^S$, $Pi^Z$ | Numerous rare alleles. $a_1$ antitrypsin deficiency, especially in homozygotes of the $Pi^Z$ allele may lead to obstructive lung disease. |
| Ceruloplasmin | CpB, CpA, CpC | Most Europeans are homozygous CpB/CpB; blacks have a gene frequency of 0.06 for CpA |
| Immunoglobulins | | |
| Gm | $Gm^{3.5}$, $Gm^{1.2}$, $Gm^1$ | This is a very complicated system with many rare haplotypes and specificities. |
| Km | $Km^1$, $Km^2$ | More alleles are known, but sera are usually not easily available |
| Complement polymorphisms C3 | $C3^F$, $C3^S$ | Apart from these two common alleles, there are a number of rare ones |

**Table 4** (continued)

| Name | Main alleles | Remarks |
|------|--------------|---------|
| Red cell enzymes | | |
| Acid phosphatases | $acP^A$, $acP^B$, $acP^C$ | An additional allele $acP^r$ has been observed in Khoisanids |
| 6-phosphogluconate dehydrogenase | $PGD^A$, $PGD^B$ | Some other, rarer alleles are known |
| Adenylate kinase | $AK^1$, $AK^2$ | Some other, rarer alleles are known |
| Peptidase A | $PepA^1$, $PepA^2$ | $PepA^2$ has a gene frequency of about 0.07 in blacks; whites have almost exclusively $PepA^1$. Rare variants are known |
| Peptidase D | $PepD^1$, $PepD^2$, $PepD^3$ | $PepD^3$ observed especially in blacks |
| Phosphoglucomutase | | |
| $PGM_1$ | $PGM_1^1$, $PGM_1^2$, | Rare alleles are known |
| $PGM_2$ | $PGM_2^1$, $PGM_2^2$, | Allele $PGM_2^2$ only in blacks other alleles very rare |
| $PGM_3$ | $PGM_3^1$, $PGM_3^2$ | Enzymes only in leukocytes, placenta and sperms. Linkage with the major histocompatibility complex (MHC). |
| Esterase D | $EsD^1$, $EsD^2$ | Rare variants are also known for esterase A |
| Adenosine deaminase | $ADA^1$, $ADA^2$ | Rare alleles $ADA^3$ and $ADA^4$ have been described |
| Some other enzyme polymorphisms | | |
| Liver acetyltransferase | Rapid and slow inactivators | |
| Serum pseudocholin-esterase | $E_1$, $E_1$, $E_1^s$ | |
| Alcohol dehydrogenase | $ADH_3^1$, $ADH_3^2$ | $ADH_2$ active in other organs; $ADH_3$ active in the intestines |
| Glutamate pyruvate transaminase | $GPT^1$, $GPT^2$ | |

species (mouse; *Drosophila*), as well. For structural proteins, for example brain proteins (Comings 1982a,b,c,d,e), the fraction of polymorphic loci appears to be much lower. According to a recent study (Klose et al. 1983; additional references in the same paper), "genetic variability occurs to quite different degrees in different classes of proteins: structure-bound proteins < soluble non-enzymatic proteins < enzymes". Moreover, quantitative variability appears to be much more important in structure-bound proteins of fibroblasts than qualitative variability as revealed, for example, by differences in speed of electrophoretic migration. In addition to mutations occurring in polymorphic frequencies, many, very rare mutants have been observed involving in part the same gene loci (Harris et al. 1974). Their frequency distribution gave rise to discussions as to whether the majority of polymorphisms were produced by natural selection, or by random processes ("neutral" mutations; for discussion see Vogel and Motulsky 1979) (see Table 4).

### 2.2.3 Transplantation Antigens

Transplantation antigens are specific structures of the cell surface of most cells (including lymphocytes) that play a major role in rejection of tissue transplants, for example skin grafts or kidneys. They can be identified using specific antibodies in appropriate assays, primarily the lymphotoxicity test (Svejgaard et al. 1979; Terasaki 1980). So far, four different, and closely linked, HLA loci have been discovered. They are located together with some genes involved in the complement part of the immune answer on the short arm of chromosome 6. The loci that have become known so far, are named HLA-A, B, C, and D – all with many different alleles (Fig. 3). A characteristic feature of the HLA types is their occurrence in "linkage disequilibrium"; i.e. the various alleles of the four loci do not occur in random combinations; rather some combinations of these alleles (haplotypes) are more common than expected, others are rarer. For the most part, this is very probably due to natural selection favoring certain haplotypes. The precise mechanisms of selection are still unknown; hints are provided by the numerous disease associations of HLA types. For example, antigen B 27 is almost 100 times more common in Bechterew's disease than other HLA types. Other disease associations involve mainly "rheumatic" and autoimmune diseases of various kinds. Since the HLA specificities are known to play a major role in the cellular and tumoral immune answer, selection might easily have worked though interaction with pathogenic agents, for example viruses (see Dorf 1981; Brown 1979; Vogel and Motulsky 1979).

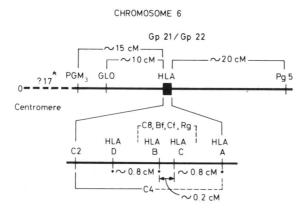

**Fig. 3.** Linkage group of MHC on chromosome 6. The HLA complex is ≈ 15 cM (centimorgans) away from the gene locus for PGM₃ and 10 cM from the locus for the enzyme glyoxalase (GLO). Within the HLA complex, the sequence D-B-C-A is the most likely. In the same region, other gene loci involved in the immune response are located, for example those determining complement factors (C₂, C₄, C₈, Bf)

Transplantion antigens have also been analyzed in other species, especially the mouse (see Götze 1977). The H2 antigen loci of the mouse are homologous with the human HLA loci. In addition to these loci, there are other genes influencing draft rejection. Since, according to some reports, spontaneous mutation rates in inbred strains as measured by draft rejection were unusually high, some attempts were made at using draft rejection as an in vivo point mutation test. Preliminary results are promising, but not yet entirely convincing (Harnasch 1982).

## 2.3 Contribution of Genes with Simple Modes of Inheritance to Complex Phenotypes, Especially Diseases

Despite the impressive number of genetic diseases with simple modes of inheritance and genetic polymorphisms, there must be many more gene loci for which mutations have not become known so far. This might have various reasons: for example, mutations involving functionally essential metabolic pathways have been observed very rarely. This can be explained by the assumptions that these mutations might not be compatible with surviving, leading to early – and undiscovered – embryonic death.

Other, recessive, mutations might be so rare that homozygotes have simply not been observed. Some of them will certainly become known when populations not covered so far by medical genetics research will be screened. China, India, and Africa are good candidates.

But there are certainly many mutations leading to non-characteristic phenotypes that are hidden in the huge heap of malformations, diseases and disease liabilities commonly classified as "multifactorial". For example, many people, especially in middle and advanced age, have high levels of blood lipids such as cholesterol. In addition to genetic factors, these lipid levels are influenced strongly by the environment, especially food habits and smoking. With increasing precision of lipid determination, however, it became possible to isolate several subtypes, some of them more or less clearly exhibiting a simple mode of inheritance. One of these subtypes, hypercholesterolemia type II a, was shown relatively early to be dominantly inherited; however, since the condition is relatively common, homozygotes were observed occasionally. Their serum cholesterol values were still higher. This anomaly increases the risk of coronary heart disease. Homozygotes normally are hit by cardiac infarction in the third decades of their life; many heterozygotes are affected during their fourth, fifth or sixth decades. Biochemical studies revealed a specific defect of cell receptors for low density lipoprotein (LDL), and cholesterol-containing plasma component. Normally, LDL-bound cholesterol enters the cell, inhibiting cholesterol synthesis. This feedback circle is interrupted (partially) by diminished (or, in homozygotes completely lacking) receptor activity (Brown and Goldstein 1976).

This example shows how combined genetic and biochemical studies succeeded in isolating a single genetic variant out of a heap of phenotypically similar conditions. Similar instances exist, for example, for common birth defects such as congenital heart disease, cleft lip and palate, or spina bifida. Studies in mouse radiation genetics give rise to the suspicion that such "hidden" single gene mutations might be much more common than hitherto expected: Parental irradiation leads to mutations causing extremely pleiotropic patterns with low penetrance and extremely variable expressivity (Selby 1978).

In humans, such major gene effects can hardly be identified for purely practical reasons.

In conclusion, hereditary diseases with simple modes of inheritance represent the tip of the iceberg of all single mutation affects in human pathology. In considerations on phenotypic effects of induced mutations, this has to be kept in mind.

## 3 Mutation Rates

A human mutation rate is defined as probability for the occurrence of a certain mutation per fertilizing germ cell per generation. This definition is different from that frequently used in microbial genetics where a mutation rate is calculated per cell division, or in *Drosophila* genetics where mutation rates are often given for groups of mutations, for example X-linked lethals. So far, mutation rates in human populations have been estimated for chromosome aberrations (see Chap. Sperling, p. 128ff.), and for some point mutations leading to clearcut phenotypes (hereditary diseases with dominant or X-linked recessive modes of inheritance). Some preliminary data are also available for mutations identifiable at the protein level.

### 3.1 General Remarks and Methods

Estimation of a mutation rate in a population requires a careful study of this population with epidemiological methods. Such a study normally starts with a determined effort to ascertain all cases and families that could belong to the disease in question in a limited population, which, however, should not be too small; 5 to 10 million people is a good choice. Such a search for cases is often very difficult logistically; it requires patience and readiness to achieve one's goal in an effort of many years on the side of the investigators, and cooperation with many doctors and other professions. Such studies were carried out mainly in the 1940's and 1950's in Denmark, the state of Michigan (USA), in Northern Ireland, and at a smaller scale also in Germany, Scandinavia, Japan, the USSR, and in some other countries. It is no mere coincidence that such activities abated with the advent of new and fascinating opportunities for laboratory research such as human cytogenetics; single cell genetics; and molecular biology. These new fields promised much faster and more impressive success. Therefore, many of the available results on human mutation rates have a somewhat outdated data base; often, more recent studies on genetic heterogeneity, based on a more thorough biochemical analysis, could not be considered. Important achievements can be expected from studies in which newly developed laboratory methods will be combined with the epidemiological approach.

Estimation of human mutation rates goes back to a classical paper of Haldane (1935), in which he estimated the mutation rate for hemophilia that has stood the test of almost 50 years. In this paper, he developed the basic rationale of such studies: if there is an appreciable amount of natural selection against a certain mutant, a genetic equilibrium will soon be established between selection and new mutation: About as many mutant genes will be lost per generation, as are added by new mutation. This equilibrium will be stable; i.a. after random deviations there will be a tendency for return to equilibrium frequencies. Therefore, by measuring the rate of selection against a certain mutation, one obtains an estimate of its mutation rate.

The "indirect method" for mutation rate estimation is founded on this rationale. The following formulas are conventionally used:

$\mu = 1/3 (1-f) x'$ for X-linked recessive mutations;   and
$\mu = 1/2 (1-f) x$ for autosomal-dominant mutations

($\mu$ = mutation rate; f = relative fertility of carriers of the mutation, average fertility of the reference population being defined f = 1; x = frequency of the condition in the reference population; for example, incidence at birth for a disease manifesting itself at birth; x′ = frequency of the condition in the population of males). Obviously, the method can be expected to give unbiased estimates only if the basic requirement of an equilibrium is fulfilled. This is not the case if selection against a certain mutation has changed recently as has occurred, for example, with hemophilia A. Only a few decades ago this disease still killed many patients, reducing their average relative fertility f to $\approx$ 1/3, whereas at present, due to factor VIII substitution, patients have a near normal life. It is an open question to which degree natural selection has been replaced in hemophilia by artificial selection (genetic counseling; voluntary reduction of reproduction). In autosomal recessive diseases, the required equilibrium between mutation and selection will almost never exist, and in any case, cannot be tested, mainly since heterozygotes are so much more common than homozygotes (Sect. 2.1.1), and small selective advantages or disadvantages that can hardly be assessed in human populations, will have a much stronger effect than even the total loss of reproductive capability in homozygotes. Therefore, mutation rate estimates for autosomal-recessive diseases are of no scientific value.

As soon as the principle was established that in dominant and X-linked conditions which appreciably diminish reproductive capacity a high fraction of affected individuals in a population should be new mutants, a still simpler approach to mutation rate estimation, the "direct method", was introduced. It simply consists of counting, for an autosomal dominant condition, all "sporadic cases", a sporadic case being defined as a patient with unaffected parents and no other affected family members (in X-linked diseases, the direct method cannot be applied). The following formula is used:

$$\mu = \frac{\text{number of sporadic cases}}{2 \times \text{reference population}} \cdot$$

The 2 in the denominator is due to the convention to denote a mutation rate per fertilizing germ cell, and not per individual.

Simple as they are, successful practical application of these methods depends on a number of preconditions that must be observed with scrupulous care. The requirement of a genetic equilibrium for the indirect method has already been mentioned; the direct method requires, for example, that all "sporadic" cases are actually caused by new mutations in the germ cell of one of the parents. But non-hereditary cases with identical or very similar phenotypes have been observed in some diseases. A major problem is genetic heterogeneity. These requirements cannot be discussed here in greater details (see Vogel and Rathenberg 1975).

## 3.2 "Classical" Mutation Rates

Table 5 gives a selection of "classical" mutation rate estimates. Almost all these estimates could be criticized on some technical accounts, for example genetic heterogeneity (see Vogel and Rathenberg 1975; Vogel and Motulsky 1979). They should not

**Table 5.** Selected "classical" mutation rates for human genes

| $\leqslant 10^{-4}$ | | $> 10^{-4} - 10^{-5}$ | | $< 10^{-5} - 10^{-6}$ | | $< 10^{-6}$ | |
|---|---|---|---|---|---|---|---|
| Trait | No of series | Trait | No of series | Trait | No of series | Trait | No of series |
| Neuro-fibroma-tosis | 1 | Achondro-plasie | 3 | Aniridia | 2 | v. Hippel-Lindau's disease | 1 |
| | | Osteo-genesis imper-fecta | 2 | Dystrophia myotonica | 2 | | |
| | | | | Retinoblas-toma | 5 | | |
| | | Polyosis intestini | 1 | Apert's syndrome | 2 | | |
| | | Polycystic Kidney disease | 1 | Tuberous slerosis | 2 | | |
| | | Haemo-philia (A) | 5 | Marfan's syndrome | 1 | | |
| | | Duchenne-type mus-cular dys-trophy | 10 | Diaphyseal aclasis | 1 | | |
| | | Incontinen-tia piamenti | 1 | Haemo-philia (B) | 2 | | |
| | | | | OFD Syn-drome | 1 | | |

be taken too strictly; but there are good reasons to assume that they give, at least, the correct order of magnitude. For some mutations, estimates are available from more than one population; for the Duchenne muscular dystrophy, no less than 10 different estimates can be compared. It is comforting to know that, wherever different estimates are available, they agree fairly well. Naturally, all estimates relate to populations living under highly civilized and industrialized conditions, and most have been carried out in Caucasoid populations. Data from Japan, however, do not deviate. It would be of high interest to have comparable figures from populations living under completely different circumstances, for example as hunters and gatherers, or in a mainly agricultural setting, such as most people in India or China.

We have good reason to assume that the few mutation rates of Table 5 are not an unbiased sample of all human mutation rates. As mentioned, mutation rate estimation requires complete ascertainment of all cases in a defined population. Such an epidemiological survey becomes unbearably cumbersome and costly when the trait is too rare and the population to be screened becomes too large. However, it is a common experience of medical geneticists that very many hereditary diseases with simple modes of inheritance, in fact, the great majority of all diseases enumerated in McKusick's catalogue, are, indeed, extremely rare. Table 5 contains, in fact, many of the most common ones. Therefore, it can reasonably be assumed that the mutation rates of Table 5 are a selection of the highest human rates for mutations leading to well-defined autosomal-dominant or X-linked diseases (see also Stevenson and Kerr 1967). Genetic heterogeneity might be another systematic bias spuriously increasing these estimates. On the other hand, the gene loci involved might be capable of mutating not only to alleles producing these genetic diseases, but to other, much more harmless alleles, as well. Obviously, this part of their mutability cannot be covered by the methods available at present. As mentioned above (Sect. 2.1.2), it is an unlucky coincidence that for almost all diseases for which reasonably reliable mutation rate estimates are available, excepting hemophilia B, the basic molecular defects are still unknown.

## 3.3 Influence of Paternal Age and Sex

The influence on the probability for nondisjunction of maternal (and, possibly, at a much smaller rate, also of paternal) age is well-known (Chaps. Sperling, p. 128ff., and Hansmann, p. 147ff.). But there is also some influence of parental (in this case exclusively or almost exclusively paternal) age on the rate of some point mutations. Penrose (1955) was the first to demonstrate convincingly an increase of paternal age for the rate of the dominant mutation leading to the most common type of achondroplasia, an anomaly affecting longitudinal growth of the long bones of limbs (and the cranial basis). Here, the risk of fathers 45 years and older having a child carrying this mutation is about six times that of fathers of the age of 20–25.

Meanwhile a paternal age effect has also been demonstrated for a number of other autosomal dominant conditions such as acroenphalosyndactyly, Marfan's syndrome, myositis ossificans, and even for maternal grandfathers of sporadic hemophilia patients (who are the main source of new mutations) and for the Lesch-Nyhan syndrome (see Vogel and Rathenberg 1975; Francke et al. 1976) (Fig. 4). More recently, an increase with paternal age has also been shown for a group of biochemically well-defined mutations – those leading to a dominant hemoglobinopathy, unstable hemoglobins (Stamatoyannopolous et al. 1981). An increase of the mutation rate only with paternal, not with maternal age is not surprising considering the general experience from molecular biology that most (not all!) gene mutations are caused by copy errors in DNA replication (see Drake 1969). The human spermatogonium undergoes many more cell divisions and, hence, DNA replication cycles during its history from early embryonic age to the mature, fertilizing sperm than the oocyte, and the number of cell divisions in the male increases with increasing age. In the female, all oocytes are ready a short time after birth, and afterwards, there are only the two meiotic divisions, irrespective of age (for a detailed discussion of germ cell kinetics in both sexes see Vogel and Rathenberg 1975).

However, some other dominant mutations fail to show a clearcut increase with paternal age, or the parental age effect was small; viz. retinoblastoma, neurofibromatosis, polyposis coli, aniridia, and osteogenesis imperfecta. In osteogenesis imperfecta, this is probably due to admixture of recessive cases, since genetic hetero-

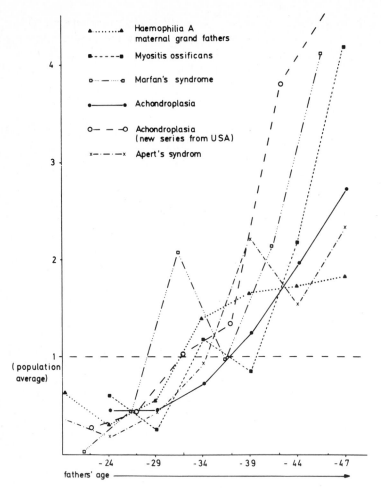

**Fig. 4.** Increase of the mutation rate with paternal age in some conditions. (For details see Vogel and Rathenberg 1975)

geneity is extensive. Interestingly enough, the three first-mentioned conditions are tumor diseases. This points to a different molecular mechanism for these mutations, possibly an insertion of a virus genome (Vogel 1980)?

Explanation of new mutation by copy errors predicts not only an increase of the mutation rate with paternal age, but also a higher rate in male than in female germ cells. This prediction was confirmed for hemophilia (see Vogel and Rathenberg 1975) and for the Lesch-Nyhan syndrome [deficiency of the enzyme hypoxathine-guanine phosphoribosyl transferase-HPRT (Francke et al. 1976)]; but not for the third X-linked disease for which appropriate data are available, Duchenne muscular dystrophy. Moreover, such a sex difference apparently exists for dominant mutations showing a strong paternal age effect (Vogel 1983). For Duchenne muscular dystrophy, the X-linked condition with the highest mutation rate, rates appear to be similar in both sexes. A special molecular mechanism – unequal crossing over involving a pseudogene – has been suspected (Winter and Pembrey 1982). Mouse data point to a higher point mutation rate in males of this animal (Searle 1972).

## 3.4 Mutations at the Molecular Level

Data on the molecular nature of human point mutations are available mainly for the hemoglobin gene families located on chromosomes 11 ($\beta$-family) and 16 ($\alpha$-family), respectively (see Shows et al. 1982). For hemoglobins in general and, specially, for molecular mechanisms see also Bunn et al. (1977), Fairbanks (1980), and Stamatoyannopolous (1981). Figure 5 gives an impression of the relative frequencies of several primary events among the mutations affecting transcribed DNA sequences of hemoglobin genes. By far the most common type of molecular change is the single base substitution but deletions, chain elongations due to mutation of a stop codon, and frameshifts have also been observed. Lepore and anti-lepore-type mutations might be caused by deletions, but might also be recombinational events linking, for example, one sequence of a Hb$\beta$ gene with another sequence of a $\delta$-gene. Such recombinations may be caused by unequal crossing over between these structurally very similar genes.

In addition, there are a number of diseases in which the rate of protein synthesis is reduced or completely abolished, leading to a group of anemias called thalassemias. Here, many deletions involving smaller or larger parts of Hb$\beta$ genes may be involved.

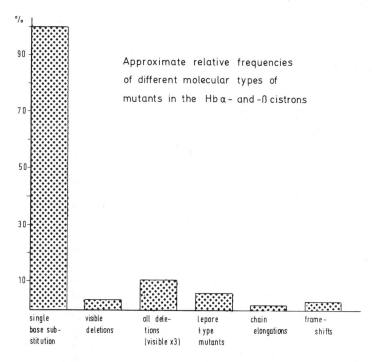

**Fig. 5.** Relative frequencies of molecular events among mutations leading to Hb variants

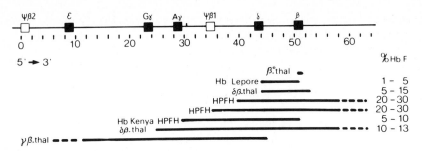

**Fig. 6.** Hbβ family. (Flavell et al. 1982)


Figure 6 shows the Hbβ gene family together with a number of deletions. Remarkable is, among others, the βo thal mutation – a deletion affecting the β gene itself. In other thal-o mutations, stop codons within the exons had been formed by a base replacement (pos. 39, and 17, respectively). In still other cases, mutations in the pre- or poststructural part of the gene might be involved. Apparently, some mutations lead to incomplete processing of mRNA or to other posttranscriptional anomalies. α-Thalassemias, on the other hand, are most often caused by a deletion of one or more Hbα-genes. Surviving of individuals carrying such an extended deletion is possible because Hbα genes are present in duplicate on one chromosome.

This brief survey of various molecular mechanisms involved in hemoglobin point mutations should remind us that apparently simple events at the level of phenotypes that are dealt with in considerations on "classical" mutation rates will very probably comprise a variety of different primary events when analyzed at the molecular level.

Attempts have been made to estimate relative probabilities for various types of base pair replacements (transitions and transversions). In principle, such inferences can be drawn even from amino acid, RNA or DNA sequences using the genetic code. They are made more difficult, however, by the all-pervading influence of natural selection. Still, there is now fairly good evidence that transitions in general may occur more often than expected, if direction of mutation would be entirely random (Vogel and Kopun 1977).

On the basis of very slender factual evidence for hemoglobin variants, attempts have been made at estimating "codon mutation rates", i.e. mutation rates for one specified base replacement (see Vogel and Rathenberg 1975; Vogel and Motulsky 1979): The estimates are in the order of magnitude of $2.5 - 5 \times 10^{-9}$; and, according to a more recent study that was based on observations of unstable Hb variants or HbM variants, between $5.9 \times 10^{-9}$ and $18.9 \times 10^{-9}$ (Stamatoyannopolous et al. 1982), not too poor an agreement with "classical" mutation rates of Table 5, assuming an average length of $\approx 1,000$ nucleotides for a structural gene.


## 3.5 Single-Cell Mutations

Since Pontecorvo's book *Trends in Genetic Analysis* was published in 1959, genetic analysis at the level of the single cell has become increasingly popular, and successful in various branches of human genetics, for example biochemical genetics, or linkage and assignment of genes to specific chromosomes. Indeed, it is fascinating to visualize a theoretical increase in discriminatory power of the order of magnitude of $10^9$ or more, from the human individual to the single cell. Attempts at utilizing single cells in work on (spontaneous and induced) mutation date back for more than 20 years. Single-cell systems for assessing gene and chromosome mutations in vitro and in vivo after treatment with mutagenic agents will be dealt with in other chapters; here, some remarks on "spontaneous" mutations might be pertinent.

Assessing a point mutation in a single cell requires a way of identifying this cell. This identification should be very effective, since mutation rates at the single-cell level can be expected to be very low. A variety of such systems have been developed (see Chu and Powell 1976). For example, mutants deficient in the enzyme HPRT can be selected since they survive poisoning with 8-azaguanine or 6-thioguanine, toxic hypoxanthine analogues that cannot be taken up into the cell due to the enzyme defect; HPRT revertants, on the other hand, can be selected by the so-called HAT medium; chloramphenicol-resistant HeLa cell mutants have been described; changes of HLA specificities were observed in single cells; and newly occurring G6PD variants were observed.

"Spontaneous" mutation rates observed with these methods range from $\approx 10^{-4}$ to $\approx 10^{-8}$ (Vogel and Rathenberg 1975). There has been considerable discussion as to the nature of these changes in cell phenotypes. Many of them appear to be genetic in origin (see the discussion in Chu and Powell 1976); but some evidence points to epigenetic modification in several of them; and the rate of "mutation" often depends on culture conditions. Therefore, the contribution of single-cell mutation to our understanding of "spontaneous" mutation in man has so far not been overwhelming.

Interestingly enough, however, an appreciable increase of the "spontaneous" mutation rate was observed in fibroblasts derived from two individuals with Bloom's syndrome compared with cells from normal controls (Warren et al. 1981). As mentioned in Sect. 2.1.1, Bloom's syndrome belongs to the chromosome instability syndromes; this observation establishes the Bloom's mutation as a human mutator gene, not only for structural chromosome changes but for gene mutations as well.

## 3.6 Point Mutation Systems for Monitoring Human Populations

Ever since Muller (1927)[3] discovered the mutagenic effect of ionizing radiation and especially, since the first chemical mutagens have been described by Oehlkers et al. (1943), Auerbach and Robson (1946), and Rapoport (1946), geneticists have warned against a possible increase of the human mutation rate due to increased exposure to mutagenic agents in our modern civilization. Muller (1950), in a paper entitled *Our load of mutation*, painted a picture of the future of our species in very dark colors[4]. It is all the more surprising that, so far, no workable rationale for monitoring entire

---

3   The ability of ionizing radiation to enhance the frequency of nondisjunction had already been discovered by Mavor (1924); but this important achievement has never been appreciated appropriately

4   Recently, the positions taken by Muller in this paper have been reconsidered in the light of new evidence by Vogel (1979)

human populations for possible changes in natural mutation rates has been developed. It has become possible to monitor relatively small population groups with especially strong exposure to mutagenic agents (for examples workers exposed to mutagenic chemically) by examining blood cells for an increase of chromosome aberrations or SCEs (see, for example, Hook 1982; Sutton 1982; Chaps. Natarajan, p. 156ff., Gebhart, p. 198ff., and Obe et al., p. 223ff.). This, however, gives little, if any, hint as to whether there is an increase of the mutation rate *in the general population* – possibly due to some undiscovered environmental agent. The main statistical problem with all possible monitoring systems is that mutagenic agents do not produce anything entirely new (contrary to some teratogenic agents such as thalidomide which produced a malformation syndrome that had almost never been observed before, and was therefore easily recognized). They only enhance the natural mutation rate, producing a somewhat higher number of events that had always occurred. Discovery of such an increase obviously requires precise knowledge of present-day incidence of such events.

diseases have been enumerated. At first glance, monitoring such diseases as "sentinel mutations" (Neel et al. 1973) seems to be relatively easy, and this approach has, indeed, often been suggested. However, it was shown by statistical model calculations that discovery, for example, of a mutation rate increase of (at least) 10% between two 10-year periods would require careful and continuous screening of a population of many millions of newborn (Vogel 1970). In theory, this is, of course, possible. But it meets with almost insuperable technical and logistic problems. For example, a team of specially trained doctors would have to examine all persons who could possibly be affected; and extensive family studies would be necessary, posing social problems, such as possible discrimination, data protection etc. So far, no society has ever attributed such high priority to the problem of mutation screening that these difficulties could be overcome. It would be much easier to monitor populations for chromosome aberrations. In many countries, an increasing fraction of the female population above the age of 35 undergoes prenatal diagnosis for chromosome aberrations; these data could easily be used for population monitoring. Moreover, there would be great immediate advantages of a newborn screening for gonosomal aberrations, since behavioral abnormalities in children having such abnormalities could probably be prevented by adequate education. Even so, monitoring programs have not been started on a larger scale.

Another possibility is the introduction of protein screening using, for example, electrophoretic methods. This approach is very attractive for a priori reasons, since effects close to the gene level are covered. Moreover, extensive experience with genetic polymorphisms (Sect. 2.2) has shown feasability of large-scale studies.

*Two Different Approaches to This Problem Are Being Developed:*

Following suggestions by J.V. Neel, an American group is developing methods for screening blood for as many different proteins as possible in order to detect qualitative and quantitative variants that could have occurred by mutation. In the USA, electrophoretic and other methods are being developed and a large-scale pilot program is under way in which feasability is being tested. The program requires collecting chord blood from newborn together with blood samples from both parents (Neel 1982).

Obviously, such a program requires a relatively large logistic structure of its own; blood sampling, storing, and transportation has to be organized. Moreover, application of all available methods to these blood specimens is relatively expensive. Now, Neel (1982) is certainly correct in asserting that the expenses of not doing proper studies are much higher; for example, due to possibly unnecessary precautions and delays in industrial development because of unjustified public fears. However, these expenses are largely indirect and not immediately felt, and their consideration will therefore always have low priority for political agencies responsible for assignment of funds. Such electrophoretic methods have been used in Japan, in children of individuals exposed to the atomic bombs in Hiroshima or Nagasaki (Satoh et al. 1982). Each child was examined for rare (= phenotype frequency less then 2%) variants of 28 proteins of blood plasma and erythrocytes, and in a part of the sample, deficiency variants of 10 erythrocyte enzymes were added. When such variants were found in the child but not in one of the parents, extensive paternity testing was added. Two enzyme variants of probable mutational origin were observed in this study among 419,666 tested gene loci in children of exposed parents and 282,848 loci in the control group, both among children of irradiated parents.

A preliminary estimate based on these figures would lead to a mutation rate similar to the "classical" rates of Table 5. This demonstrates the amount of work necessary for monitoring a population based on protein variants. Attempts at drawing conclusions regarding mutations from more indirect evidence, such as frequency differences of electrophoretic variants between different populations (Neel and Rothmann 1978, 1981) are unsatisfactory, since such differences could also have other causes.

The main source of expense in the monitoring system described above is not the laboratory work but the organization necessary for collecting blood samples. It was therefore an obvious question whether these expenses could be saved by coupling a monitoring program to screening programs already under way in human populations and more immediately useful for public health. In such a monitoring programm, logistics should not be adapted to the requirements of laboratory techniques, but laboratory techniques should be adjusted to pre-existing logistics.

Based on this "jump on the bandwagon"-principle, methods were developed for utilizing part of the dried blood spots collected in most populations of developed countries for testing for phenylketonuria (PKU) and other inherited metabolic diseases (Vogel and Altland 1982; Altland 1982). It has now been shown that most hemoglobin variants can be identified in minute amounts of dried blood by a very inexpensive variety of isoelectric focusing. The only logistic requirement for practical application is that the units for PKU screening do not put their abandoned microtiter plates (or postcards with blood spots) into the dust bin, but keep them in the refrigerator until they are collected by a staff member of the protein laboratory. In a successful test run with 25,000 blood specimens in Hessen, Germany, 51 Hb variants were found; one of them was identified by extensive paternity testing as a fresh mutant. Already now, frequency and distribution of the other, familial variants pose interesting population genetic problems (Altland 1982). Presently, the method is being extended to include several additional proteins.

Both approaches – the Neel approach and the Vogel-Altland approach – have their advantages and disadvantages. It is a big advantage of the Neel approach (examining child-mother-father triplets for as many as possible different variants) that a relatively comprehensive pattern of mutational events can be covered in one individual. In addition to the smaller number of families to be included, this is also useful since studies on spontaneous and induced mutations in the mouse have shown that the "spontaneous" mutation rate does not necessarily indicate the degree of susceptibility of a certain gene locus to a mutagenic agent. Therefore, limitation of a test system to only very few genes could easily lead to under- or overestimation of a mutation rate trend if inferences are drawn for the entire genome.

On the other hand, it is the advantage of the Vogel-Altland approach that almost no special organization needs to be established. This together with the highly rationalized laboratory procedure, makes the method much less expensive, allowing, in principle, application to very large populations with realistic expenditure. The disadvantage of generalized conclusions from too limited a set of genes will be diminished up to a certain degree by inclusion of some additional proteins.

In conclusion, both approaches have their merits and limitations. Probably, the Neel approach will sometime be used for monitoring of limited, high-risk populations whereas the Vogel-Altland approach will be applicable for long-term screening of very large populations. Both together may help to protect the human gene pool as far as possible from mutational damage.

# References

Altland K, Kaempfer M, Forssbohm M, Werner W (1982) Monitoring for changing mutation rates using blood samples submitted for PKU screening. In: Human Genetics, part A: The unfolding genome, p 277–287. Liss, New York

Auerbach C, Robson JM (1946) Chemical production of mutations. Nature 137:302

Bartholomé K (1979) Genetics and biochemistry of phenylketonuria-present state. Hum Genet 51:241–245

Bora KG, Douglas GR, Nestmann ER (eds) (1982) Chemical mutagenesis, human population monitoring and genetic rsik assessment. Progress in Mutation Research, vol 3. Elsevier, Amsterdam

Brown WE (1979) HLA and disease: a comprehensive review. CRC, Boca Raton

Brown MS, Goldstein JL (1976) New directions in human biochemical genetics: understanding the manifestation of receptor deficiency states. Prog Med Genet (new series) I:103–119

Bunn HF, Forget BS, Ranney HM (1977) Human hemoglobins. Saunders, Philadelphia London Toronto

Cantz M, Gehler J (1976) The mucopolysaccharidoses: inborn errors of glycosaminoglycan catabolism. Hum Genet 32:233–255

Carter CO (1977) Monogenic disorders. J Med Genet 14:316–320

Chu EHY, Powell SS (1976) Selective systems in somatic cells genetics. In: Harris H, Hirschhorn K (eds) Adv Hum Genet, pp 189–258. Plenum, New York London

Comings DE (1982) Two-dimensional gel electrophoresis of human brain hypothesis. I. Technique and nomenclature of proteins. Clin Chem 28/4:782–789

Comings DE, Garraway NG, Pekkula-Flagan A (1982) II. Specific proteins and brain subfractions. Clin Chem 28/4:790–797

Comings DE (1982) III. Genetic and non-genetic variations in 145 brains. Clin Chem 28/4:798 to 804

Comings DE (1982) IV. Disorders of glial proliferation and a polymorphism of glial fibrillary acidic protein – GFAP Duarte. Clin Chem 28/4:805–812

Comings DE, Pekkula-Flagan A (1982) V. Non-equilibrium gel electrophoresis with detection of a myelin basic protein mutation – MBL Duarte. Clin Chem 28/4:813–818

Drake JW (1969) The molecular basis of mutation. Holden Day, San Francisco

Dorf ME (1981) The role of the major histocompatibility complex in immunobiology. Wiley, Chichester

Fairbanks VF (1980) Hemoglobinopathies and thalassemias. Thieme, Stuttgart

Flavell RA, Bud H, Bullman H, Busslinger M, deBoer E, deKeine A, Golden L, Groffen J, Grosveld FG, Mellor AL, Moschones N, Weiss E (1982) The structure and expression of mammalian gene clusters. In: Bonné-Tamir B (ed) The unfolding genome (Human Genetics, part A). Liss, New York

Francke U, Felsenstein J, Gartler SM, Migeon BR, Dancis J, Seegmiller JE, Bakay F, Nyhan WL (1976) The occurrence of new mutants in the X-linked recessive Lesch-Nyhan disease. Am J Hum Genet 28:123–137

Francois J, deBie S, Verriest G, Matton MT (1972) Conceptions actuelles sur l'heredité des d'eficiences congenitales de la vision des couleurs. Acta Genet Med Gemellol 21:233–256

Garrod AE (1902) The incidence of alcaptonuria: A study in chemical individuality. Lancet II: 1616–1620

Gartler SM, Francke U (1975) Half chromatid mutations: Transmission in humans? Am J Hum Genet 27:218–223

German J, Archibald R, Bloom D (1965) Chromosomal breakage in a rare and probably genetically determined syndrome of man. Science 148:506

German J (1974) Bloom's syndrome. II. The prototype of genetic disorders predisposing to chromosome instability and cancer. In: German J (ed) Chromosomes and cancer. New York, pp 601–618

Götze D (ed) (1977) The major histocompatibility system in man and animals. Springer, Berlin Heidelberg New York

Haldane JBS (1935) The rate of spontaneous mutation of a human gene. J Genet 31:317–326

Harnasch D (1982) Untersuchungen über die Möglichkeiten der in vivo Mutagenitätsprüfung mit Hilfe von Histokompatibilitäts-Reaktionen bei Inzuchtmäusen. Dissertation, Freiburg

Harris H, Hopkinson DA (1972) Average heterozygosity per locus in man: an estimate based on the incidence of enzyme polymorphisms. Ann Hum Genet 36:9–20

Harris H, Hopkinson DA (1976) Handbook of the enzyme electrophoresis in human genetics. North-Holland, Amsterdam Oxford New York

Harris H, Hopkinson DA, Robson EB (1974) The incidence of rare alleles determining electrophoretic variants: data on 43 enzyme loci in man. Ann Hum Genet 37:237–253

Hecht F, Koler RD, Rigas DA, Dahuke GS, Case MP, Tisdale V, Miller RW (1966) Leukemia and lymphocytes in ataxia-teleangiectasia. Lancet II:1193

Hecht F, Kaiser McCaw B (1977) Chromosome instability syndromes. In: Mulvihill JJ, Miller RW, Fraumeni JF Jr (eds) Genetics of human cancer. Raven, New York, pp 105–123

Hook RB (1982) Contribution of chromosome abnormalities to human morbidity and mortality and some comments upon surveillance of chromosome mutation rates. In: Bora KC et al. (eds) Progress in mutation research, vol 3. Elsevier, Amsterdam Oxford New York, pp 9–38

Klose J, Willers I, Singh S, Goedde W (1983) Two-dimensional electrophoresis of soluble and structure-bound proteins from cultured human fibroblasts and hair root cells: qualitative and quantitative variation. Hum Genet 63:262–267

Lenz W (1961) Medizinische Genetik, 1. Aufl. Thieme, Stuttgart

Lenz W (1975) Half chromatid mutations may explain incontinentia pigmenti in males. Am J Hum Genet 27:690–691

Lyon MH (1968) Chromosomal and subchromosomal inactivation. Ann Rev Genet 2:31–52

Mavor JW (1924) The production of nondisjunction by X-rays. J Exp Zool 39:381–432

McKusick VA (1978) Mendelian inheritance in man, 5th edn. The Johns Hopkins University Press, Baltimore

McKusick VA (1982) The human genome through the eyes of a clinical geneticist. Cytogenet Cell Genet 31:7–23

McKusick VA (1983) A neo-Vesalian view of the past and the future of clinical genetics. In: Bonné-Tamir B (ed) Medical aspects. Liss, New York (Human Genetics, part B)

Mohn G, Würgler FE (1972) Mutator genes in different species. Hum Genet 16:49–58

Motulsky AG, Vogel F, Buselmaier W, Reichert W, Kellermann G, Berg P (eds) (1978) Human genetic variation in response to medical and environmental agents: pharmacogenetics and ecogenetics. Hum Genet [Suppl] 1

Mourant AE, Kopeć AC, Domaniewska-Sobczak K (1978) Blood groups and diseases. Oxford University Press, London

Muller HJ (1927) Artificial transmutation of the gene. Science 66:84–87

Muller HJ (1950) Our load of mutation. Am J Hum Genet 2:111–176

Mulvihill JJ (ed) (1977) Genetics of human cancer. Raven, New York

Neel JV, Tiffany TO, Anderson NG (1973) Approaches to monitoring human populations for mutation rates and genetic disease. In: Hollaender A (ed) Chemical mutagens, vol 3. Plenum, New York, pp 105–150

Neel JV (1982) Chairman's comments: symposium on environmental mutagenesis. In: Bonné-Tamir B (ed) The unfolding genome. Liss, New York, pp 263–266 (Human genetics, part A)

Neel JV, Rothman ED (1978) Indirect estimates of mutation rates in tribal Amerindians. Proc Natl Acad Sci USA 75:5585–5588

Neel JV, Rothman ED (1981) Is there a difference among human populations in the rate with which mutation produces electrophoretic variants? Proc Natl Acad Sci USA 78:3108–3112

Neufeld EF (1974) The biochemical basis of mucopolysaccharidoses and mucolipidoses. Prog Med Genet 10:81–101

Oehlkers F (1943) Die Auslösung von Chromosomenmutationen in der Meiosis durch Einwirkung von Chemikalien. Z Induktiven Abstammungs-Vererbungslehre 81:313–341

Penrose LS (1955) Parental age and mutation. Lancet II:312

Pontecorvo G (1959) Trends in genetic analysis. Columbia University Press, New York

Prokop O, Göhler W (1976) Die menschlichen Blutgruppen, 4. Aufl. Fischer, Stuttgart New York

Putnam FW (1975) The plasma proteins, 2 vols. Academic, New York

Race RR, Sanger R (1975) Blood groups in man. Blackwell, Oxford

Rapoport IA (1946) Carbonyl compounds and the chemical mechanism of mutation. CR Acad Sci USSR 54:65

Satoh C, Awa AA, Neel JV, Schull WJ, Kato H, Hamilton HB, Otake M, Goriki K (1982) Genetic effects of atomic bombs. In: Bonné-Tamir B (ed) The unfolding genome. Liss, New York, pp 267–276 (Human genetics, part A)

Schempp W, Meer B (1983) Cytologic evidence for three human X-chromosomal segments escaping inactivation. Hum Genet 63:171–174

Schroeder TM, Anschütz F, Knopp A (1964) Spontane Chromosomenaberrationen bei familiärer Panmyelopathie. Hum Genet I:194–196

Schroeder TM (1982) Genetically determined chromosome instability syndromes. Cytogenet Cell Genet 33:119–132

Scriver CR, Clow CL (1980) Phenylketonuria and other phenylalanine hydroxylation mutants in man. Ann Rev Genet 14:179–202

Searle AG (1972) Spontaneous frequencies of point mutations in mice. Hum Genet 16:33–38

Selby PB, Selby PR (1978) Gamma-ray-induced dominant mutations that cause skeletal abnormalities in mice. II. Description of proved mutations. Mutat Res 51:199–236

Shows TB, Sakaguchi AY, Naylor SL (1982) Mapping the human genome, cloned genes, DNA polymorphisms and inherited disease. In: Harris H, Hirschhorn K (eds) Advances in Human Genetics, vol 12. Plenum, New York, pp 341–454

Stamatoyannopoulos G, Nienhuis AW (1981) Organization and expression of globin genes. Liss, New York

Stamatoyannopoulos G, Nute PE, Miller M (1981) De novo mutations producing unstable Hbs or HbsM. I: Establishment of a depository and use of data to test for an association of de novo mutations with advanced parental age. Hum Genet 58:396–404

Stamatoyannopoulos G, Nute PE (1982) De novo mutations producing unstable Hbs or HbsM. II. Direct estimates of minimum nucleotide mutation rates in man. Hum Genet 60:181–188

Stanbury JB (1974) Inborn errors of the thyroid. In: Steinberg AG, Burn AG (eds) Progress in Medical Genetics, vol X, pp 55–80

Stanbury JB, Wyngaarden JB, Fredrickson DS (1978) The metabolic basis of inherited disease, 4th edn. McCraw-Hill, New York Toronto London

Stevenson AC (1959) The load of heredity defects in human populations. Radiat Res (Suppl I): 306–325

Stevenson AC, Kerr CB (1967) On the distribution of frequencies of mutation in genes determining harmful traits in man. Mutat Res 4:339–352

Sutton HE (1982) Somatic mutation in man. In: Bonné-Tamir B (ed) The unfolding genome. Liss, New York, pp 289–298

Svejgaard A et al. (1979) The HLA system, 2nd edn. Karger, Basel New York

Swift M, Sholman L, Perry M, Chase C (1976) Malignant neoplasms in the families of patients with ataxia-teleangiectasia. Cancer Res 36:209–215

Swift M, Chase C (1979) Cancer in families with Xeroderma pigmentosum. JNCI 62:1415–1421

Terasaki PI (ed) (1980) Histocompatibility testing 1980. University of California, Los Angeles

Thalhammer O, Havelec L, Knoll E, Wehle E (1977) Intellectual level (I.Q.) in heterozygotes for phenylketonuria (PKU). Hum Genet 38:285–288

Trimble BK, Doughty JH (1974) The amount of hereditary disease in human populations. Ann Hum Genet 38:199–223

Vogel F (1970) Monitoring of human populations. In: Vogel F, Röhrborn G (eds) Chemical mutagenesis in mammals and man. Springer, Berlin Heidelberg New York, pp 445–452

Vogel F (1979) Our load of mutation – reappraisal of an old problem. Proc R Soc Lond 205: 77–90

Vogel F (1980) The influence of paternal age on some human mutation rates and its significance for comparison of mutation rates in female germ cells. Conf on struct pathology in DNA and aging, Freiburg, 1979. Boldt, Boppard

Vogel F (1983) Mutation in man. In: Emery AE (ed) Principles and practices of medical genetics

Vogel F, Altland K (1982) Utilization of material from PKU-screening programs for mutation screening. In: Chemical mutagenesis, human population monitoring, and genetic risk assessment. Progress in Mutation Research, vol 3. Elsevier, Amsterdam Oxford New York, pp 143–159

Vogel F, Helmbold W (1972) Blutgruppen – Populationsgenetik und Statistik. In: Becker PE (ed) Humangenetik. Thieme, Stuttgart, S 120–557

Vogel F, Kopun M (1977) Higher frequencies of transitions among point mutations. J Mol Evol 9:159–180

Vogel F, Motulsky AG (1979) Human genetics – problems and approaches. Springer, Berlin Heidelberg New York

Vogel F, Rathenberg R (1975) Spontaneous mutation in man. Adv Hum Genet 5:223–318

Warren ST, Schultz RA, Chang C-C, Wade MH, Trosko JE (1981) Elevated spontaneous mutation rate in Bloom syndrome fibroblasts. Proc Natl Acad Sci USA 78:3133–3137

Watkins WM (1980) Biochemistry and genetics of the ABO, Lewis, and P Blood group systems. In: Harris H, Hirschhorn K (eds) Advances Human Genetics, vol 10. Plenum, New York London, pp 1–136

Wendt GG, Drohm D (1972) Fortschritte der allgemeinen und klinischen Humangenetik, vol IV: Die Huntingtonsche Chorea. Thieme, Stuttgart

Wettke-Schäfer R, Kantner G (1983) X-linked dominant inherited diseases with lethality in hemizygous males. Hum Genet

Winter RM, Pembrey ME (1982) Hypothesis: does unequal crossing over contribute to the mutation rate in Duchenne muscular dystrophy? Am J Med Genet 12:437–441

# Frequency and Origin of Chromosome Abnormalities in Man

K. SPERLING[1]

## 1 Introduction

During the past 25 years human cytogenetics has changed from a highly specialized academic discipline restricted to only a few institutions to a rapidly expanding area of research and routine application involving hundreds of laboratories. World-wide several millions of individuals have now been karyotyped, perhaps more than from any other species. This immense progress was based on at least three important methodological developments:

1. the introduction of the lymphocyte culture technique as an easily accessible source of mitotic cells (Nowell 1960),
2. the improvement in preparing metaphase spreads by combination of the hypotonic solution pretreatment (Hsu 1952), and the air-drying technique (Rothfels and Siminovitch 1958), and
3. the development of various differential staining techniques which allowed the identification of each chromosome pair and the accurate delineation of structural rearrangements (see Schwarzacher 1976; Comings 1978; Schweizer 1981 for review).

Moreover, the discovery of heteromorphic chromosome regions makes it possible to trace the transmission of individual chromosomes from one generation to the next and to determine the origin of a number of structural and numerical chromosomal anomalies (see Jacobs 1977; Schnedl 1978; Erdtmann 1982 for review). This methodical progress is illustrated in Fig. 1.

## 2 Overall Rate of Chromosome Abnormalities in Recognized Conceptions

The total rate of chromosome abnormalities in man should theoretically be derived from the analysis of human preimplantation embryos. This approach, however, has severe limitations as early human embryos are difficult to obtain and even after in vitro fertilization considerable problems exist in making clear chromosomal preparations (Edwards 1977; Wramsby et al. 1982). It would therefore be premature to

---

1 Institut für Humangenetik der FU Berlin, Heubnerweg 6, 1000 Berlin 19, FRG

Mutations in Man, ed. by G. Obe
© Springer-Verlag Berlin Heidelberg 1984

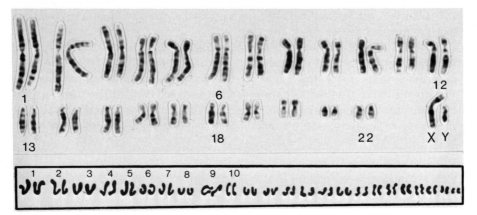

**Fig. 1.** *Above* Human G-banded lymphocyte metaphase arranged pairwise from 1 to 22,XY; *Below* Human oogonial metaphase. (Andres and Vögel 1936). It was reproduced in C. Stern's *Principles of Human Genetics* (1st ed, 1949) and has been characterized in a review article by Heberer (1940) as follows: without question we are here near the limit of the form-analytical possibility attainable

make any exact calculation though the frequency of abnormal zygotes appears high: two of three 8-cell embryos completely analyzed were chromosomally abnormal (Angell et al. 1983). This lack of information is highly regrettable, as perhaps almost 50% of all conceptuses are lost within the first two weeks of development, i.e. before the pregnancies were recognized (Fig. 2). This number has been calculated by Leridon (1977) and agrees fairly well with the recent data of Edmonds et al. (1982) based on the detection of fertilization by analysis of urinary HCG in women trying to conceive.

In contrast to this, extensive data exist on the incidence of chromosomal anomalies in clinically recognized pregnancies, based on the analysis of spontaneous abortion

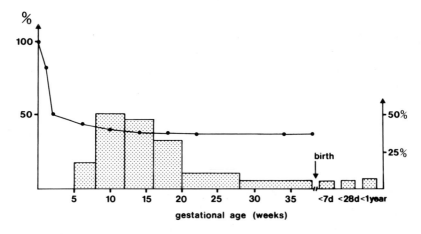

**Fig. 2.** Proportion of cytogenetic abnormalities in spontaneous abortions at different gestational intervals and in neonatal deaths (based on the data presented by Hook 1982) in comparison to intrauterine mortality as proposed by Leridon (1977; modified) starting with 100% fertilized ova

and stillbirth[2], the material obtained from prenatal diagnosis and from several large
unselected series on newborn.

Since the first description of a genome mutation in a macerated fetus (Penrose
and Delhanty 1961) and the pioneering work of Carr (see Carr 1970 for review) a
considerable number of cytogenetic surveys were performed on spontaneously aborted embryos, more than 4,000 have now been analyzed using differentially stained
chromosomes. As can be seen from Table 1 the total number of chromosome anomalies is on an average at least 50 times higher in spontaneous abortions than in live
births.

**Table 1.** Chromosome abnormalities in spontaneous abortions and live births. (Data from several
authors in Warburton et al. 1980a)

| Population | | Total exam- ed | Total abnor- mal | Type of abnormality | | | | |
|---|---|---|---|---|---|---|---|---|
| | | | | Structur- al | Mono- somy | Tri- somy | Poly- ploidy | Other |
| Sponta- neous abor- tions[1] | N | 3,714 | 1,499 | 47 | 311 | 779[a] | 289 | 73 |
| | % | 100 | 40.4 | 1.3 | 8.4 | 21.0 | 7.8 | 1.9 |
| Live births[2] | N | 59.452 | 364 | 124 | 2 | 177 | 1 | 60 |
| | % | 100 | 0.61 | 0.21 | – | 0.30 | – | 0.10 |

[a]   including mosaics

However, the proportion of cytogenetic abnormalities among early abortions varies
considerably between different studies, though the relative contribution of the individual chromosome anomalies is fairly constant. As was pointed out by Warburton
et al. (1980) these differences can only partly be explained by technical artifacts
(i.e. the different success rates in karyotyping or contamination of the fetal with
maternal tissue, the problem of discriminating between spontaneous and induced
abortions) or by biological differences in the incidence rate between different populations or the proportion of chromosomal anomalous conceptuses being aborted. The
most important factor seems to be the gestational age at which the specimens were
collected. Thus, the frequency of chromosomal anomalies is low in early abortions,
reaches a maximum (Fig. 2) between the 8th and 15th week with about 50%, and
then diminishes to an overall rate of about 5.7% in stillbirths, which is about the
same rate as that in infant deaths (Sutherland et al. 1978).

The relatively low incidence of chromosome anomalies in early abortions can be
explained by the long retention of the dead embryo in utero until abortion takes
place (Boué and Boué 1981a,b).

---

2  The term "stillbirth" is applied to fetal death that occurred subsequently to the 28th week of
pregnancy

Hook (1981, 1982) has tentatively estimated the proportion of chromosomally affected conceptuses lost at different stages of gestation. His calculation based on five abortion studies with a total of 3,364 specimens cytogenetically analysed, and the overall frequency of spontaneous abortions derived from a prospective study on more than 30,000 women who were recruited at their first antenatal visit to a medical center in Northern California. As can be seen from Fig. 3, 15.5% of all recognized pregnancies did not survive to term, about 30% of them had a cytogenetic abnormality, that is 4.5%. Among live births this rate has dropped to 0.6%, corresponding to a frequency of 0.5% (84.5%/0.6%) for all recognized pregnancies. Thus, the overall incidence of chromosome anomalies among *recognizable pregnancies* is about 5% (4.5% + 0.5%). This estimate is still tentative as it depends upon several putative assumptions and extrapolations. The error may be in the range of at least ±1% but it seems to be the most exact calculation as yet available.

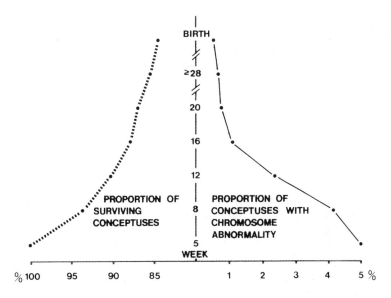

**Fig. 3.** Life-curve estimates for conceptuses from age 5 weeks to term, based on the tabulated data from Hook (1981)

It should, however, be pointed out that the frequency of cytogenetic defects *at conception* is significantly higher, as the enormous loss of preimplantation embryos (Fig. 2) is at least partially be due to lethal chromosome abnormalities (see Sect. 4). Their level seems to be an order of magnitude higher than in any other mammalian species (Chandley 1981, Chap. Hansmann, p. 147ff.).

# 3 Frequency and Origin of Different Types of Chromosome Anomalies

## 3.1 Classification of Chromosomal Abnormalities

Chromosome anomalies can be roughly divided into structural aberrations which can either be balanced (reciprocal translocations; Robertsonian translocations; inversions) or unbalanced (partial deletions and duplications) and in numerical anomalies. The latter can be divided in aneuploids (e.g. monosomies: $2n - 1$; trisomies: $2n + 1$) and polyploids (e.g. triploids: $3n$; tetraploids: $4n$).

In addition, the existence of two or more chromosomally distinct cell lines in one individual can be either due to mosaicism or – very rarely – to chimaerism. According to Anderson et al. (1951) a mosaic is formed of the cells of a single zygote lineage and goes back to an abnormal mitotic event (anaphase lag; non-disjunction), while a chimaera "is an organism whose cells derive from two or more distinct zygote lineages". They can originate either through two separate acts of syngamy or from the association of cells derived from two independent zygotes (see Ford 1969 for further details).

The relative distribution of these various chromosomal anomalies in spontaneous abortions and unselected live births is completely different (Fig. 4). Thus, monosomies and polyploids which are almost absent in live births, account for about 40% of the abnormalities in the abortions, while balanced structural rearrangements clearly predominate among live births.

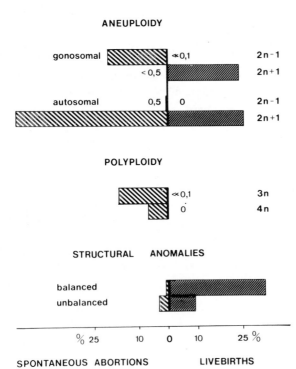

Fig. 4. Relative frequency of different types of chromosome anomalies in sponataneous abortions and live births. (For references, see legends to Tables 1 and 6)

These differences can principally be explained by different incidences in their origin or by different selective forces operating against the various chromosome anomalies. To decide between both possibilities, it is essential to first get insight into their mechanisms of origin. This can be done with the aid of cytogenetical and biochemical polymorphisms.

## 3.2 Tracing the Origin of Chromosome Abnormalities by Genetic Polymorphisms

According to Harris (1975), a polymorphic locus can be defined as one for which at least two loci exists with frequencies above 1% in the particular population. In principle, this definition can also be applied to the more than 20 different heteromorphic chromosome regions which have already been described. They vary in size, position, or staining properties and are apparently without phenotypic effect. As they are inherited according to Mendel's law they are ideal markers to trace the origin of specific chromosomes or whole chromosome sets (Fig. 5).

It is, however, almost impossible to exactly estimate their real frequency as most of them show a quantitative variation which is, from a methodical point of view, most difficult to classify objectively. It is therefore preferred to use the descriptive term "heteromorphism" instead of "polymorphism".

The mode of assignment of the timing of the segregational error by these heteromorphisms is illustrated in Fig. 6a. However several biasses, including undetected false paternity, have to be kept in mind:

1. the ascertainment of second meiotic division non-disjunction is favored as not all families are informative (Langenbeck et al. 1976),
2. crossing-over between the heteromorphism and the centromere may lead to incorrect assignment of the meiotic error, and
3. second division errors cannot be distinguished from mitotic errors in the zygote (leading to a trisomic individual from a diploid zygote) as well as from undetected mosaicism in the parents.

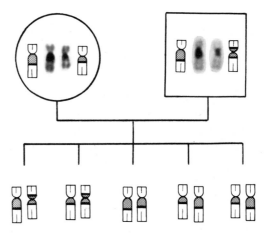

**Fig. 5.** Inheritance of C-band polymorphisms of chromosome 9 in a family, in which 3 of the 4 parental chromosome 9 can be distinguished. Note that all combinations are observed among the offspring. (Karkut, Inaug. Dissert., Berlin 1982)

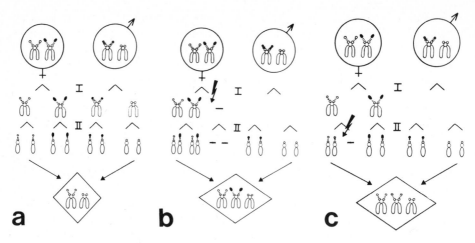

**Fig. 6a–c.** Schematic representation of the inheritance of chromosomal polymorphism in a kindred, in which all four parental chromosomes 21 can be distinguished. a Normal situation: segregation of homologous chromosomes takes place in the first meiotic division (I), separation of the homologous chromatids in the second (II). After fertilization the resulting zygote contains two chromosomes 21, one from each parent. b Kindred with trisomy 21; due to non-disjunction (↯) in the first oogonial division, the child has inherited both homologous 21 from its mother. c Kindred with trisomy 21; due to non-disjunction in the second oogonial division, the child has inherited two identical chromosomes 21 from its mother (in all cases crossing-over has been neglected). (Modified from Sperling 1983)

Bias 1 is of less importance as in case of chromosome 21 for example, almost 80% of all families are informative if both Q-band and Ag-staining polymorphisms are carefully registered. It can be further reduced by using a maximum likelihood method based on complete ascertainment of all relevant family members including those with non-informative markers (Jacobs and Morton 1977). Bias 2 will only marginally influence the results as the majority of human polymorphisms is located at or near to the centromere. Bias 3 which is due to non-disjunction in early cleavage division is perhaps negligible as most of these events will lead to easily detectable mosaicism. It could be of some relevance with respect to parental mosaicism, as perhaps several percent of all parents of children with Down's syndrome may be "hidden mosaics" (Rodewald et al. 1981).

Generally, as most ascertainment biasses are in favor of second division errors the tabulated frequencies of non-disjunction in meiosis I may be an underestimate.

In case of the human X, for which no reliable cytogenetic markers are available, the biochemical heterogeneity at the Xg locus can be used instead. The gene frequency of the dominant allele Xg (a+) among northern europeans which corresponds to the phenotype frequency in males is 0.66, that for the recessive allele Xg (a–) 0.34, respectively (Sanger et al. 1977). The phenotype frequency for females is then Xg (a+) 88% and Xg (a–) 12%.

## 3.3 Aneuploidy

The most common type of chromosome abnormality in both spontaneously aborted fetuses and among newborn is aneuploidy, making up about 70% or 50%, respectively (Fig. 4). It arises predominantly by non-disjunction during gametogenesis (Fig. 6b,c), a segregational error, which should produce equal ratios of trisomic and monosomic embryos. However, the latter are extremely rare, with the exception of the 45,XO status, and are expected to be eliminated before implantation (see Sect. 4).

Among trisomies significant differences are observed in the frequencies of the various chromosomes involved. Among live births only three autosomal trisomies are found with any appreciable frequency (trisomy 13; 18; and 21 with a rate of 0.05; 0.15; and 1.15 per 1,000) while in abortions trisomies for all autosomes except no. 1 have been observed (Fig. 7). Trisomy 16 is by far the most common, making up 30% of all human trisomies. Interestingly, those trisomies which are compatible with live are also frequently observed in spontaneous abortions. Thus, about three quarters of all trisomy 21 embryos die before term.

The parent and meiotic stage of origin for the extra chromosome 21 in trisomy 21 patients and for six autosomal trisomies in spontaneous abortions is given in Tables 2 and 3. It can be seen that in the majority of all cases the error occurs in the first maternal meiotic division. This is in accordance with the clear association of most trisomies with increasing maternal age which will be discussed in the chapter of Hansmann, p. 147ff.

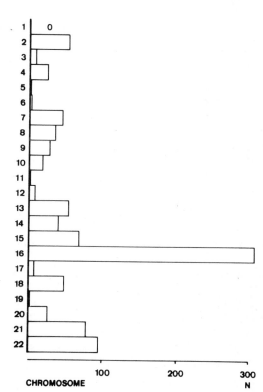

**Fig. 7.** Frequency of different trisomies among 950 human spontaneous abortions. (Data from 8 surveys presented by Chandley 1981)

**Table 2.** Parental origin of extrachromosome 21 in patients with Down's syndrome. (Data from an European collaborative study, presented by Jongbloet et al. 1982)

|                  | Abs. | %    |     |
| ---------------- | ---- | ---- | --- |
| Maternal I       | 158  | 55   |     |
| Maternal II      | 43   | 15   | 79% |
| Maternal I or II | 27   | 9    |     |
| Paternal I       | 33   | 12   |     |
| Paternal II      | 18   | 6    | 21% |
| Paternal I or II | 8    | 3    |     |
| Total            | 287  | 100% |     |

**Table 3.** Origin of extrachromosomes in human trisomic spontaneous abortions (131 cases analyzed – origin determined in 40%). (Data from Jacobs and Hassold 1980, incl. the data on chromosome 16 from Meulenbroek and Geraedts 1982)

|                  | Chromosome |    |    |    |    |    |    |     |
| ---------------- | ---------- | -- | -- | -- | -- | -- | -- | --- |
|                  | 13 | 14 | 15 | 16 | 21 | 22 | N  |     |
| Maternal I       | 3  | 1  | 2  | 15 | 4  | 11 | 36 |     |
| Maternal II      |    |    |    | 1  | 1  |    | 2  | 92% |
| Maternal I or II | 1  |    | 2  | 1  | 3  | 3  | 10 |     |
| Paternal I       |    |    |    | 2  |    |    | 2  |     |
| Paternal II      | 1  |    |    | 1  |    |    | 2  | 8%  |
| Paternal I or II |    |    |    |    |    |    | 0  |     |
| Total            | 5  | 1  | 4  | 20 | 8  | 14 | 52 |     |

Mosaic trisomies may constitute approximately 5% of all trisomies among spontaneous abortions (Hassold 1982) and 1%–2% of all liveborn children with Down syndrome (Mikkelsen 1977). They usually originate from a trisomic zygote (Fig. 10a) followed by a mitotic error leading to a normal diploid line.

The detection of mosaicism depends on the number of cells and tissues studied and on the developmental stage at the time of analysis (Table 4). Due to in vivo selection against aneuploid cells, they may gradually become eliminated and can even completely disappear (Taysi et al. 1970; Zankl and Rodewald 1977). Thus, it is practically impossible to exactly determine the frequency of mosaicism in man.

Among sex chromosome aneuploidy there is little or no selective loss of either 47,XYY or 47,XXX embryos but a yet unexplained loss of 47,XXY fetuses (Warburton et al. 1980). As all 47,XYY males go back to a 24,YY sperm, their frequency of 0.04/1,000 live births directly reflects the frequency of non-disjunctional events of the Y chromosome at second paternal division.

The parental origin of the extra X in Klinefelter's syndrome has been estimated from the distribution of the Xga blood group antigen (Sanger et al. 1977). Almost one third of all informative cases contained an extra paternal X which must be due

**Table 4.** Mosaic status in a fetus with 46,XX/47,XX,+17 cell lines which was diagnosed prenatally. After interruptio mosaicism could be confirmed in different extraembryonal and fetal tissues. Note that the non-disjunctional event must have occurred at least in the morula stage and that the underrepresentation of the aneuploid line is most probably due to selective deaths of these cells. The length of the fetus was below the 3rd percentile. (Data from Struck and Wieczorek personal communication)

|                        | Tissue          | 46,XX | 47,XX,+17 |
|------------------------|-----------------|-------|-----------|
| 1. Amniocentesis       | Amniotic        | 55    | 5         |
| 2. Amniocentesis       | Fluid cells     | 52    | 8         |
| Extraembryonal tissue  | Amnion          | 199   | 1         |
|                        | Placenta        | 72    | 1         |
| Fetal tissue           | Achilles tendon | 100   | 3         |
|                        | Diaphragm       | 177   | 2         |
|                        | Heart           | 84    | 2         |
|                        | Liver           | 330   | 0         |
|                        | Lung            | 76    | 1         |
|                        | Lymphocytes     | 100   | 0         |
|                        | Peritoneum      | 50    | 0         |
|                        | Pleura          | 157   | 1         |
|                        | Skin            | 156   | 0         |
|                        | Umbilical cord  | 199   | 1         |

to non-disjunction in meiosis I (Fig. 8b). In Figure 8c the Xg data of a 48,XXYY proband indicate a segregational error in both paternal meiotic divisions.

Most striking among sex chromosome aneuploidies is the rare occurrence of the clinically harmless 45,XO status among newborn (0.06/1,000) compared to its high frequency among spontaneous abortions (Fig. 4, Table 1). Thus, only one of 150 to 200 recognized 45,XO embryos survive to term. In 77% of all individuals with Turner

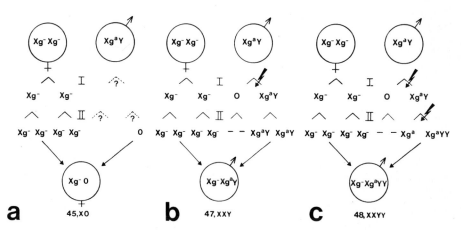

**Fig. 8a–c.** Schematic representation of the timing of non-disjunction in case of gonosomal aneuploids: 45,XO (**a**), 47,XXY (**b**), and 48,XXYY (**c**), based on the Xg antigen status (see text for further information). (Modified from Sperling 1983)

syndrome the single X was of maternal origin (Sanger et al. 1977), and consequently no paternal gonosome was present (Fig. 8a). Moreover, in contrast to most trisomies, the maternal age in XO abortions is not increased but even lower than expected (Boué and Boué 1973; Warburton et al. 1980).

These observations can be explained in that a substantial portion of XO zygotes is attributed to a postzygotic sex chromosome loss (see Ford 1981 for discussion). In accordance to this, mosaics with a 45,XO cell line accompanied by either a 46,XX or 46,XY line are the most common. Occasionally monozygotic twins with 46,XX/45,XO mosaicism have been found of whom one develops as a male, the other as a female with symptoms of Turner's syndrome (see Hata et al. 1982 for review).

## 3.4 Polyploidy

Polyploids are practically unknown among live births but comparatively frequent among spontaneous abortions (Table 1). More than two third are triploid, the remaining tetraploid (Fig. 4).

Among triploids about 60% have the sex chromosome constitution 69,XXY, while 36% are 69,XXX, and only 4% are 69,XYY. The parental origin of the extra haploid genome can be determined by the distribution of the chromosomal heteromorphisms and by biochemical markers. In about 75% of all informative cases it was of paternal origin (Table 5).

**Table 5.** Parental origin of triploidy in spontaneous abortions (173 cases analyzed – origin determined in 70%). (Data from several authors in Meulenbroek and Geraedts 1982; incl. the data from Jacobs et al. 1982)

|                          | Abs. | %    |      |
| ------------------------ | ---- | ---- | ---- |
| Maternal I               | 16   | 13   |      |
| Maternal II              | 16   | 13   | 28%  |
| Maternal I or II         | 2    | 2    |      |
| Paternal I or dispermy   | 23   | 19   |      |
| Paternal II or dispermy  | 7    | 6    | 72%  |
| Dispermy                 | 57   | 47   |      |
| Total                    | 121  | 100% |      |

Principally, triploidy can result from fusion of a haploid ovum with two haploid sperm (dispermy) or from fertilization between a normal haploid gamete and a diploid gamete (Fig. 9). Here, the extra set can be due to a failure of segregation at first or second paternal or maternal meiotic division (diandry or digyny). Formally, fertilization by a diploid sperm cannot be distinguished from dispermy, as by chance both may contain the same polymorphisms. The probabilities for each event can however be estimated by a maximum-likelihood method (Jacobs and Morton 1977). As can be seen from Table 5, all five mechanisms seem to play a role in the origin of triploidy, of which dispermy is the dominating one.

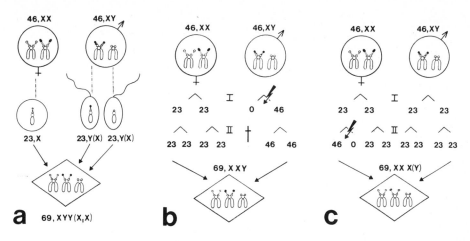

**Fig. 9a–c.** Schematic representation of the origin of triploidy. **a** dispermy; **b** fertilization of a normal ovum by a diploid sperm; **c** fertilization of a diploid ovum by a normal sperm. Note that **a** and **b** cannot be distinguished on the basis of a single polymorphism, but only statistically (maximum-likelihood method) if a number of different polymorphism are present. (Modified from Sperling 1983)

The unique occurrence of a malformed triploid acardiac monster with 69,XXX karyotype together with a normal 46,XY male twin was explained by independent fertilization of a normal haploid ovum and its diploid first polar body. Both developed then independently within the same chorion (Bieber et al. 1981).

Theoretically, such an event could also lead to a chimaeric individual. Indeed, several patients with diploid/triploid cell lines have already been described. In one case with 46,XX/69,XXY constitution, chromosomal heteromorphisms proved that the extra haploid set was of paternal origin, indicating double fertilization (Dewald et al. 1975).

A 48,XXYY and a 71,XXXYY cell line was revealed in a child with multiple malformations (Schmid and Vischer 1967). The most plausible explanation was the fertilization of a normal 23,X ovum by a 25,XYY sperm, and after first cleavage division fusion of one of the two cells with the second polar body.

All tetraploid specimens observed so far had either the sex chromosome constitution 92,XXXX or 92,XXYY, indicating failure of the first cleavage division as the most probable mode of origin. However, in one case the conceptus had one maternal and three haploid paternal sets and may have arisen by trispermy (Sheppard et al. 1982).

These examples have illustrated that anomalies at fertilization leading to chromosomally unbalanced embryos affect more than 1% of all recognized pregnancies in our species and occasionally even two abnormal events can coincide.

### 3.5 Chimaerism, Androgenesis, and Gynogenesis

Most of human individuals with 46,XX/46,XY karyotype originate from fusion of the haploid ovum nucleus and the haploid nucleus of a polar body with one sperm

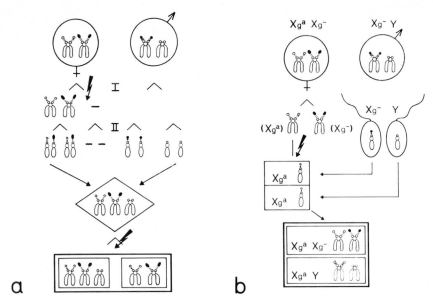

**Fig. 10a,b.** Schematic representation of the origin of mosaicism and chimaerism (for further information, see text). (Modified from Sperling 1983)

each and therefore represent chimaeras (Fig. 10b). Both, polar body and egg may be of equal size, if the anaphase spindle is displaced to the center of the egg or orientated parallel instead of vertical to the egg surface. After fertilization of each half with a 23,X and a 23,Y sperm, respectively, the resulting individual is composed of two genetically distinct cell lines of about equal frequency (Sperling et al. 1974).

Experimentally, chimaeras can be produced by fusion of 2-cell embryos, an event which has also been suggested to occur in vivo in man. The best known types of chimaeras result from exchange of cells between dizygotic twins through placental cross-circulation. There is also evidence for maternal-fetal cell exchange (see Ford 1969 for review).

As yet, no evidence exists in man that parthenogenetic development of a haploid ovum has lead to a recognized pregnancy. However, the very common benign ovarian teratomas which are composed of mature histologic structures of ectodermal, meso-dermal, and endodermal origin are of gynogenetic origin. They consistently had a 46,XX constitution and originate from oocytes that had undergone their first meiotic division, but failed to pass the second division (Linder 1969; Linder and Power 1970; Linder et al. 1975). Consequently, the teratomas can be homozygous for loci for which the mother was heterozygous, or also heterozygous if crossing-over has occurred during meiotic prophase (Fig. 11a).

Recently, androgenetic origin has been proven in complete hydatidiform moles (Kajii et al. 1977; Jacobs et al. 1980). This abnormal pregnancy has no visible fetus and is characterized by its grossly swollen chorionic villi and hydropic degeneration of placental villi. Usually it shows a 46,XX karyotype but is entirely homozygous for all paternal markers (Fig. 11b). Thus, these moles originate from fertilization of

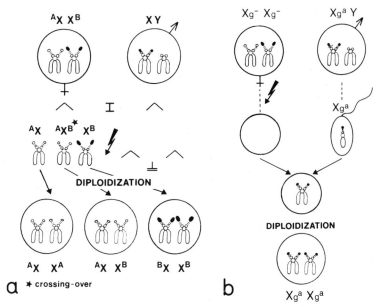

**Fig. 11a,b.** Schematic representation of the origin of an ovarian teratoma (**a** gynogenesis) and a hydatidiform mole (**b** androgenesis) (for further information, see text). (Modified from Sperling 1983)

an anucleate egg by a normal haploid sperm and its subsequent duplication. The very rare 46,XY hydatidiform moles are most probably the result from penetration of an "empty egg" by two haploid spermatozoa (Ohama et al. 1981).

### 3.6 Structural Abnormalities

The frequency of structural chromosome abnormalities ascertained through several studies on spontaneous abortions, fetuses at amniocentesis, and in live births is given in Table 6. While balanced aberrations predominate among live births, unbalanced are more frequent in spontaneous abortions, which clearly reflect their different selective values. As can be seen from Table 6 most of the balanced mutations and a considerable fraction of the unbalanced are familial. These had to be omitted in all calculations on the germinal mutation rate. According to Jacobs (1981) the mutation rate for all structural rearrangements is approximately $1 \times 10^{-3}$. The rate for balanced rearrangements was estimated to $2.2 \times 10^{-4}$, for unbalanced to $6.96 \times 10^{-4}$. Clearly, these data are still preliminary. They are, however, of general importance, as this class of aberration is sensitive to a number of environmental hazards and is therefore a good candidate for population monitoring of mutagens. For further details on this topic the reader is referred to the articles of Jacobs (1981), Jacobs et al. (1981), Hook (1981, 1982), Boué and Boué (1981a,b), Warburton (1982), and Hook et al. (1983).

**Table 6.** Frequency of structural chromosome abnormalities that are de novo, familial or not known, in spontaneous abortions, fetuses at amniocentesis, and live births

| Population | | Total examined | Total struct. abnormal | Type of structural abnormality | | | | | | | | | | | | | | |
|---|---|---|---|---|---|---|---|---|---|---|---|---|---|---|---|---|---|---|
| | | | | Balanced | | | | | | Unbalanced | | | | | | | | |
| | | | | De novo | | | | Fam. | Un-known | De novo | | | | Fam. | Un-known | | | |
| | | | | Rob. Tr. | Invers. | Rec. Tr. | Total | | | Rob. Tr. | Super-num. | Others | Total | | | | | |
| Spontaneous abortions[1] | N | 5,726 | 104 | – | – | 4 | 4 | 10 | 2 | 20 | – | 16 | 36 | 30 | 22 | | | |
| | o/oo | 1,000 | 18.1 | | | | 0.7 | 1.7 | 0.4 | | | | 6.3 | 5.2 | 3.8 | | | |
| Fetuses at amnio-centesis[2] | N | 27,225 | 61 | 6 | 2 | 9 | 17 | – | 9 | 2 | 7 | 7 | 16 | – | 19 | | | |
| | o/oo | 1,000 | 0.22 | | | | 0.06 | | 0.03 | | | | 0.06 | | 0.07 | | | |
| Live births[3] | N | 59,452 | 135 | 4 | 1 | 13 | 18 | 73 | 22 | 2 | 2 | 3 | 7 | 9 | 6 | | | |
| | o/oo | 1,000 | 0.23 | | | | 0.03 | 0.12 | 0.04 | | | | 0.01 | 0.02 | 0.01 | | | |

1 Data from several authors in Jacobs (1981)
2 Data from: the New York State Chromosome Registry and United States Interregional Chromosome Register System presented by Hook et al. (1983)
3 Data from several authors in Jacobs (1978)

Rob. Tr. (Robertsonian translocation); Invers. (inversion); Rec. Tr. (reciprocal translocation); Fam. (familiar); Supernum. (supernumericals)

# 4 Estimation of the Frequency of Chromosomal Anomalies at Conception

It has already been pointed out (see Sect. 2) that an estimated 5% of all clinically recognized pregnancies have a chromosomal abnormality. If one considers that trisomics account for about 50% of all anomalies among spontaneous abortions and live births as well, the total frequency of trisomics is about 2.5% in all recognized conceptions. As almost all of these anomalies go back to meiotic non-disjunction an equivalent number of monosomies would be expected, which obviously die before implantation and therefore contribute to the massive embryonic loss seen in Fig. 2. However, also trisomics could become eliminated in early pregnancy which may explain their completely different frequencies among spontaneous abortions (Fig. 7). If all chromosomes had the same non-disjunction rate as no. 16, then about one in every two conceptions should be aneuploid (Boué and Boué 1973). Though this assumption is not very plausible, it cannot be excluded from the beginning. A considerable step forward in this respect is the new technique of direct karyotype analysis of male gametes, based on the in vitro fertilization of hamster eggs with human sperm (Rudak et al. 1978).

In a large series on the chromosome complements in 1,000 sperm from 33 normal men, the frequency of chromosomal abnormalities was 8.5% (Martin et al. 1983). Interestingly, the predominant type of aberration was aneuploidy (5.2%), with about the same frequency of monosomies and trisomies and all chromosome groups represented. The sample size is still too small to detect significant differences among the chromosomes involved. There was a tendency for G-group chromosomes to be over-represented: 4 hyperhaploid and 6 hypohaploid among 52 aneuploids. The remaining 3.3% had a structural aberration, most of them a chromatid break.

No reliable data exist on the frequency of abnormalities in human oocytes. If one considers that about 4/5 of all trisomics are due to maternal non-disjunction, the proportion of aberrant gametes in the female should be significantly higher than in males with 8.5%. It is therefore realistic to assume that at least 1 of every 10 zygotes, perhaps even 2 or 3 are chromosomally aberrant. Most of them are lost before recognition of pregnancy.

This is in accord with the findings in the tobacco mouse, *Mus poschiavinus*, which has several balanced translocations and hence a high rate of meiotic mal-segregation. While monosomies and trisomies were of about the same frequency in preimplantation embryos, the first are lost very early in development (Gropp et al. 1976; Epstein and Travis 1979).

Due to the massive selective death of cytogenetically defective conceptuses the rate of chromosome anomalies drops to 0.6% in live births of whom 0.2% to 0.3% have a clinically significant abnormality. This still represents a considerable genetic burden. If more is known about the etiology of chromosome anomalies in man, more criteria will be detected for identifying families with high risks of cytogenetical abnormalities. This may then help to reduce this genetic load, as an efficient tool is now available for the detection of any chromosome abnormality in early pregnancy: prenatal diagnosis.

# References

Anderson D, Billingham RE, Lampkin GH, Medawar PB (1951) The use of skin drafting to distinguish between monozygotic and dizygotic twins in cattle. Heredity 5:379–397

Angell RR, Hitken RJ, Look van PFA, Lumsden MA, Templeton AA (1983) Chromosome abnormalities in human embryos after in vitro fertilization. Nature 303:336–338

Bieber FB, Nance WE, Morton CC, Brown JA, Redwine FO (1981) Genetic studies of an acardiac monster: evidence of polar body twinning in man. Science 213:775–777

Boué J, Boué A (1973) Chromosomal analysis of two consecutive abortuses in each of 43 women. Humangenetik 19:275

Boué J, Boué A (1981a) Genetic causes for human fetal wastage. In: Cortes-Prieto J, Campos da Paz A, Neves-e-Castro M (eds) Research on fertility and sterility. MTP Press, Lancaster, pp 393 to 405

Boué J, Boué A (1981b) Chromosome structural rearrangements and reproductive failure. In: Bennett JD, Bobrow M, Hewitt GM (eds) Chromosomes today, vol 7. Allen and Unwin, London, pp 281–290

Carr DH (1970) Chromosome abnormalities and spontaneous abortions. In: Jacobs PA, Price WH, Law P (eds) Human population cytogenetics. Edinburgh University Press, Edinburgh, pp 103 to 118

Chandley AC (1981) The origin of chromosomal aberrations in man and their potential for survival and reproduction in the adult human population. Ann Genet 24:5–11

Comings DA (1978) Mechanism of chromosome banding and implications for chromosome structure. Annu Rev Genet 12:25–46

Dewald G, Alvarez MN, Cloutier MD, Kelalis PP, Gordon H (1975) A diploid-triploid human mosaic with cytogenetic evidence of double fertilization. Clin Genet 8:149–160

Edmonds DK, Lindsay KS, Miller JF, Williamson E, Wood P (1982) J Fert Steril 38:447–453

Edwards RG (1977) In: Phillipp EE, Barnes J, Newton M (eds) Scientific foundations of obstetrics and gynaecology. Heinemann Medical, London, pp 175–252

Epstein CJ, Travis B (1979) Preimplantation lethality of monosomy for mouse chromosome 19. Nature 280:144–145

Erdtmann B (1982) Aspects of evaluation, significance, and evolution of human C-band heteromorphism. Hum Genet 61:281–294

Ford CE (1969) Mosaics and chimaeras. Br Med Bull 25:104–109

Ford CE (1981) Nondisjunction. In: Burgio GR, Fraccaro M, Tiepolo L, Wolf U (eds) Trisomy 21. Springer, Berlin Heidelberg New York, pp 103–144

Gropp A, Putz B, Zimmermann U (1976) Autosomal monosomy and trisomy causing developmental failure. In: Gropp A, Benirschke K (eds) Developmental biology and pathology. Curr Top Pathol 62:117

Hassold T (1982) Mosaic trisomies in human spontaneous abortions. Hum Genet 61:31–35

Harris H (1975) The principles of human biochemical genetics, 2nd edn. North-Holland, Amsterdam Oxford

Hata A, Suzuki Y, Matsui J, Kuroki Y (1982) Ring 18 mosaicism in identical twins. Hum Genet 62:364–367

Hook EB (1981) Prevalence of chromosome abnormalities during human gestation and implications for studies of environmental mutagens. Lancet II:169–172

Hook EB (1982) Contribution of chromosome abnormalities to human morbidity and mortality and some comments upon surveillance of chromosome mutation rates. In: Bora KC, Douglas GR, Nestmann ER (eds) Chemical mutagenesis, human population monitoring and genetic risk assessment. Elsevier Biomedical, Amsterdam Oxford New York, pp 9–38

Hook EB, Schreinemachers DM, Willey AM, Cross PK (1983) Rates of mutant structural chromosome rearrangements in human fetuses: data from prenatal cytogenetic studies and associations with maternal age and parental mutagen exposure. Am J Hum Genet 35:96–109

Hsu TC (1952) Mammalian chromosomes in vitro. I. The karyotype of man. J Hered 43:172

Jacobs PA (1977) Human chromosome heteromorphisms. In: Steinberg AG, Bearn AG, Motulsky AG, Childs B (eds) Progress in medical genetics, new series, vol 2, pp 251–274

Jacobs PA (1978) Population surveillance: a cytogenetic approach. In: Morton NE, Chung CS (eds) Genetic epidemiology. Academic, New York San Francisco London, pp 463–481

Jacobs PA (1981) Mutation rates of structural chromosome rearrangements in man. Am J Hum Genet 33:44–54

Jacobs PA, Morton NE (1977) Origin of human trisomics and polyploids. Hum Hered 27:59–72

Jacobs PA, Hassold TJ (1980) In: Porter JH, Hook EB (eds) Human embryonic and fetal death. Academic, New York London, pp 289–298

Jacobs PA, Wilson CM, Sprenkle JA, Rosenstein NB, Migeon BR (1980) Mechanism of origin of complete hydatidiform moles. Nature 286:714–716

Jacobs PA, Funkhouser J, Matsuura J (1981) In: Hook EB, Porter IH (eds) Population and biological aspects of human mutation. Academic, New York London, pp 135–145

Jacobs PA, Szulman AE, Funkhouser J, Matsuura JS, Wilson CC (1982) Human triploidy: relationship between parental origin of the additional haploid complement and development of partial hydatidiform moles. Am J Hum Genet 46:223–231

Jongbloet PH, Mulder AM, Hamers AJ (1982) Seasonality of pre-ovulatory non-disjunction and the aetiology of Down syndrome. A european collaborative study. Hum Genet 62:134–138

Kajii T, Ohamek (1977) Androgenetic origin of hydatidiform mole. Nature 268:633–634

Langenbeck U, Hansmann J, Hinney B, Hönig V (1976) On the origin of the supernumerary chromosome in autosomal trisomies, with special reference to Down's syndrome. Hum Genet 33:89–102

Leridon H (1977) Human fertility. The University of Chicago Press, Chicago

Linder D (1969) Gene loss in human teratomas. PNAS 63:699–704

Linder D, Power J (1970) Further evidence for postmeiotic origin of teratomas in the human female. Am J Hum Genet 34:21–30

Linder D, McCaw BK, Hecht F (1975) Parthenogenetic origin of benign ovariant teratomas. New Eng J Med 292:63–66

Martin RH, Balkan W, Burns K, Rademaker AW, Lin CC, Rudd NL (1983) The chromosome constitution of 1,000 human spermatozoa. Hum Genet 63:305–309

Meulenbroek GHM, Geraedts JPM (1982) Parental origin of chromosome abnormalities in spontaneous abortions. Hum Genet 62:129–133

Mikkelsen M (1977) Down's syndrome: cytogenetic epidemiology. Hereditas 86:45–59

Nowell PC (1960) Phytohemagglutinin: an initiator of mitosis in cultures of normal human leukocytes. Cancer Res 20:462–466

Ohama K, Kajii T, Okamoto E, Fukuda Y, Imaizumi K, Tsukahara M, Kobayashi K, Hagiwara K (1981) Dispermic origin of XY hydatidiform moles. Nature 292:551

Penrose LS, Delhanty JDA (1961) Triploid cell cultures from macerated foetus. Lancet I:1261

Rodewald A, Zang KD, Zankl H, Zankl M (1981) In: Burgio GR, Fraccaro M, Tiepolo L, Wolf U (eds) Trisomy 21. Springer, Berlin Heidelberg New York, pp 41–56

Rothfels KH, Siminovitch L (1958) An air-drying technique for flattening chromosomes in mammalian cells grown in vitro. Stain Technol 38:73–77

Rudak E, Jacobs PA, Yanagimachi R (1978) Direct analysis of the chromosome constitution of human spermatozoa. Nature 274:911–913

Sanger R, Tippett P, Gavin J, Teesdale P, Daniels GL (1977) Xg groups and sex chromosome abnormalities in people of northern European ancestry: an addendum. J Med Genet 14:210 to 213

Schmid W, Vischer D (1967) A malformed boy with double aneuploidy and diploid-triploid mosaicism 48,XXYY/71,XXXYY. Cytogenetics 6:145–155

Schnedl W (1978) Structure and variability of human chromosomes analyzed by recent techniques. Hum Genet 41:1–9

Schwarzacher HG (1976) Chromosomes in mitosis and interphase. Springer, Berlin Heidelberg New York

Schweizer D (1981) Counterstain – enhanced chromosome banding. Hum Genet 57:1–14

Sheppard DM, Fisher RA, Lawler SD, Povey S (1982) Tetraploid conceptus with three paternal contributions. Hum Genet 62:371–374

Sperling K, Kaden R, Gillert K-E, Weise W (1974) Ein Fall von XX/XY-Chimärismus mit normalem männlichem Habitus. Humangenetik 21:237–244

Sperling K (1983) Häufigkeit und Entstehung von Chromosomenanomalien beim Menschen. Biuz 13:144–152

Sutherland GR, Carter RF, Bauld R, Smith II, Bain AD (1978) Chromosome studies at the pediatric necropsy. Ann Hum Genet 42:173–181

Taysi K, Kohn G, Mellmann WJ (1970) Mosaic mongolism. II. Cytogenetic studies. J Pediatr 76: 880–885

Warburton D (1982) In: Willey AM, Carter TP, Kelly S, Porter JH (eds) Clinical genetics: problems in diagnosis and counseling. Academic, New York London, pp 63–75

Warburton D, Stein Z, Kline J, Susser M (1980a) Chromosome abnormalities in spontaneous abortion; data from the New York City study. In: Porter IH, Hook EB (eds) Human embryonic and fetal death. Academic, New York, pp 261–287

Wramsby H, Hansson A, Liedholm P (1982) Chromosomal preparations from in-vitro inseminated human oocytes. In: Hafez ESE, Semm K (eds) World Conf on embryo transfer, in vitro fertilization and instrumental insemination. Kiel 1980, MTP Press, Lancaster, pp 263–271

Zankl H, Rodewald A (1977) Diagnostische Probleme beim Mosaik-Down Syndrom. Klin Pädiatr 189:430–439

# Aspects of Nondisjunction

I. HANSMANN[1]

## 1 Introduction

Aneuploidy is a significant cause of pregnancy wastage, perinatal death, malformation and mental retardation in man (see Chap. Sperling, p. 128ff.). Much information is available on the incidence of aneuploidy at birth and at various gestational stages after implantation, in addition to the association of several trisomies with maternal age which has been known for many years (Ford 1981; Sankaranarayanan 1982, for review). Many studies with mammals have convincingly shown that physical and chemical factors may cause aneuploidy in exposed germ cells (Hansmann 1983). However, 25 years after the first cytogenetical description of trisomy in man (Léjeune et al. 1959; Jacobs and Baikie 1959; Ford et al. 1959) we can still only speculate as to the mechanisms involved in chromosomal nonsegregation during meiosis.

Retrospective studies, with human newborn or with specimens from abortions, on the association of aneuploidy and preconceptional exposure to chemicals (Yamamoto et al. 1982) or to irradiation (Sankaranarayanan 1982, for review) will hardly contribute to an understanding of the mechanisms of nondisjunction. That conclusion is valid because the aneuploid specimen at any of the ontogenetic stages analysed cytogenetically developed as far for two reasons (1) nondisjunction within one of the parental gametes (2) survival of critical steps during gestation, not with standing selection against chromosomal imbalance.

Cytogenetic studies with human germ cells (Jagiello et al. 1976; Rudak et al. 1978) are therefore important not only with respect to measurement of the incidence of aneuploidy at conception but also to provide a better understanding of the mechanisms involved in nondisjunction.

Martin et al. (1983) recently analysed the chromosomal complement of 1,000 human sperm after in vitro fertilization of zona-free hamster eggs. They found that hyperhaploid and hypohaploid metaphases from 33 semen donors occurred with frequencies of 2.4 and 2.7, respectively. Such results are relevant to several aspects of nondisjunction. Firstly, the above data show that chromosomes from all groups may be involved in nondisjunction; the question as to whether some chromosomes are preferentially involved cannot yet be answered. The observation that all groups of chromosomes are involved indicates that the predominance of X-chromosomal

1 Institut für Humangenetik, Universität Göttingen, Nikolausberger Weg 5a, 3400 Göttingen, FRG

Mutations in Man, ed. by G. Obe
© Springer-Verlag Berlin Heidelberg 1984

aneuploidy and trisomy 13, 18, and 21 at late developmental stages (Boué and Boué 1973; Hassold et al. 1978) and at birth (Sankaranarayanan 1982, for review) is either due to their preferential survival and/or to their origin from maternal germ cells. The finding of Martin et al. (1983) of double aneuploidy corroborates findings in spontaneous abortions (Kaji et al. 1980) and demonstrates that several bivalents may be involved in nondisjunction. Future studies will show whether the frequency of double aneuploidy is solely an addition of probabilities for two independent events. There is tentative evidence to suggest that the frequency of aneuploidy in sperm samples may vary between donors (Martin et al. 1983). Further investigations with this technique in areas such as the occupational or therapeutic exposure to mutagens, the association of aneuploidy with paternal age, with chromosomal heteromorphisms or with heterozygosity for varios genetic diseases or even with HLA-types will provide us with important information on the events of chromosomal segregation during meiosis. With the advent of in vitro fertilization in man, studies on nondisjunction in human oocytes or early cleavage embryos will become more feasable. First results give evidence of a surprisingly high frequency of chromosome anomalies in such conceptions after in vitro fertilization (Testart et al. 1983; Angell et al. 1983). These abnormalities, however, should possibly be related to the methods applied (Edwards 1983).

In the near future we shall be able to evaluate hypotheses formulated after investigation of causes and mechanisms of nondisjunction in experimental mammals. Hence, experiments on the mechanisms of spontaneous and induced nondisjunction with appropriate animal models should also receive further attention. These studies, advantageously, can measure aneuploidy directly in germ cells in parallel with biochemical and endocrinological measurements in the surrounding somatic cells of the reproductive system.

One further aspect which is responsible for the generation of aneuploidy in newborns will be discussed briefly. The frequency of chromosomal imbalance at birth appears to be approx. 1/10 to 1/20 of that estimated at conception (Ford 1981; Sankaranarayanan 1982). Most of the monosomic and trisomic conceptions are aborted at a more or less specific stage of gestation (Boué and Boué 1970), by mechanisms which are still unknown. Recent data suggest that the determining factors which decide whether a trisomic embryo is aborted or whether it develops until parturition may change with maternal age (Aymé and Lippman-Hand 1982). According to the work by Aymé and Lippman-Hand (1982) these determining factors affecting the survival of chromosomally unbalanced conceptions are most stringent in mothers around 30 years of age. These factors should be characterized experimentally because changes in the incidence of aneuploidy in the population might also be caused by modulation of the efficiency of prenatal selection.

## 2 Causes of Meiotic Nondisjunction

### 2.1 Irradiation

Results from retrospective studies on the association of Down's syndrome with preconceptional X-ray exposure are inconsistent. Some authors found an association, others not (Sankaranarayanan 1982, for review). No evidence for an increase of aneuploidy has been found in the population of Hiroshima and Nagasaki (Schull et al. 1982).

Since the first experimental cytogenetic study by Uchida and Lee (1974) which demonstrated that irradiation during the preovulatory phase of oogenesis significantly increased the incidence of hyperploid mouse oocytes, several studies have been published supporting the hypotheses that X-rays induce nondisjunction during male or female germ cell development. Several authors found a rather moderate increase of nondisjunction in mouse (Uchida and Freeman 1977; Hansmann et al. 1982; Tease 1982) and Chinese hamster oocytes (Mikamo 1980), as well as in spermatocytes from mice (Hansmann et al. 1979) and *Microtus oeconomus* (Tates 1979).

Whether ageing oocytes are in fact more susceptible to radiation-induced nondisjunction, as was first suggested by Yamamoto et al. (1973), is not yet proven because data have been published both in favor (Uchida and Freeman 1977) and against (Max 1977; Strausmanis et al. 1978; Tease 1982) this hypothesis. The only evidence so far for a dose-dependency of radiation-induced aneuploidy comes from a study by Tease (1982) on mouse pronuclei. As no banding techniques were used to identify chromosomal aneuploidy, it is difficult to exclude the involvement of radiation-induced translocations in the generation of nondisjunction.

Several authors, although using mice and similar experimental schemes (target cells analyzed, germ cell stage irradiated, exposure conditions) found no effect of X-rays on chromosome segregation (de Boer and Tates 1983; Hansmann 1983, for review).

The findings of Nijhoff and de Boer (1980) as well as of Hansmann et al. (1983) of a decreased incidence of aneuploidy after irradiation are somewhat puzzling. Both studies were performed in mouse strains with a high level of aneuploidy in nonirradiated animals. Whether these data indicate a protecting capability of irradiation at certain dose levels remains unknown.

Considering all data on the exposure of mammalian germ cells it is reasonable to suggest that X-rays may have three different effects on chromosome segregation:

1. A moderate increase in the incidence of nondisjunction,
2. a non measureable effect, and
3. a reducing/protecting effect.

## 2.2 Chemical Agents

A variety of chemicals were tested for their ability to induce nondisjunction in either spermatogenesis or oogenesis (Hansmann 1983, for review). Compounds that induced nondisjunction are Trenimon, Cyclophosphamide, Methotrexate, MMS, EMS, Natulan, Bleomycin, the amino acid analogue 4-Fluoro-*dl-β* phenylalanine (PFPA), the benzimidazol derivative Carbendazim (MBC), Cadmium, Colchicin, Vincristine, and Ethanol. No statistically significant increase of aneuploidy was found after treatment with Mitomycin C, TEM, IMS, Hycanthone, Benomyl, Nocodazol, 6-Mercaptopurine, Trypaflavin, and Isoniazid (Hansmann 1983, for review). The data obtained from studies in ovulated oocytes, in spermatocytes, or even in $F_1$-embryos do not, however, provide sufficient information on the mechanisms involved.

The utilization of compounds known to interfere with specific cellular functions, e.g. inhibitors of the spindle apparatus should prove useful in future studies. Agents such as Colchicin or Vincristine could be used to answer questions such as the preferential involvement of specific bivalents in nondisjunction. Ethanol which concomitantly activates oocytes parthenogenetically (Kaufman 1982) would seem to be a suitable compound for the investigation of mechanisms of nondisjunction during meiosis II.

It is nevertheless rather unlikely that the above-mentioned chemical agents, or environmental pollutants, together with the exposure to medical or environmental irradiation are responsible for the considerable burden of aneuploidy in man. In addition to a possible involvement of a genetically determined prevalence for aneuploidy (Alfi et al. 1980; Hansmann and Jenderny 1983), other factors have to be considered here, too. Evaluation in more detail of the influence of our reproductive behavior and, more importantly, the involvement of alterations of gonadal functions, as main causes for aneuploidy in man are essential.

## 2.3 Reproduction

The postulate that our reproductive behavior can influence the frequency of aneuploidy goes back to German (1968) who tried to explain the association of Down's syndrome with maternal age by altered sexual behavior at advanced ages. The method of birth control practiced and hence also sexual behavior is also thought to be responsible for a higher incidence of Down's syndrome in Catholic populations (Jongbloet et al. 1978; Mulcahy 1978). There is now sufficient experimental evidence to accept that overripeness of ovulated oocytes may cause aneuploidy in the $F_1$-generation (Szollosi 1975; Adachi and Ingalls 1976; Yamamoto and Ingalls 1972). The effects of sperm ageing before fertilization (Martin-Deleon and Boice 1982) is in the author's opinion still open. It cannot be excluded that the observed chromosome anomalies are caused by physiological alterations in the ligated ducts (Martin-Deleon et al. 1973).

Whether delayed fertilization really contributes significantly to aneuploidy in man is doubtful, since it has been shown by the use of chromosomal polymorphisms that at least 60%–80% of all trisomies in live-born children and abortions are derived from nondisjunction during meiosis I (Hansmann 1983, for review and Chap. Sperling, p. 128ff.). Apparently, this stage of germ cell development appears to be independent of alterations introduced by sexual behavior.

Jongbloet et al. (1982) published evidence in favor of their "seasonal preovulatory overripeness ovopathy" hypotheses. They found a clustering around May–July of those trisomy-21 conceptions with an error of chromosome segregation at maternal meiosis I. Conceptions of trisomy 21 patients with either maternal meiosis II or paternal errors were found to be not clustered in the same way. These data do not, however, prove that nondisjunction occurs more frequently during the transitional phases of decreasing and increasing ovulation rate. Such a clustering may also be caused secondarily by e.g. an altered prenatal survival of specific types (meiosis I versus meiosis II) of trisomic conceptions. Preovulatory overripeness may nevertheless cause aneuploidy (Butcher and Fugo 1967) (monosomy, trisomy, and polyploidy) in rat embryos after a delay of ovulation induced by pentobarbital treatment (Mikamo et al. 1975; Kamiguchi et al. 1979). Further studies have still to elucidate, however, the influence of pentobarbital itself on chromosome segregation.

More evidence for the hypothesis that alterations in germ cell development are the primary cause for nondisjunction has been obtained from two studies on the effects of oral contraceptives on chromosome segregation in mouse oocytes. Röhrborn and Hansmann (1974) analysed oocytes after spontaneous ovulation from mice treated with 1 or 10 mg Norethisteroneacetate daily for 4 weeks. Mice that received the high and nonphysiological dose of gestagen ovulated significantly more hyperploid oocytes, both soon after cessation of treatment and after a longer interval of 6 weeks without hormonal uptake. Becker and Schöneich (1982) detected in another study significantly more diploid oocytes and still more importantly, more trisomic pronuclei in mice treated with ethinylestradiol-sulfonate, daily with 0.63 mg for 5 weeks.

The current widespread use of such hormones in birth control requires further experimental studies on the effect of various oral contraceptives on chromosome segregation in oocytes from several species.

In a retrospective cytogenetic study with human abortions it was reported that hormonally induced ovulation is apparently associated with a higher incidence of trisomic and polyploid spontaneous abortions (Lazar et al. 1974). It is not known whether the hormones used to stimulate follicular maturation and ovulation are responsible for this higher incidence of chromosomally abnormal abortions or whether this is due to the gynecological disturbance requiring such hormonal therapy. A third possibility is that these above mentioned circumstances may exert a favorable influence on the development of aneuploid conceptions up to a gestational stage where the aborted specimens are accessible for cytogenetic diagnosis.

## 3 Nondisjunction by Failures of the Endocrine Control of Meiosis

Most of the proposed hypotheses on the mechanisms of nondisjunction are based upon morphological properties and are confused to mechanical events within the germ cell. Prominent examples are the hypotheses on nucleolus persistance (Evans 1967), the generation-line hypotheses (Henderson and Edwards 1968), spindle degeneration (Mikamo 1968), and precocious chromatid separation in ovulated oocytes (Rodman 1971). Experimental evidence in favor and against these hypotheses have recently been discussed in more detail (Hansmann 1983). The recently proposed chiasma-hormonal hypothesis (Crowley et al. 1979) considers only events expected to be associated with an increased maternal age. Nondisjunction in oocytes from young females, however, or even in spermatocytes is not considered by these hypotheses.

Hansmann et al. (1980), influenced by the tremendous progress made during the last years in the understanding of follicular and oocyte maturation (Masui and Clarke 1979; Moor et al. 1980), and of the multiplicity of intercellular communication which is required for the maturation of the oocyte (Tarkowsky 1982; Brower and Schultz 1982; Schultz et al. 1983), proposed a model for the hormonal control of meiosis (Hansmann et al. 1980). The hypothesis of "nondisjunction caused by alterations of neuroendocrine control of germ cell maturation" is based on observations made during recent years with ovulated oocytes from Djungarian hamsters of our own colony. Gonadotropic hormones (GTH) used to stimulate follicular maturation and ovulation cause hyperploidy in oocytes from these females. Females from other species studied so far (Hansmann and El Nahass 1979; Hansmann and Probeck 1979) did not respond to GTH-treatment with the ovulation of hyperploid oocytes. A significant increase in nondisjunction after GTH-treatment was recently observed, however, in females from two $F_1$-hybrid mouse strains, but not in females from their parental strains (one of these parental strain is the sensitive NMRI/Han line, females from that line ovulate diploid oocytes after GTH-stimulation) (Hansmann and Jenderny 1983). Such an increase of hyperploidy in cross-breeds of the $F_1$-generation is an indication of a favorable influence of heterozygosity on nonsegregation during meiosis.

The frequency of hyperploidy in oocytes from Djungarian hamsters depends entirely on the injected dose of GTH. This is in approximately 10% after the highest doses injected. Several bivalents are affected by nondisjunction at higher doses and the probability of involvement in that process is not the same for each bivalent. The large metacentric chromosomes as well as the subtelocentric chromosome 5, which, very interestingly, carries a nucleolus organizer on its short arm, are most often involved.

Similar observations were made with oocytes from prepubertal females after the same GTH-treatment (Theuring 1982). Hence, the aneuploid oocytes are not derived from degenerated follicles of the preceding estrus cycle. Furthermore, preliminary data on measurements of ovulated oocytes demonstrate that these oocytes are not ovulated before they have reached their full size, as seen in large mature follicles.

Since it appears to be very unlikely that the gonadotropins themselves would interfere with oocyte maturation, we became interested in the effects of steroids on germinal vesicle break down (GVBD) and chromosome segregation. Dihydrotestosterone potently induced nondisjunction in a dose dependent manner in GTH-primed females during the first meiotic division. Preliminary data also provided evidence that Gn-releasing hormone, i.e. a synthetic analogue, may enhance chromosomal nonsegregation in oocytes from GTH-primed females. It was concluded that the primary cause of the creation of an intracellular milieu rendering oocytes susceptible to nondisjunction is derived from outside the germ cell and causes an imbalance in the normal pathway in the final process of follicular and oocyte interaction. Such alterations in normal follicular function may result from various failures at any level of the hypothalamic-pituitary-gonadal axis.

The hypotheses of neuroendocrine control failures as a primary cause of nondisjunction may also be valid for spermatogenesis. Experimental evidence in favor of that suggestion may be derived from the finding of an increased frequency of hyperploid spermatocytes in hypopituitary Snell-dwarf mice (Wauben-Penris and van Buul-Offers 1982).

The intracellular mechanisms causing nondisjunction are still unclear. It is assumed, however, that the error manifests itself biochemically. One of our assumptions is that the altered endocrine control creates desynchronized processes of nuclear and cytoplasmic differentiation, causing a premature arrest – directly or indirectly – of chromosome segregation. These bivalents therefore, which start anaphase I very late, should be those which are most likely and most often involved in nondisjunction.

Hence, chromosomal properties like length, position of the centromere, number of chiasmata, pairing forces (e.g. by quantity/quality of the synaptonemal complexes) presence of nucleolus organizers, as well as the number of active copies of ribosomal DNA and the position of the bivalent within the spindle, will all provide an individual risk for each bivalent to be, more or less, passively involved in the process of nondisjunction. It would not be at all surprising if the influence of single chromosomal properties outlined above changes during aging.

The hypotheses on the mechanisms of nondisjunction are transferable also to events occurring during the second meiotic division and during mitosis in early cleavage stages. It is hoped that it may stimulate further experimental work in this field and that we shall soon arrive at a level where we are not only satisfied intellectually, but also in a position where we are able to recognize and protect individuals in our population who risk aneuploidy in their offspring.

*Acknowledgements.* Most of the experimental work with the Djungarian hamsters was done during the last years in collaboration with Christel Düls, Jutta Jenderny and Manni Theuring.

The secreterial and editorial help by Marita Heinze-Sprotte and by David Cooper is highly acknowledged. The author's experimental work is supported by grants from the Deutsche Forschungsgemeinschaft (DFG Ha 747/5), the Bundesministerium für Forschung und Technologie (CMT 40) and from the European Commission (206-BIO D).

# References

Adachi K, Ingalls TH (1976) Ovum aging and pH imbalance as a cause of chromosomal anomalies in the hamster. Science 194:946–948

Alfi OS, Chang R, Azen SP (1980) Evidence for genetic control of nondisjunction in man. Am J Hum Genet 32:477–483

Angell RR, Aitken RJ, van Look PFA, Lumsden MA, Templeton AA (1983) Chromosome abnormalities in human embryos after in vitro fertilization. Nature 303:336–338

Aymé S, Lippman-Hand A (1982) Maternal-age effect in aneuploidy: does altered embryonic selection play a role? Am J Hum Genet 34:558–565

Becker K, Schöneich J (1982) Expression of genetic damage induced by alkylating agents in germ cells of female mice. Mutat Res 92:447–464

Boué J, Boué A (1970) Les abérrations chromosomiques dans les avortements spontanés humains. La Presse médical 14:635–645

Brower PT, Schultz RM (1982) Intercellular communication between granulosa cells and mouse oocytes: existence of possible nutritional role during oocyte growth. Dev Biol 90:144–153

Butcher RL, Fugo NW (1967) Overripeness and mammalian ova. II. Delayed ovulation and chromosome anomalies. Fertil Steril 18:297

Crowley PH, Gulati DK, Hayden TL, Lopez P, Dyer R (1979) A chiasma-hormonal hypothesis relating Down's syndrome and maternal age. Nature 280:417–418

de Boer P, Tates AD (1983) Radiation-induced nondisjunction. In: Ishihara T, Sasaki, MS (eds) Current topics of cytogenetics. Liss, New York, p 299

Edwards RG (1983) Chromosomal abnormalities in human embryos. Nature 303:283

Evans HJ (1967) The nucleolus, virus infection and trisomy in man. Nature 214:361–463

Ford CE (1981) Nondisjunction. In: Burgio GR, Fraccaro M, Tiepolo L, Wolf U (eds) Trisomy 21. Springer, Berlin Heidelberg New York, p 103

Ford CE, Jones KW, Miller GJ (1959) The chromosomes in a patient showing both mongolism and the Klinefelter syndrome. Lancet I:709

German J (1968) Mongolism, delayed fertilization and human sexual behaviour. Nature 217: 516–518

Hansmann I (1983) Factors and mechanisms involved in nondisjunction and X-chromosome loss. In: Sandberg A (ed) The cytogenetics of the mammalian X-chromosome. Liss, New York, pp 131–170

Hansmann I, El Nahass E (1979) Incidence of nondisjunction in mouse oocytes. Cytogenet Cell Genet 24:115–121

Hansmann I, Jenderny J (1983) Genetic basis of nondisjunction. Increased incidence of hyperploidy in oocytes from $F_1$-hybrids but not from their parental mice. Hum Genet 65:56–60

Hansmann I, Probeck HD (1979) Chromosomal imbalance in ovulated oocytes from Syrian Hamsters (Mesocricetus auratus) and Chinese hamsters (Cricetulus griseus). Cytogenet Cell Genet 23:70–76

Hansmann I, Jenderny J, Probeck HD (1980) Mechanisms of nondisjunction: Gonadotrophininduced aneuploidy in oocytes from Phodopus sungorus. Eur J Cell Biol 22:24

Hansmann I, Jenderny J, Probeck HD (1982) Nondisjunction and chromosome breakage in mouse oocytes after various X-ray doses. Hum Genet 61:190–192

Hansmann I, Jenderny J, Probeck HD (1983) Low doses of X-rays decrease the risk of diploidy in mouse oocytes. Mutat Res 109:99–110

Hansmann I, Zmarsly R, Probeck HD, Jenderny J, Schäfer J (1979) Aneuploidy in mouse foetuses after paternal exposure to X-rays. Nature 280:228–229

Hassold UJ, Matsuyama A, Newlands IM, Matsuura JS, Jacobs PA, Mannel B, Tsuei J (1978) A cytogenetic study of spontaneous abortions in Hawaii. Ann Hum Genet 41:443–454

Henderson SA, Edwards RG (1968) Chiasma frequency and maternal age in mammals. Nature 218:22–28

Jacobs PA, Baikie AG, Court-Brown WM, Strong JA (1959) The somatic chromosomes in mongolism. Lancet I:710

Jagiello G, Ducayen M, Fang JS, Graffeo J (1976) Cytogenetic observations in mammalian oocytes. In: Pearson PL, Lewis KR (eds) Chromosomes today, no. 5. Wiley, New York, pp 43–63

Jongbloet PH, Mulder AM, Hamers AJ (1982) Seasonality of preovulatory nondisjunction and the aetiology of Down-syndrome. A European collaborative study. Hum Genet 62:134–138

Jongbloet PH, Poestkoke AJM, Hamers AJH, van Erkelens-Zwets JHJ (1978) Down Syndrome and religious groups. Lancet II:1310

Kajii T, Ferrier A, Niikawa N, Takahara H, Ohama K, Avirachan S (1980) Anatomic and chromosomal anomalies in 639 spontaneous abortuses. Hum Genet 55:87–98

Kamiguchi Y, Funaki K, Mikamo K (1979) Chromosomal overripeness of the primary oocyte. Proc Jap Acad 55:398

Kaufman MH (1982) The chromosome complement of single-pronuclear haploid mouse embryos following activation by ethanol treatment. J Embryol Exp Morphol 71:139–154

Lazar P, Gueguen S, Boué J, Boué A (1974) Epidémiologie des avortéments spontanés precoces: à propos de 1469 avortéments caryotypes. In: Boué A, Thibault C (eds) Chromosomal errors in relation to reproductive failure. INSERM, Paris, pp 317–339

Léjeune J, Gautier M, Turpin R (1959) Le mongolism, premier éxemple d'abérration autosomique humaine. CR Acad Sci (Paris) 248:1721–1722

Martin RH, Balkan W, Burns K, Rademaker AW, Lin CC, Rudd NL (1983) The chromosome constitution of 1,000 human spermatozoa. Hum Genet 63:305–309

Martin-Deleon PA, Boice ML (1982) Sperm aging in the male and cytogenetic anomalies. An animal model. Hum Genet 62:70–77

Martin-Deleon PA, Shaver El, Gammal EB (1973) Chromosome abnormalities in rabbit blastocysts resulting from spermatozoa aged in the male tract. Fertil Steril 24:212–219

Masui Y, Clarke HJ (1979) Oocyte maturation. In: Bourne GH, Danielli JF, Jeon KW (eds) International review of cytology, vol 57. Academic, New York, pp 186–282

Max C (1977) Cytological investigation of embryos in low-dose X-irradiated young and old female inbred mice. Hereditas 85:199–206

Mikamo K (1968) Mechanism of nondisjunction of meiotic chromosomes and degeneration of maturation spindles in eggs affected by intrafollicular overripeness. Experientia 24:75–78

Mikamo K (1980) Cytogenetic effects of radiation on developing ovarian oocytes. Proc Jpn Acad 82:174

Mikamo K, Hamaguchi H (1975) Chromosomal disorder caused by preovulatory overripeness of oocytes. In: Blandau RJ (ed) Aging gametes. Karger AG, Basel, pp 72–97

Moor RM, Polge C, Willadsen SM (1980) Effect of follicular steroids on the maturation and fertilization of mammalian oocytes. J Embryol Exp Morphol 56:319–335

Mulcahy MT (1978) Down syndrome and parental coital rate. Lancet II:895

Nijhoff JH, de Boer P (1980) Radiation-induced meiotic autosomal nondisjunction in male mice. The effects of low doses of fission neutrons and X-rays in meiosis I and II of a Robertsonian translocation heterozygote. Mutat Res 72:431–446

Rodman TC (1971) Chromatid disjunction in unfertilized aging oocytes. Nature 233:191

Röhrborn G, Hansmann I (1974) Oral contraceptives and chromosome segregation in oocytes of mice. Mutat Res 26:535–544

Rudak E, Jacobs PA, Yanagimachi R (1978) Direct analysis of the chromosome constitution of human spermatozoa. Nature 274:911–913

Sankaranarayanan K (1982) Genetic effects of ionizing radiation in multicellular eukaryotes and the assessment of genetic radiation hazards in man. Elsevier, Amsterdam

Schull WJ, Otake M, Neel JV (1981) Genetic effects of the atomic bombs: a reappraisal. Science 213:1220–1227

Schultz RM, Montgomery RR, World-Bailey PF, Eppig J (1983) Regulation of oocyte maturation in the mouse: possible roles of intracellular communication, c AMP and Testosterone. Dev Biol 95:294–304

Strausmanis R, Henrikson IB, Holmberg M, Rönnbäck C (1978) Lack of effect on the chromosomal nondisjunction in aged mice after low dose X-irradiation. Mutat Res 49:269–274

Szollosi D (1975) Mammalian eggs aging in the fallopian tubes. In: Blandau RJ (ed) Aging gametes. Karger AG, Basel, pp 98–121

Tarkowsky A (1982) Nucleo-cytoplasmic interactions in oogenesis and early embryogenesis in the mouse. In: Embryonic development. A: Genetic aspects. Liss, New York, pp 407–416

Tates AD (1979) Microtus oeconomus (Rodentia), a useful mammal for studying the induction of sex-chromosome nondisjunction and diploid gametes in male germ cells. Environ Health Perspect 31:151–159

Tease C (1982) Similar dose-related chromosome nondisjunction in young and old female mice after X-irradiation. Mutat Res 95:287–296

Testart J, Lasalle B, Frydman R, Belaisch JC (1983) A study of factors affecting the success of human fertilization in vitro. II. Influence of semen quality and oocyte maturity on fertilization and cleavage. Biol Reprod 28:425–431

Theuring F (1982) Der Einfluß von Gonadotropinen und des LH-Releasing Hormons auf die Ovarfunktion und den Ablauf der Meiose beim russischen Hamster (Phodopus sungorus). Diplomarbeit der Math. Nat. Fakultät, Göttingen

Uchida IA, Freeman CPV (1977) Radiation induced nondisjunction in oocytes of aged mice. Nature 265:186–187

Uchida IA, Lee CPV (1974) Radiation-induced nondisjunction in mouse oocytes. Nature 250:601–602

Wauben-Penris PJJ, van Buul-Offers SC (1982) Meiotic nondisjunction in male Snell dwarf mice. J Hered 73:365–369

Yamamoto M, Ingalls TH (1972) Delayed fertilization and chromosome anomalies in the hamster embryo. Science 176:518

Yamamoto M, Endo A, Watanabe G (1973) Maternal age dependence of chromosome anomalies. Nature 241:141

Yamamoto M, Ito T, Watanabe M, Watanabe G (1982) Causes of chromosome anomalies suggested by cytogenetic epidemiology of induced abortions. Hum Genet 60:360–364

# Origin and Significance of Chromosomal Alterations

A.T. NATARAJAN[1]

## 1 Introduction

The spontaneous frequency of chromosomal changes (structural and numerical aberrations) in humans is in the order of 6 in 1,000 newborn. Chromosomal analysis of spontaneous abortuses indicate that about 50% of all spontaneous abortions are chromosomally abnormal. Populations exposed to ionizing radiations (atom bomb survivors) or chemical mutagens (e.g., workers occupationally exposed to vinyl chloride or benzene) show increased frequencies of chromosomal aberrations in their peripheral blood lymphocytes. Many types of human cancer are associated with specific or nonspecific chromosomal aberrations. Several human recessive diseases, such as ataxia telangiectasia (A-T), Fanconi's anemia (FA) and Bloom's syndrome (BS) are associated with increased frequencies of chromosomal aberrations. However, no detectable increase in the frequency of spontaneous point mutations in human populations exposed to ionizing radiations or chemical mutagens has been demonstrated so far. These observations point to the importance of understanding the mechanism involved in the origin of chromosomal alterations and their significance.

## 2 Origin of Chromosomal Aberrations

Most of the known mutagenic carcinogens induce chromosomal aberrations and it is generally agreed that DNA is the main target for the action of most of the chromosome breaking agents. A number of different types of lesions is induced in DNA depending on the type of mutagenic agent employed. These include DNA strand breaks (ionizing radiations, UV, bleomycin), pyrimidine dimers (short wave UV), base alkylation (alkylating agents), inter- and intra-strand cross links (polyfunctional alkylating agents, psoralen and long wave UV) and intercalation (acridine, proflavin, etc.) (Fig. 1). These lesions are subjected to cellular repair and unrepaired lesions or misrepaired lesions lead to chromosomal aberrations. The repair of DNA

1 Department of Radiation Genetics and Chemical Mutagenesis, University of Leiden, Sylvius Laboratories, Wassenaarseweg 72, 2333 AL Leiden, The Netherlands and J.A. Cohen Institute, Interuniversity Institute of Radiopathology and Radiation Protection, Rijnsburgerweg 86, 2333 AD Leiden, The Netherlands

Mutations in Man, ed. by G. Obe
© Springer-Verlag Berlin Heidelberg 1984

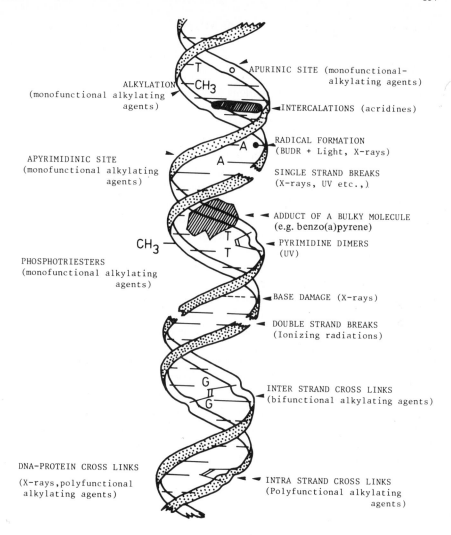

**Fig. 1.** Primary DNA lesions in DNA following treatment with mutagens. [Obe and Natarajan (1982) In: Müller D et al. (ed) Sister chromatid exchange test. Georg Thieme, Stuttgart]

lesions is processed by several ways. There are mainly three kinds of repair recognized, namely, (1) photoreactivation, (2) excision repair, and (3) post replication repair (for detailed description of these repair pathways, see Chap. van Zeeland, p. 35ff.). Establishing direct correlations between these repair processes and chromosomal alterations is difficult at present.

# 3  Lesions Leading to Chromosomal Alterations

## 3.1  Ionizing Radiations

Quantitative and qualitative correlations between primary DNA lesions and chromosomal alterations can be attempted to some extent. For example, ionizing radiations induce single strand breaks (SSB), double strand breaks (DSB), base damage of different kinds in the DNA, as well as DNA-protein cross links. With X-rays, base damage and single strand breaks are induced in similar frequencies, while 10 to 20 SSB's are induced for every DSB. The relative importance of the involvement of these lesions in the origin of chromosomal aberrations has been a subject of investigation for many years. Based on the available evidence, such as high RBE for high LET radiations (which induce relatively high proportion of DSB's to SSB's) in the DNA, it was proposed that DSB's should be responsible for the origin of chromosomal aberrations (Bender et al. 1974). Direct experimental support for this conclusion has come from experiments involving posttreatment with *Neurospora* endonuclease, of X-irradiated Chinese hamster ovary cells (CHO). *Neurospora* endonuclease is specific for cleaving single stranded DNA. The rationale behind these experiments is that X-ray induced SSB's and gaps can be converted into DSB's and if DSB's are indeed responsible for the origin of aberrations, then there should be an increase in the frequencies of induced aberrations (Natarajan and Obe 1978). All classes of aberrations, namely gaps, breaks and exchanges, increase by post treatment with NE. Parallel measurement of DNA DSB's induced by X-rays, by neutral sucrose gradient analysis indicated a doubling in the frequencies of DSB's induced by X-rays due to post incubation with NE with concentrations of the enzyme with which cytological experiments were carried out (Natarajan et al. 1980). Similar experiments with 0.5 MeV neutrons in CHO cells did not yield any increase in the frequencies of induced aberrations due to post treatment with NE indicating single strand breaks/gaps are not available for conversion to DSB's by NE, following neutron irradiation. Chemical mutagens which induce DNA strand breaks, like bleomycin, also respond with increased frequencies of aberrations following NE treatment.

Monofunctional alkylating agents induce several types of DNA lesions including DNA single strand breaks. Normally, lesions such as single strand breaks, are repaired quickly and do not lead to chromosomal aberrations, whereas base damages are repaired slowly. Thus, treatment of $G_2$ cells does not lead to any chromatid aberrations in the forthcoming mitosis. However, if $G_2$ cells are treated with methyl methanesulfonate and post treated with NE, there is a dramatic increase in the frequencies of aberrations, indicating that some DNA SSB's are available for conversion to DSB's by NE (Natarajan and Obe 1978). Such a response has been observed, following treatment with several monofunctional agents, including dimethyl sulfate, ethyl methane sulfonate, methyl nitrosoguanidine and methyl nitrosourea (Natarajan unpublished results). Short wave UV irradiation, as well as treatment with chemicals such as mitomycin C, trenimon, diepoxybutane do not respond to NE treatment with increased aberrations (Natarajan et al. 1980, Nowak 1983). This indicates that these treatments, even if they induce SSB's to some extent, do not respond to NE treatment, most probably, a strand break is not enough for NE to act, but a larger stretch such as a gap is needed. However, when cells which have been grown in medium containing 5-bromodeoxy uridine (BrdUrd) are treated with short wave UV, there is an immediate induction of chromatid aberrations in $G_2$ (S-independent) and the frequencies of aberrations can be increased by NE post treatment. This is due to the types of lesions induced by UV in unsubstituted and BU-substituted DNA-predominantly pyrimidine dimers and DNA strand breaks, respectively (Obe et al. 1982). From these results, it can be concluded that DNA DSB's are mainly responsible for chromosomal aberrations induced by ionizing radiations.

There are other hypotheses implying DNA SSB's and base damage as possible lesions for the origin of chromosomal aberrations. The involvement of SSB's in the aberration formation can be accommodated as in this model, SSB's are converted to DSB's by an endonuclease, and this has been shown to be feasible with the experiments involving *Neurospora* endonuclease (Craig 1980; Natarajan and Obe 1978). Arguments favoring involvement of DNA base damage induced by ionizing radiations in the origin of chromosomal aberrations have been put forward by Preston (1980). These are based on the observation that when X-irradiated $G_0$ lymphocytes are treated with cytosine arabinoside (ara C, an inhibitor of DNA polymerase alpha) the frequencies of aberrations increase with increase in incubation time up to 3 h. In earlier studies, it has been found that when cells are irradiated first and allowed to repair for 2 h and then challenged with ara C, there was an accumulation of strand breaks, indicating that a slow repair process goes on, probably of the base damage, as one expects that most of the strand breaks are repaired by 2 h. Thus, the increase in the frequencies of aberrations should be due to mis-repair of base damage. There are at least three observations which do not support this argument. (1) In their studies, the maximum increase in the frequencies of aberrations due to ara C treatment occurred during the first 1 h of treatment, a period when most of the strand breaks are repaired (Preston 1980; Bender and Preston 1982). (2) When irradiated $G_0$ lymphocytes are challenged with ara C after 1 or 2 h incubation in normal medium, the potentiating effect is greatly reduced (Natarajan and Darroudi unpublished). (3) While there is a profound oxygen effect for X- or gamma-ray induced chromosomal aberrations and DNA strand breaks, the oxygen effect is minimal for DNA base damage (Paterson and Setlow 1972). Thus, contribution of base damage towards the production of chromosomal aberrations following X-irradiation should be minimal.

Ionizing radiations are poor inducers of SCE's. However, if BrdUrd-substituted cells are irradiated in the S phase, there is an increase in the frequencies of SCE's at high doses (Perry and Evans 1975). This is perhaps due to the higher radiosensitivity of BrdUrd-substituted DNA. By manipulating the irradiation atmosphere (air or nitrogen), one can alter the proportion of strand breaks to base damage. If cells are irradiated in nitrogen and scored for SCE's and chromosomal aberrations, the ratio of SCE to chromosomal aberrations was much higher when compared to this ratio obtained following irradiation in air (Andersson et al. 1981; Uggla and Natarajan 1983). These results indicate that while DNA strand breaks induced by ionizing radiations lead to chromosomal aberrations, and induced base damage leads to SCE's.

## 3.2 Short Wave UV

Most of the biological effects induced by short wave UV (254 nm) irradiation has been attributed to pyrimidine dimers. These include cell killing, induction of chromosomal aberrations, SCE's in eukaryotic cells, point mutations in pro and eukaryotic cells and tumors in fishes (Natarajan et al. 1980a; Griggs and Bender 1973; van Zeeland et al. 1980; Hart et al. 1977). Specific involvement of pyrimidine dimers in these effects can be demonstrated in many ways. When CHO cells are irradiated with UV of different wavelengths, namely 254 and 290 nm, the frequencies of SCE's increase linearly to the induction of dimers (detected as endonuclease sensitive sites), irrespective of the wavelength of UV employed (Reynolds et al. 1979). More direct evidence comes from studies involving eukaryotic cells which possess photoreactivating enzyme. This enzyme monomerizes the cyclobutane dimers when UV irradiated cells are exposed to light with a wave length between 300–600 nm. This enzyme is present in lower vertebrates and plants. Griggs and Bender (1973) first demonstrated that UV induced chromosomal aberrations can be reduced by photoreactivation (PR) in *Xenopus* cells. Wolff (1978) found that in chicken embryonic fibroblasts, the frequencies of SCE's induced by short wave UV are not reduced by PR. However, in a detailed analysis, it has been demonstrated in chicken embryonic fibroblasts, both chromosomal aberrations and SCE's induced by UV can be drastically reduced by PR (Natarajan et al. 1980a). In a further study this phenomenon has been analyzed employing fibroblasts from *Xenopus laevis* (van Zeeland et al. 1980). In this study, the induction and disappearance of dimers were measured on UV irradiation and photoreactivation, respectively. In addition, several biological effects were measured. These included cell killing, chromosomal aberrations, SCE's and point mutations (resistance to ouabaine, Table 1). There was a clear photoreactivation effect for all the end points studied.

**Table 1.** Influence of photoreactivating light on the frequencies of short wave UV induced pyrimidine dimers, chromosomal aberrations, sister chromatid exchanges and point mutations. (UV dose 10 J/m$^2$ and PR light 60 min)

| Treatment | Dimers/10$^8$ M.W. DNA | Chromatid exchanges/ 100 cells | SCE's/cell | Mutation frequency $\times$ 10$^{-6}$ |
|-----------|-------------------------|--------------------------------|------------|----------------------------------------|
| UV        | 18.4                    | 8.3                            | 28.7       | 26.2                                   |
| UV + PR   | 0.92 (95)               | 1.6 (80.7)                     | 8.9 (69.0) | 1.3 (96.5)                             |

Figures in parantheses indicate reduction due to PR. Resistance to ouabaine was used as mutational system. (Data from van Zeeland et al. 1980)

In cells derived from patients of xeroderma pigmentosum, which are deficient in the excision of pyrimidine dimers, it has been shown that these cells respond with increased cell killing, increased frequencies of chromosomal aberrations, sister chromatid exchanges and point mutations following UV irradiation in comparison to normal human cells (de Weerd-Kastelein et al. 1977; McCormick and Maher 1978).

These observations point out but do not prove that UV induced pyrimidine dimers are involved in bestowing biological effects.

The influence of density inhibition on several biological parameters in mouse and human cells – two types of cells with great difference in their capacity to excise dimers – has been studied (Nagasawa et al. 1982). This approach gives further insight into the relation of excision repair to UV induced biological effects. About 60% of the dimers were excised in human cells in 5 h and 30% of dimers were excised in mouse cells in 24 h. After density inhibition, the frequencies of UV induced chromosomal aberrations declined in human cells in comparison to mouse cells indicating residual dimers present in the DNA lead to chromosomal aberrations. The kinetics of reduction of SCE's during density inhibition, however, showed a pattern similar in both types of cells, with a slow decline of SCE's in a biphasic manner up to 48 h, a pattern similar to protein DNA cross link repair following UV irradiation of these cells. This suggests that protein-DNA cross links would be involved in the origin of SCE's. Photoreactivation experiments do not suggest this, unless protein DNA cross links can also be photoreactivated.

## 3.3 Chemical Mutagens

The number of types of DNA lesions induced by chemical mutagens is very large. All the different types of adducts formed in the DNA due to treatment even with well-defined chemical mutagens have not been identified and quantified. Using an indirect approach, Vogel and Natarajan (1979a,b, 1982) selected a series of monofunctional alkylating agents which differ in their reactivity to nucleophiles (the Swain-Scott substrate constant-S) and showed that at equal toxicity levels, agents with low S-values, were more mutagenic in Drosophila than those with high S-values. Consequently, agents with high S-values induced more chromosomal aberrations than those with low S-values. Since alkylating agents with low S-values alkylate $O^6$ guanine more efficiently in comparison to alkylating agents with high "S", it was concluded that $O^6$ alkylation must be an important lesion leading to point mutations and other lesions such as $N^7$ guanine may lead to chromosomal aberrations and SCE's (Vogel and Natarajan 1982). Data from other eukaryotic systems, such as barley and mammalian cells in vitro, confirm these conclusions. In a comparative study involving five monofunctional alkylating agents, namely ethyl methanesulfonate (EMS), methyl methanesulfonate (MMS), ethyl nitrosourea (ENU), methyl nitrosourea (MNU), and dimethyl sulfate (DMS), differing in their S-values, Natarajan et al. (1984) investigated the relationship between induced cell killing, chromosomal aberrations, sister chromatid exchanges and point mutations in CHO cells. While there was a good correlation between SCE's, chromosomal aberrations and cell killing, induction of point mutations was not correlated with any of these end points, indicating that cytotoxic effects and mutagenic effects arise from different DNA lesions.

Loveless (1969) was the first to postulate that $O^6$ alkylation of guanine is the most important lesion for the origin of point mutations. Claims have been made that $O^6$ alkylations of guanine also lead to SCE's (Wolff 1982) and that SCE's are direct indicators of point mutations in mammalian cells (Carrano et al. 1978). However, if one compares monofunctional alkylating agents with different S-values (and thus with different ratios of $O^6$ to $N^7$ alkylation), the ones with low S-values are potent inducers of point mutations, while those with high S-values are potent inducers of SCE's (Natarajan et al. 1984). In a comparative study of the induction of point mutations and SCE's in CHO cells by EMS and ENU, Heflich et al. (1982) found a good correlation between $O^6$ alkylation and point mutations, but not for SCE's, and SCE induction cannot be a good quantitative predictor for mutagenicity. The number of $O^6$ ethylations needed for the induction of point mutations by EMS and ENU is similar, whereas for induction of SCE's, the difference between these two agents is by a factor of 2.5 (van Zeeland et al. 1981). It is difficult to establish 1 to 1 relationship between biological effects and DNA lesions. For example, to increase the frequencies of spontaneously occurring SCE's by 1 in CHO cells, one would have to induce 20,000 pyrimidine dimers per genome (Reynolds et al. 1979). Similarly, 16,000 ethylations per genome are needed to increase the frequencies of SCE's by 1, following treatment with EMS. This indicates that not all lesions lead to biological effects. For induction of SCE's, the location of the lesion in DNA (e.g. between replicon clusters) may be important.

## 3.4 Interaction of X-Ray Induced Lesions with Lesions Induced by Other Agents to Produce Chromosomal Aberrations

Results from interaction experiments can throw some light on the types of lesions responsible for the induced chromosomal aberrations. However, results from experiments from combined treatment with X-rays and chemicals are difficult to interpret as the mechanisms involved are different. While ionizing radiation induce aberrations in an S-independent manner, most of the chemical mutagens such as alkylating agents (as well as UV) induce aberrations in S-dependent manner, meaning that the cells have to pass through an S-phase before the aberrations can be visualized.

When a combined treatment of 4 NQO and X-rays was given to $G_1$ lymphocytes, a synergistic effect was observed. This increase was interpreted as due to interaction of lesions induced by 4 NQO and X-rays (Preston 1982). This increase was more dramatic when the cells were post treated with ara C. The question as to whether this increase is due to real interaction between lesions induced by these two agents or due to some other effects on the repair of X-ray induced lesions by 4 NQO treatment is open. Recently, Holmberg and Strausmanis (1983) have presented very interesting data on the frequencies of dicentrics induced by combined treatment of UV and X-rays in human $G_0$ lymphocytes. In these experiments, 6 $J/m^2/s$ of UV did not induce any dicentrics and 150 rad of X-rays induced about 0.24 dicentrics per cell, UV pretreatment increased the frequencies of dicentrics to 0.55/cell. This potentiation effect lasted at least till 90 min. However, if UV irradiation was given as a post treatment, the frequencies of dicentrics were 0.45 at the start and dropped to control level by 40 min, thus decreasing with a half life of 20 min at 20 °C. This means that X-ray induced lesions giving rise to enhanced frequencies of dicentrics in combined treatments are repaired within 20–30 min at 20 °C, which in turn indicates a few minutes at 37 °C. This short life is not according to well-documented estimates of rejoining time derived from fractionation experiments, which is in the order of 90 min to 4 h. It is obvious in dose fractionation experiments, that one is studying the dose square term of the equation for yield of chromosome exchange aberrations, namely $Y = a D + b D^2$. The repair phenomenon studied by these authors are those related to aberrations having the same origin as those giving rise to the first linear term in the equation, namely interaction between lesions from the same ionization track, meaning short-lived DNA lesions. These results can be easily explained by an elaboration of the existing models (Natarajan and Ahnström 1968; Bender et al. 1974; Kihlman 1977). The following assumptions can be made: (1) X-ray induced aberrations arise from interactions between short lived DNA lesions (repair half life of few minutes at 37 °C) as well as by long lived lesions (repair half life of several minutes to a few hours), former arising predominantly from the interaction of lesions within the same ionizing track and the latter by interaction between lesions from different tracks. (2) An aberration results from competition between association of damaged sites and DNA repair of the lesions before the association. The association of damaged DNA regions is faster for lesions which are in close proximity immediately after irradiation (i.e., in the same ionization track).

This model can explain the temperature effect on the yield of aberrations, namely a higher yield at 37 °C when compared to 20 °C. This temperature effect is most marked at lower doses where the contribution of linear term in the equation predominates (Lloyd et al. 1975). It is also interesting to note that the frequencies of chromosomal aberrations are higher in X-irradiated lymphocytes from ataxia telangiectasia patients, in comparison to those from the normal individuals, at lower doses than at higher doses, indicating the importance of the linear term in the equation on the yield of aberrations (Taylor 1978).

In these UV + X-ray experiments, the increased yield of aberrations may not be due to interaction between UV induced lesions and X-ray induced lesions, but to the interference in the repair of X-ray induced lesions by UV pretreatment, which switches on the excision repair system. This type of interpretation can also be valid for the results obtained from the combination treatment of 4 NQO and X-rays. Moreover, when chemical mutagens, such as bleomycin, which directly break DNA, are used along with X-rays true interaction between the lesions would occur (Preston 1982). It is important to generate dose-response curves over a wide range, to interpret data obtained in such experiments.

## 3.5  Are Two Lesions Necessary to Induce an Exchange?

There are two hypotheses for the formation of chromosomal aberrations. The breakage first hypothesis of Sax (1941) envisages that X-rays break chromosomes directly, most of the breaks restitute. Unrestituted ones remain as breaks or rejoin illegitimately with another break to give rise to exchange aberrations. Exponential dose-response curves, as well as results from fractionation experiments indicate interaction between two breaks. Revell (1974) proposed the exchange hypothesis, according to which all aberrations arise due to exchange events, and the primary events are not direct breaks. The primary lesions heal or start an exchange initiation process if a second primary lesion had occurred in the near vicinity. Since the chromatin is a highly condensed structure, there will be a large number of loops in the interphase nucleus and these exchange events occur only in these loops, and abortive exchanges will lead to breaks.

Though this hypothesis can accommodate chemically induced breaks very well (van Kesteren-van Leeuwen and Natarajan 1980), Sax's hypothesis seems to be most appropriate for X-ray induced aberrations (Natarajan and Obe 1978).

Based on a reciprocal recombination model put forward by Resnick (1976), attempts have been made to explain exchange aberrations arising from a single DNA double strand break (Chadwick and Leenhouts 1978). This problem has been investigated in Vicia faba root meristematic cells, with differential substitution of BrdUrd in the chromatids, namely TT-TB (T=thymine, B=Bromouracil). These chromatids can be differentially stained and exchanges between TT-TT, TB-TB, and TT-TB chromatids can be identified and qualified. Such an analysis of exchanges following $G_2$ X-irradiation favoured the idea that two lesions are necessary for an exchange formation (Kihlman et al. 1978).

Recently, carbon ultra soft X-rays, which produce within the cell only secondary electrons of very small energy and range have been shown to be efficient in inducing cell killing, point mutations and chromosomal aberrations (Thacker et al. 1982). Since this type of radiation deposits a small amount of highly localized energy, 280 eV (14 ionizations) within 7 nm, it can be argued that a single lesion is enough to induce an exchange event, as these X-rays are unlikely to traverse two chromosome stands. However, if the chromosomes involved in the exchange are lying very close together in the interphase, these soft X-rays will be able to induce lesions in both the chromosomes. More knowledge on the organization of the interphase nucleus in $G_0$ cells is necessary to arrive at a definitive interpretation.

## 3.6 Influence of Inhibitors of DNA Synthesis/
## Repair on the Biological Effects Induced by Mutagens

There are several inhibitors of DNA synthesis/repair which have been studied for their ability to potentiate the frequencies of induced chromosomal aberrations (Natarajan and Obe 1983; Kihlman and Natarajan 1984). These include (1) inhibitors of deoxyribonucleotide synthesis (5-fluorodeoxy uridine, $2'$ deoxy adenosine, hyroxyurea), (2) inhibitors of DNA polymerase alpha (aphidicolin, ara C), (3) inhibitors of poly (ADP-ribose) synthetase (benzamide, 3 aminobenzamide), and (4) caffeine. All these inhibitors increase the frequencies of X-ray induced chromosomal aberrations, though there are qualitative differences. Inhibitors of the classes 1 and 2 increase in $G_2$ only the frequencies of breaks and not exchanges, while inhibitors of group 3 and caffeine increase all classes of aberrations including exchanges. Some information on the mechanisms of DNA and chromosomal repair has emergerd from studies involving these inhibitors. Inhibitors such as aphidicolin and ara C, which are considered as specific inhibitors of DNA polymerase alpha (i.e., related to replicative synthesis) do affect repair synthesis (designated to DNA polymerase beta) indicating that some of the repair synthesis may be semi-conservative and be affected by these inhibitors. Aphidicolin inhibits repair replication in $G_0$ lymphocytes after UV irradiation, but not in exponentially growing cells (Natarajan et al. 1982; van Zeeland et al. 1982). The influence of aphidicolin on X-ray induced chromosome aberrations is greater in $G_0$ lymphocytes in comparison to synchronized $G_1$ CHO cells. The influence of inhibitors on DNA repair depends on the metabolic status and the type of cells under study.

### 3.7 Studies on the Influence of DNA Repair Inhibitors in $G_2$ Stage of Cell Cycle of Mutagen-Treated Cells

When X-irradiated $G_2$ human lymphocytes are post treated with hydroxyurea, there is a potentiation for induced aberrations. The maximum potentiation is observed when HU is given in late $G_2$, rather than in early $G_2$. Similarly, if CHO cells are X-irradiated and post treated with *Neurospora* endonuclease in 30-min pulses, the frequencies of aberrations increase. While immediate post treatment increases all classes of aberrations, treatment just prior to fixation increases dramatically only chromatid breaks (Natarajan et al. 1980). These observations indicate that a large number of lesions (single strand gaps) persist until mitosis, and their repair can be influenced even at that stage. The observation that there was no potentiation for chromatid exchanges by NE treatment in the last 30 min is due to the spatial organization of the chromatids, just prior to mitosis.

When human lymphocytes in $G_0$ or synchronized CHO cells are treated with an S-dependent agent such as short wave UV or alkylating agents and challenged in $G_2$ with hydroxyurea, aphidicolin or caffeine (3 or 2 h prior to fixation, respectively), the frequencies of induced aberrations increase (Hansson et al. 1982; Palitti et al. 1983). These results indicate that though the cells have passed through the S phase following mutagenic insult, there are unrepaired primary or secondary lesions persisting till the onset of mitosis and their repair can be influence even at this stage. These observations could be interpreted as due to a post replication repair or a proof reading and editing of the replicated DNA which go on in $G_2$. The other possibility is that there is a residual DNA synthesis which goes on in $G_2$ and misreplication of these regions due to the presence of lesions can lead to chromosomal aberrations. The division of cell cycle into $G_1$, S, and $G_2$ is only operational and according to the accepted definition, $G_2$ stage begins when it is no longer possible to detect with autoradiographic techniques the incorporation of tritiated thymidine in chromosomal DNA. This technique is not sensitive enough to detect small amounts of DNA synthesis. In HeLa and CHO cells, we have been able to measure a small but significant incorporation of $H^3$ TdR in the DNA during the last hour, prior to mitotic shake-off, and this incorporation was inhibited by HU (Zwanenburg and Natarajan unpublished). Treatment of control cells with ara C, an inhibitor of DNA synthesis, in $G_2$ increases the frequencies of spontaneous chromosomal aberrations in human lymphocytes and CHO cells, which indicate the existence of residual DNA synthesis in $G_2$ stage (Preston 1982; Natarajan et al. 1982). Whichever mechanism may be involved – post replication repair or residual DNA synthesis – the available data indicate that a large proportion of lesions is repaired just prior to mitosis and that this repair can be inhibited.

The mechanism by which caffeine potentiates mutagen induced chromosomal aberrations is not well understood (Kihlman et al. 1982). Potentiation by caffeine in the $G_2$ stage is most remarkable. There are at least two unrelated mechanisms by which caffeine can potentiate the frequencies of induced aberrations. (1) Caffeine can interfere with the DNA repair taking place in $G_2$, and (2) caffeine can counteract $G_2$ delay induced by physical and chemical mutagens. By reducing $G_2$ delay, caffeine would shorten the time available for repair and cells would arrive at mitosis with higher frequencies of aberrations (Kihlman and Natarajan 1984).

Poly (ADP-ribose) synthetase is a nuclear enzyme credited for maintaining the integrity of chromatid during replication and repair of DNA. When DNA strand breaks are induced in the cell, there is a depletion in the level of NAD with concomitant increase in the activity of poly (ADP-ribose) synthetase indicating their involvement in the repair of DNA strand breaks. Inhibitors of this enzyme, such as 3-aminobenzamide, have been shown to inhibit the ligation step in excision repair, especially involving DNA ligase II (Shall 1983). These inhibitors do not induce any chromosomal aberrations by themselves, even when present for several cell cycles. However, in cells grown in nicotinamide starved medium, they increase the frequencies of aberrations. In AT and FA fibroblasts, 3-aminobenzamide increases the frequencies of aberrations.

These inhibitors also potentiate X-ray induced chromosomal aberrations (Natarajan et al. 1981). They are also efficient inducers of SCE's by themselves. Their ability to induce SCE's is directly correlated with their ability to inhibit the activity of poly (ADP-ribose) synthetase (Oikawa et al. 1980). They induce SCE's efficiently only when BU containing DNA is replicated (a condition necessary for the differentiation of the sister chromatids) in their presence (Natarajan et al. 1981). This increase is directly proportional to the amount of BU substitution in the DNA, 100% substitution yielding maximum increase in the frequencies of SCE's (Natarajan and Csukas 1982). It is interesting to note that cells from BS patients, which are characterized by increased frequencies of spontaneously occurring SCE's, generate most of the SCE's during the second cell cycle, when BU containing DNA is replicated, indicating that these cells are not able to repair efficiently lesions which arise due to incorporated BU (Shirashi et al. 1982). Though this observation may suggest that BS cells may have low levels of poly (ADP-ribose) synthetase activity, no difference has been found so far between these and normal human cells (Shirashi et al. 1983).

Though it is generally believed that all agents which induce SCE's also induce point mutations, and SCE's and point mutations are directly correlated (Carrano et al. 1978), inhibitors of poly (ADP-ribose) synthetase do not induce point mutations by themselves under conditions, in which they increase the baseline SCE frequencies by a factor of 10. When CHO cells are treated with monofunctional alkylating agents and post incubated for two cycles in a medium containing BrdUrd, there is a synergistic increase in the frequencies of induced SCE's, but not mutations for 6-thioguanine resistance (Natarajan et al. 1983). Neither 2 nor 24 h treatment with ethylmethanesulfonate in the presence of 3-aminobenzamide leads to an increase in the frequency of induced mutations, though there is an increased cytotoxicity due to 3 AB treatment (Fig. 2) (Natarajan et al. 1983). It has also been reported that in mouse L 1210 cells treated with methyl nitrosourea and post treated with 5-methylnicotinamide, another inhibitor of poly (ADP-ribose) synthetase, for 10 days, covering the entire expression time, did not increase but decrease the frequency of induced mutations to thioguanine resistance, though there was decreased survival due to post treatment with 5-methylnicotinamide (Durrant et al. 1981). Therefore, it appears that there is no direct cor-

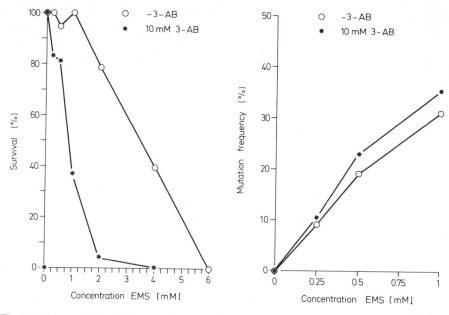

**Fig. 2.** Influence of 3-aminobenzamide on cell killing and mutations (HGPRT⁻) in CHO cells induced by ethyl methanesulfonate. The treatment was given together for 24 h

relation between induction of mutations and SCE's, under these experimental conditions.

# 4  Chromosomal Alterations in Human Diseases

There are several human recessive diseases associated with chromosomal instability. The exact defects in these diseases which lead to the chromosomal instability are not clear. Xeroderma pigmentosum (XP), a disease characterized by increased sensitivity to sun light and UV, is not characterized by increased frequencies of spontaneously occurring chromosomal aberrations. The increased sensitivity of XP has been attributed to the lack of UV endonuclease function in these cells. This enzyme is needed to initiate repair processes to excise pyrimidine dimers. Fanconi's anemia (FA) and ataxia telangiectasia (A–T) are associated with increased spontaneous frequencies of chromosomal aberrations, but not sister chromatid exchanges. FA is sensitive to DNA cross-linking agents in all cases studied so far, while the sensitivity to X-rays in this syndrome is variable. The increased sensitivity to cross-linking agents such as mitomycin C and diepoxybutane has been used with success to diagnose Fanconi's anemia. A–T cells are extremely sensitive to ionizing radiations, such as X-rays, neutrons as well as soft β-rays from incorporated tritium. In addition, they are also sensitive to DNA cross-linking agents which induce DNA strand breaks, as well as agents which produce directly DNA strand breaks like bleomycin, and neocarcinostatin (for references see Natarajan and Obe 1982). The defect in A–T cells appears to be inefficient repair of DNA strand breaks, although all efforts to directly quantify and demonstrate convincingly a reduced repair in A–T cells when compared to normal cells have not been successful so far. This may be due to the insensitivity of the biochemical techniques employed to quantify DNA strand breaks. Bloom's syndrome cells exhibit a very high frequency of spontaneously occurring SCE's and chromosomal aberrations. These cells have also been found to mutate spontaneously at higher rate when compared to normal cells. BS cells are more sensitive to near UV light in comparison to normal cells. The molecular basis for all these observations in BS cells is not yet clear.

FA cells and BS cells are sensitive to oxygen free radicals, whereas A–T cells are not, which may indicate that these cells are susceptible to DNA damaging agents which operate at least in part via direct action, i.e., the intermediate formation of active oxygen species (superoxide radicals, singlet oxygen, hydroxyl radicals, hydrogen peroxide, etc.). These are chemical mutagens which in part act indirectly, such as bleomycin, adriamycin, streptonigrin, mitomycin C. Thus, part of the sensitivity of these cells might arise from their susceptibility to active free radicals (Cerutti 1982). When BS cells are cocultivated with Chinese hamster ovary cells, the frequencies of SCE's in BS cells are reduced. This is due possibly to some diffusible factor from normal cells which can correct the defect in BS cells to some extent (van Buul et al. 1978). Similarly, clastogenic factors, i.e., components which can induce chromosomal aberrations and SCE's in normal cells, have been obtained from the serum of A–T patients as well as BS patients. Superoxide dismutase has been shown to inhibit the activity of the clastogenic factor from BS cells, indicating that superoxide radicals ($O_2$) may be involved in the induction of chromosomal alterations (Cerutti 1982). Increased oxygen tension induces higher frequencies of chromosomal aberrations in FA cells in comparison to normal cells (Joenje et al. 1981). The role of oxygen radicals in bestowing sensitivity to mutagens of these human disorders is important and calls for further research in this area.

All these recessive disorders are associated with increased predisposition to cancer. There are several dominantly inherited diseases with increased chromosomal instability. Included in this group of diseases are: familial polyposis coli, Peutz-Jagher's syndrome, Gardner's syndrome, familial colon cancer, etc. In this group of disorders, a single mutation is expressed dominantly and may initiate neoplasm in a particular organ at a particular stage of development as the sole manifestation of the inherited mutation (Sasaki 1982). No specific defect at DNA level has yet been identified in these syndromes. The identification of such defects and the study of the relationships between molecular defects leading to chromosome instability and the development of cancer in these diseases constitute a potential area of research. Gardner's syndrome and familial polyposis coli exhibit increased frequencies of non-constitutional chromosomal aberrations, as well as polyploidy in cultured skin fibroblasts. In addition, the frequencies of mitomycin C induced SCE's in these cells are higher than that induced in normal cells (Gardner et al. 1981).

## 5 Detection of Heterozygotes for Recessive Diseases with Increased Chromosomal Instability

The frequency of A–T homozygotes in American populations is estimated to be 1 in 40,000. If an estimate for heterozygotes is made according to Swift (1982), this could be in the order of 1% to 2% in the normal population. When one takes into consideration other recessive disorders, such as Xeroderma pigmentosum, Fanconi's anemia and Bloom's syndrome, the frequencies of heterozygotes for all these disorders will be 3% to 4% of the normal population. It is important to devise techniques to identify heterozygotes of these different diseases. In the literature, claims have been made that heterozygotes for A–T can be detected by their intermediate X-ray sensitivity between normal and A–T homozygotes. These results have been obtained on A–T lymphoblastoid cell lines. It is known that transformed cells change their characteristics easily and it will be difficult to transform every cell line with Epstein Bar virus in heterozygote detection studies. It will be simpler if experiments can be done with normal lymphocytes. In our laboratory, we have successfully identified some A–T heterozygotes, based on X-ray sensitivity for the induction of chromosomal aberrations (Natarajan et al. 1982b). Since there are several complementation groups known for A–T, it may be that only some of the complementation groups can be detected by this technique. Claims have been made of successful detection of heterozygotes of FA by studying the response to diepoxybutane (Auerbach et al. 1979). This observation is controversial, as no other laboratory has been able to confirm it. Detection of heterozygotes for these diseases is important for genetic counseling and occupational counseling of such heterozygous individuals. Further attempts should be made to standardize reliable techniques for detection of heterozygotes for such recessive diseases.

Most of the chromosomally unstable dominant diseases are autosomal with 100% penetrance, which means 50% of the progeny are liable to get the disease. The age of onset of these diseases varies very much. The same is true for the dominant disease Huntington chorea. Identification of individuals prone to diseases at a young age will be of great help in terms of adaptation to the type of life as well as medical management.

## 6 Cytogenetic Screening of Human Populations for Possible Exposure to Mutagenic Carcinogens

Since most of the mutagenic carcinogens induce chromosomal aberrations in experimental organisms or cell cultures in vitro, one would expect that cells from individuals occupationally or accidentally exposed to mutagens will have higher incidence of chromosomal aberrations. Peripheral blood lymphocytes are ideally suited for such studies, as they are easy to sample and to culture in vitro. In human populations exposed to ionizing radiations in Hiroshima and Nagasaki, there was a significant increase in the frequencies of chromosomal aberrations in a dose-dependent manner. Chromosomal aberrations in lymphocytes due to ionizing radiations are formed immediately on the production of DNA strand breaks and their frequencies cannot be influenced by further repair. Though stable aberrations such as translocations and unstable aberrations such as dicentrics are formed roughly in a proportion of 1:1, this proportion will change in favour of stable aberrations over the years. This trend was observed in the exposed populations of Hiroshima and Nagasaki, as well as in patients with ankylosing spondylitis subjected to partial body X-irradiation. This change can be due to several factors such as population dilution of the peripheral lymphocytes, stable aberrations produced in the bone marrow stem cells persisting and forming clones, and unstable aberrations formed in the bone marrow being eliminated during differentiation. Workers exposed to chemical mutagens occupationally respond with increased frequencies of aberrations (e.g., vinyl chloride). However, when industrial hygiene is improved, the exposure levels are low. The frequencies of aberrations usually found in these workers are at the control level or just above the control level. If one does not find an increase in the frequencies of chromosomal aberrations, does it mean that it is safe for the workers? This question remains to be solved.

## 7 Chromosomal Aberrations in Smokers and Alcoholics

Several laboratories have presented data indicating that smoking increases the frequencies of chromosomal aberrations and sister chromatid exchanges in the lymphocytes of smokers. The frequencies of chromosomal aberrations have also been shown to depend on the number of cigarettes smoked per day (see Chap. Obe and Beek, p. 177ff.). Similarly, chronic alcoholics have been shown to have elevated frequencies of chromosomal aberrations. Smoking and alcohol consumption increase the frequencies of chromosomal aberrations in a synergistic way (Obe et al. 1980). A part of the population working in chemical industries also smokes and drinks and one has to take into account the influence of these habits in the monitoring of workers from chemical or nuclear industries. This is important especially if synergistic effects could be involved in these combinations as well.

Most of the chemically induced aberrations are formed during the DNA synthetic period of the cell cycle. As circulating lymphocytes are in a dormant stage ($G_0$), the lesions induced by chemicals will be transformed to aberrations only during the oncoming DNA synthesis, which will be when the cells are transformed in vitro by a mitogen. In a typical occupational exposure situation, there is a low constant exposure over a period of months during which new lesions are induced in the DNA of the lymphocytes of the individuals, as well as a constant repair process is going on eliminating some of the lesions. The residual unrepaired lesions would give rise to the chromosomal aberrations. Thus, the frequencies of aberrations expected would be rather low, unless the exposure has been high. To get statistically significant data, one has to score a large number of cells from several individuals in order to compare with a "control" population (Natarajan and Obe 1980).

To obtain a comparable control population in humans is a difficult problem. Controls should be matched for sex, age, smoking, and drinking habits, as close as possible to the exposed population. Standardized protocols should be employed. The first cell division following the initiation of cultures (usually 48 h) should be used. When these conditions are met, one can generate reliable and meaningful data. Any investigation undertaken without proper control populations studied under identical conditions will not be of much value.

# 8 Inter-Individual Variability in Human Populations

There is ample evidence in the literature to show that in experimental organisms such as *Drosophila* and mouse, there are differences in the repair capacities of different genotypes, and there are mouse strains differing in their capacity to produce aryl-hydrocarbon hydroxylase (AHH), an enzyme necessary to metabolically activate carcinogens such as benzo(a)pyrene. The incidence of tumors due to benzo(a)pyrene treatment of mice depends on their ability to activate this carcinogen and this characteristic is inherited in a Mendelian fashion. Variability for the induction of this enzyme AHH is also known in humans and there are results which suggest a possible relationship between the inducibility of this enzyme and lung cancer among cigarette smokers (Kellermann 1977).

For induction of chromosomal aberrations it is known that Down's syndrome (trisomy 21) patients, as well as others who carry extra chromosomes respond with higher frequencies following X-irradiation than the normal individuals. In a survey done in our laboratory, among 40 normal non-smoking individuals, the frequencies of aberrations induced by X-rays varied within a factor of 2 between individuals. Post treatment with caffeine, which is known to increase the frequencies of chromosomal aberrations induced by X-rays, led to extreme variability with regard to caffeine modification. A group of individuals was insensitive to caffeine, and another group responded with 60% to 100% increase, while two individuals responded with about 400% increase (Natarajan et al. 1982a). We have found similar differences between individuals diagnosed as having aplastic anemia, some of these patients proving extremely sensitive to ionizing radiations.

In a population monitoring programme, one has to be aware of the extent of variability present among individuals. It is therefore always difficult to evaluate individual risk, but the risk due to occupational exposure has to be estimated on a population basis.

# 9 Future Approaches for Population Monitoring

While studies of chromosomal aberrations and sister chromatid exchanges are the only available techniques to monitor human populations, at present one has to accept that this is a time-consuming and tedious procedure. Alternate techniques are being developed aiming at point mutations, namely HGPRT⁻ mutations in peripheral blood lymphocytes (Albertini 1980) and hemoglobin mutants in erythrocytes. The technique employed to detect HGPRT⁻ mutations in lymphocytes involves growing isolated lymphocytes in medium containing 6 thioguanine (TG) and tritiated thymidine. TG resistant cells (HGPRT⁻) will incorporate thymidine in their DNA and these cells can be detected by autoradiography. The spontaneous frequencies of these mutations are several fold higher than found in human fibroblasts. This increase arises due to the varying proportions of non-mutant cycling cells present in circulating blood, which are also labeled in the presence of TG. Freezing and thawing of isolated lymphocytes seem to selectively kill these cycling cells. It has now been possible to clone human lymphocytes and propogate them for a long time by supplementing with T cell growth factor. Using this technique, the frequencies of spontaneously occurring HGPRT⁻ mutants in lymphocytes have been estimated and these values are similar to those obtained with the human fibroblasts (Albertini et al. 1982; Vijayalaxmi and Evans 1984). This technique can now be reliable employed in population monitoring studies. Efforts are also being made to monoclonal antibodies for hemoglobin mutations and these can be useful in the future for such population monitoring purposes.

## 10 Significance of Chromosomal Aberrations

Human syndromes which are chromosomally unstable are associated with increased predisposition to malignant neoplasms. Most human tumors are associated with specific or unspecific chromosomal abnormalities, though it is not known whether these aberrations are the cause or outcome of the transformation. Occupational exposure to vinyl chloride has been shown to induce chromosomal aberrations as well as angiosarcomas in exposed individuals at a rate higher than found in normal populations. Atom bomb survivors from Hiroshima and Nagasaki have been found to have increased frequencies of chromosomal aberrations in their lymphocytes as well as increased frequencies of leukemias and solid tumors. Smoking and alcohol consumption are shown to increase the frequencies of chromosomal aberrations in a synergistic way and this combination also increases the frequencies of cancer of the upper pharynx and esophagus in a synergistic way, among these individuals. Though these two events may not be directly connected, it is obvious that increased chromosomal aberrations in an individual may be indicative of possible predisposition to cancer. It will be important to follow up normal individuals with increased frequencies of chromosomal aberrations, to determine whether an early and increased onset of cancer is prevalent among these individuals in comparison to a parallel control who had no chromosomal aberrations. The recent findings on the involvement of specific chromosomal translocations in etiology of malignancy further point to the importance of chromosomal aberrations in the origin of cancer.

*Acknowledgements.* This work was financially supported by the Koningin Wilhelmina Fonds (project 81-91) and Euratom contract with the Leiden University.

# References

Albertini RJ (1980) Drug-resistant lymphocytes in man as indicators of somatic cell mutation. Teratogenesis, Carcinogenesis, and Mutagenesis 1:25–48

Albertini RJ, Castle KL, Bonchending WR (1982) T-cell cloning to detect the mutant 6-thioguanine resistant lymphocytes present in human peripheral blood. Proc Natl Acad Sci USA 79:6617–6621

Andersson HC, Kihlman BA, Palitti F (1981) Production of sister chromatid exchanges by X-rays under aerobic and anaerobic conditions. Hereditas 94:41–44

Auerbach AD, Warburton D, Bloom AD, Chaganti RSK (1979) Prenatal detection of Fanconi's anaemia gene by cytogenetic methods. Am J Human Genet 31:77–81

Bender MA, Griggs HG, Bedford JS (1974) Mechanisms of chromosomal aberration production III. Chemical and ionizing radiation. Mutat Res 23:197–212

Bender MA, Preston RJ (1982) Role of base damage on aberration formation. Interaction of aphidicolin and X-rays. In: Natarajan AT, Obe G, Altmann H (eds) DNA repair, chromosome alterations and chromatin structure. Elsevier Biomedical, Amsterdam, pp 37–46

Buul PPW van, Natarajan AT, Verdegaal-Immerzeel E (1978) Suppression of the frequencies of sister chromatid exchanges in Bloom syndrome fibroblasts by cocultivation with chinese hamster cells. Hum Genet 44:187–189

Carrano AV, Thompson LH, Lindl PA, Minkler (1978) Sister chromatid exchange as indicator of mutagenesis. Nature 271:551–553

Cerutti P (1982) Abnormal oxygen metabolism in Bloom syndrome? In: Natarajan AT, Obe G, Altmann H (eds) DNA repair, chromosome alterations and chromatin structure. Elsevier Biomedical, Amsterdam, pp 203–214

Chadwick KH, Leenhouts HP (1978) The rejoining of DNA double strand breaks and a model for formation of chromosomal rearrangements. Int J Radiat Biol 33:517–529

Craig AG (1980) The formation of chromosome aberrations from single strand damage in irradiated DNA. J Theor Biol 82:633

de Weerd-Kastelein EA, Keizer W, Rainaldi W, Bootsma D (1977) Induction of sister chromatid exchanges in xeroderma pigmentosum cells after exposure to ultra violet light. Mutat Res 45:253–261

Durrant LG, Margison GP, Boyle JM (1981) Effects of 5-methylnicotinamide on mouse L1210 cells exposed to N-methyl-N-nitrosourea: mutation induction, formation and removal of methylated products in DNA, and unscheduled DNA synthesis. Carcinogenesis 2:1013

Gardner EJ, Woodward SR, Burt RW (1981) Chromosomal diagnosis of polyploid diseases and cancer in colonectrum. Proc Int Human Genetics Congress, Jerusalem, p 312

Griggs HG, Bender MA (1973) Photoreactivation of ultraviolet-induced chromosomal aberrations. Science 179:86–88

Hansson K, Palitti F, Kihlman BA, Karlsson MB (1982) Potentiation of X-ray and streptonigrin induced chromosomal aberrations in human lymphocytes by post-treatment with hydroxyurea and caffeine. Hereditas 97:51–58

Hart RW, Setlow RB, Woodhead AD (1977) Evidence that pyrimidine dimers in DNA can give rise to tumors. Proc Natl Acad Sci USA 74:5574–5578

Heflich RH, Baranek DT, Kodell RL, Morris SM (1982) Induction of mutations and sister chromatid exchanges in Chinese hamster ovary cells by ethylating agents. Relationship to specific DNA adducts. Mutat Res 106:147–161

Holmberg M, Strausmanis R (1983) The repair of chromosome aberrations in human lymphocytes after combined irradiation with UV irradiation (254 nm) and X-rays. Mutat Res 120:45–50

Joenje H, Arwerr F, Ericksson AW, de Koning H, Oostra AB (1981) Oxygen dependence of chromosomal aberrations in Fanconi's anaemia. Nature 290:142–143

van Kesteren-van Leeuwen AC, Natarajan AT (1980) Localization of 7–12 dimethylbenz(a)-enthracene induced chromatid breaks and sister chromatid exchanges in chromosomes 1 and 2 of bone marrow cells of rats in vivo. Chromosoma 81:473–481

Kellermann G (1977) Hereditary factors in human cancer. In: Hiatt HH, Watson JD, Winstein JA (eds) Origin of human cancer, book B. Mechanisms of carcinogenesis. Cold Spring Harbor, New York, pp 837–845

Kihlman BA (1977) Caffeine and chromosomes. Elsevier Biomedical, Amsterdam, p 1

Kihlman BA, Hansson K, Palitti F, Andersson HC, Hartley-Asp B (1982) Potentiation of induced chromatid-type aberrations by hydrocyureau and affeine in $G_2$. In: Natarajan AT, Obe G, Altmann H (eds) DNA repair, chromosome alterations and chromatin structure. Elsevier Biomedical, Amsterdam, pp 11–24

Kihlman BA, Natarajan AT (1984) Potentiation of chromosomal alterations by inhibitors of DNA repair. In: Collins A et al. (eds) DNA repair and its inhibition. Dekker, Basel New York (in press)

Kihlman BA, Natarajan AT, Andersson HC (1978) Use of 5-bromodeoxyuridine labelling technique for exploring mechanisms invilved in the formation of chromosomal aberrations. I. $G_2$ experiments with root tips of *Vicia faba*. Mutat Res 52:181–198

Lloyd DC, Purrott RJ, Dolphin GN, Bolton D, Edwards AA, Corp MJ (1975) The relationship between chromosome aberrations and low LET dose to human lymphocytes. Int J Radiat Biol 28:75–90

Loveless A (1969) Possible relevance of $O^6$ alkylation of deoxyguanosine to the mutagenicity and carcinogenicity of nitrosamines and nitrosamides. Nature 223:206–207

McCormick JJ, Maher VM (1978) Mammalian cell mutagenesis as a biological consequence of DNA damage. In: Hanawalt PC, Friedberg EC, Fox CF (eds) DNA repair mechanisms. Academic, New York, pp 739–749

Nagasawa H, Forence AJ, Little JB (1982) Relationship of enhanced survival during confluent holding recovery in ultra violet irradiated human and mouse cells to chromosome aberrations, sister chromatid exchanges and DNA repair. Radiat Res 92:483–496

Natarajan AT, Ahnström G (1968) Cytogenetical effects of inorganic pyrophate and 5-fluorodeoxyuridine. Hereditas 59:229–241

Natarajan AT, Csukas I (to be published 1984) Mechanisms of sister chromatid exchanges. In: Critical evaluation of mutagenicity tests, German Ministry of Health (in press)

Natarajan AT, Csukas I, Degrassi F, van Zeeland AA, Palitti F, Tanzarella C, de Salvia R, Fiore M (1982) Influence of inhibition of repair enzymes on the induction of chromosomal aberrations by physical and chemical agents. In: Natarajan AT, Obe G, Altmann H (eds) DNA repair, chromosome alterations and chromatin structure. Elsevier Biomedical, Amsterdam, pp 47–59

Natarajan AT, Csukas I, van Zeeland AA (1981) Contribution of incorporated 5-bromodeoxyuridine in DNA to the frequencies of sister chromatid exchanges induced by inhibitors of poly(ADP-ribose)polymerase. Mutat Res 84:125–132

Natarajan AT, Meijers M, van Rijn JLS (1982a) Individual variability of human cells in induction of chromosomal alterations by mutagens. In: Sorsa M, Vainia H (eds) Mutagens in our environment. Liss, New York, pp 75–88

Natarajan AT, Meijers M, van Zeeland AA, Simons JWIM (1982b) Attempts to detect ataxia-telangiectasia (A–T) heterozygotes by cytogenetical techniques. Cytogenet Cell Genet 33:145–151

Natarajan AT, Obe G (1978) Molecular mechanisms involved in the production of chromosomal aberrations. I. Utilization of Neurospora endonuclease for the study of aberration production in $G_2$ stages of cell cycle. Mutat Res 52:137–149

Natarajan AT, Obe G (1980) Screening human populations for mutations induced by environmental pollutants: use of human lymphocyte system. Ecotoxicol Environ Safety 4:468–481

Natarajan AT, Obe G (1982) Mutagenicity testing with cultured mammalian cells. Cytogenetic assays. In: Heddle JA (ed) Mutagenicity – new horizons in genetic toxicology. Academic, New York, pp 171–213

Natarajan AT, Obe G (1984) Influence of DNA-repair on radiation induced chromosomal aberrations in human peripheral lymphocytes. In: Ishihara T, Sasaki MS (eds) Radiation induced chromosome damage in man. Liss, New York (in press)

Natarajan AT, Simons JWIM, Vogel EW, van Zeeland AA (1984) Relationship between cell killing, chromosomal aberrations, sister chromatid exchanges and point mutations induced by monofunctional alkylating agents in CHO cells. Mutat Res (in press)

Natarajan AT, Obe G, van Zeeland AA, Palitti F, Meijers M, Verdegaal-Immerzeel EAM (1980) Molecular mechanisms involved in the production of chromosomal aberrations. II. Utilization of Neurospora endonuclease for the study of aberration production by X-rays in $G_1$ and $G_2$ stages of the cell cycle. Mutat Res 69:293–305

Natarajan AT, van Zeeland AA, Verdegaal-Immerzeel EAM, Filon AR (1980a) Studies on the influence of photoreactivation on the frequencies of UV-induced chromosomal aberrations, sister chromatid exchanges and pyrimidine dimers in chicken embryonic fibroblasts. Mutat Res 69:307–317

Natarajan AT, van Zeeland AA, Zwanenburg TS (1983) Influence of inhibitors of poly(ADP-ribose)polymerase on DNA repair and chromosomal alterations. In: Miwa M, Hayashi O, Shall S, Smulson S, Sugimura T (eds) ADP-ribosylation, DNA repair, and cancer. Japan Scientific Society Press, Tokyo, pp 227–242

Nowak C (1983) Thesis. Free University, Berlin

Obe G, Göbel D, Engeln H, Herha J, Natarajan AT (1980) Chromosomal aberrations in peripheral blood lymphocytes of alcoholics. Mutat Res 73:377–386

Obe G, Natarajan AT, Palitti F (1982) Role of DNA double-strand breaks in the formation of radiation-induced chromosomal aberrations. In: Natarajan AT, Obe G, Altmann H (eds) DNA repair, chromosome alterations and chromatin structure. Elsevier Biomedical, Amsterdam (Progress in Mutation Research, vol 4, pp 1–9)

Oikawa A, Thoda H, Kanai M, Miwa M, Sugimura T (1980) Inhibitors of poly(adenosine diphosphate ribose)polymerase induced sister chromatid exchanges. Biochem Biophys Res Commun 97:1311–1316

Palitti F, Tanzarella C, Degrassi F, de Salvia R, Fiore M, Natarajan AT (1983) Formation of chromatid type of aberrations in $G_2$ stage of cell cycle. Mutat Res

Paterson MC, Setlow RB (1972) Endonucleolytic activity from *Micrococcus luteus* that acts on X-ray induced damage in plasmid DNA of Escherichia coli. Proc Natl Acad Sci USA 69:2927 to 2931

Perry P, Evans HJ (1975) Cytological detection of mutagen-carcinogen exposure by sister chromatid exchange. Nature 258:121–225

Preston RJ (1980) The effects of cytosine arabinoside on the frequency of X-ray induced chromosome aberrations in normal human leukocytes. Mutat Res 69:71–79

Preston RJ (1982) DNA repair and chromosome aberrations. Interactive effects of radiation and chemicals. In: Natarajan AT, Obe G, Altmann H (eds) DNA repair, chromosome alterations, and chromatin structure. Elsevier Biomedical, Amsterdam, pp 25–35

Resnick MA (1976) The repair of DNA double strand break: a model involving recombination. J Theor Biol 59:97–106

Revell SH (1974) The breakage and reunion theory and the exchange theory of chromosomal aberrations induced by ionizing radiations. A short history. In: Lett JT, Adler H, Zelle M (eds) Adv Radiat Biol 4:367–416

Reynolds RJ, Natarajan AT, Lohman PHM (1979) Micrococcus luteus UV endonuclease sensitive sites and sister chromatid exchanges in chinese hamster ovary cells. Mutat Res 64:353–356

Sasaki MS (1982) Dominantly expression procancer mutations and induction of chromosome rearrangements. In: Natarajan AT, Obe G, Altmann H (eds) DNA repair, chromosome alterations and chromatin structure. Elsevier Biomedical, Amsterdam, pp 75–84

Sax K (1941) Types and frequencies of chromosomal aberrations induced by X-rays. Cold Spring Harbor Symp Quant Biol 9:93–103

Shall S (1983) ADP-ribosylation, DNA repair cell differentiation and cancer. In: Miwa M, Hayashi O, Shall S, Smulson M, Sugimura T (eds) ADP-ribosylation DNA repair and cancer. Japan Scientific Societies Press, Tokyo, pp 3–25

Shiraishi Y, Tanaka Y, Kato M, Miwa M, Sugimura T (1983) Effect of poly(ADP-ribose)polymerase inhibitors on the frequency of sister chromatid exchanges in Bloom syndrome cells. Mutat Res 122:223–228

Shiraishi Y, Yosida TH, Sandberg AA (1982) Analysis of single and twin sister chromatid exchange in endoreduplicated normal and Bloom syndrome B. Lymphoid cell. Chromosoma 87:1–8

Swift M (1982) Disease predisposition of ataxia telangiectasia heterozygotes. In: Bridges BA, Harnden DG (eds) Ataxia-telangiectasia. Wiley, Chichester, p 355

Taylor AMR (1978) Unrepaired DNA strand breaks in irradiated ataxia-telangiectasia lymphocytes suggested from cytogenetic observations. Mutat Res 50:407–418

Thacker J, Goodhead DT, Wilkinson RE (1982) The role of localized single track events in the formation of chromosome aberrations in cultured mammalian cells. 8th Symp on Microdosimetry. Jülich, Sept. 1982 (in press)

Uggla AH, Natarajan AT (1983) X-ray induced SCEs and chromosomal aberrations in CHO cells. Influence of nitrogen and air during irradiation in different stages of cell cycle. Mutat Res

Vijayalaxmi, Evans HJ (1984) Measurement of spontaneous and X-irradiation induced 6-thioguanine resistant blood lymphocytes using a T-cell cloning technique. Mutat Res 125:87–94

Vogel E, Natarajan AT (1979a) Comparative studies on the relation between reaction kinetics and mutagenic action of monofunctional alkylating agents in Drosophila (in vivo) and chinese hamster cells (in vitro). Mutat Res 53:279–280 (Abstr)

Vogel E, Natarajan AT (1979b) The relationship between reaction kinetics and mutagenic action of monofunctional alkylating agents in higher eukaryotic systems I and II. Mutat Res 62:51 to 123

Vogel E, Natarajan AT (1982) The relationship between reaction kinetics and mutagenic action of monofunctional alkylating agents in higher eukaryotic systems: Interspecies comparisons. In: Hollaender A, de Serres FJ (eds) Chemical mutagens, vol VII. Plenum, New York, pp 295 to 336

Wolff S (1978) Relationship between DNA repair, chromosome aberrations and sister chromatid exchanges. In: Hanawalt PC, Friedberg EC, Fox CF (eds) DNA repair mechanisms ICN-UCLA Symposia on molecular and cellular biology, vol IX. Academic, New York, pp 751–760

Wolff S (1982) Chromosome aberrations, sister chromatid exchanges and the lesions that produce them. In: Wolff S (ed) Sister chromatid exchanges. Wiley, pp 41–57

Zeeland AA van, Bussmann CJM, Degrassi F, Filon AR, van Kesteren-van Leeuwen AC, Palitti F, Natarajan AT (1982) Effects of aphidicolin on repair replication and induced chromosomal aberrations in mammalian cells. Mutat Res 92:379–392

Zeeland AA van, Mohn GR, Tates AD (1981) A quantitative comparison of genetic effects of ethylating agents on the basis of the DNA ethylation lavel. In: 3rd Intern Conference on Environmental Mutagens, Japan, p 56 (abstract)

Zeeland AA van, Natarajan AT, Verdegaal-Immerzeel EAM, Filon AR (1980) Photoreactivation of UV induced cell killing, chromosome aberrations, sister chromatid exchanges, mutations and pyrimidine dimers in *Xenopus laevis* fibroblasts. Mol Gen Genet 180:495–500

# Human Peripheral Lymphocytes in Mutation Research

G. OBE[1] and B. BEEK[2]

## 1 Introduction

Human peripheral lymphocyte (PL) cultures are a widely used test system for the
analysis of chromosomal aberrations and sister chromatid exchanges (SCE), induced
either in vivo or in vitro. The analysis of chromosomal aberrations and SCE's in PL
of exposed individuals still represents the only feasible method for the determination
of mutagenic damage induced in man in vivo. The methodology and various aspects
of "the human leukocyte test system" have been reviewed extensively (Buckton and
Evans 1982; Crossen 1982; Gebhart 1982; Obe and Beek 1982a; Obe and Madle
1981; Natarajan and Obe 1980, 1982). In the following section we shall focus only
on some special aspects and recent advantages concerning the "human leukocyte test
system", namely:

2. Stimulation of peripheral lymphocytes in vitro.
3. Inter-individual variabilities in induced chromosomal alterations and in repair
   capacities.
4. Intra-individual variabilities – the problem of subpopulations.
5. Fragile sites in human chromosomes.

## 2 Stimulation of Peripheral Lymphocytes in Vitro

Most of the human peripheral lymphocytes (PL) are in a presynthetic resting phase
of the cell cycle, which is called the $G_0$ phase. Antigenic stimuli induce these cells to
enter the autosynthetic cell cycle and to undergo mitoses. There are many reviews of
the process of stimulation (Cantor 1981; Ling and Kay 1975; Obe and Beek 1982a;
Oppenheimer and Rosenstreich 1976; Trepel 1975, 1976).

In vitro the PL can be stimulated by fusion with cycling cells (Harris 1968, 1970)
and by a variety of mitogenic factors, like the T-cell mitogen phytohemagglutinin
(PHA) (Nowell 1960), which is the most widely used mitogen in mutation research
with PL.

1  Institut für Genetik, Freie Universität Berlin, Arnimallee 5–7, 1000 Berlin 33, FRG
2  Umweltbundesamt Berlin, Bismarckplatz 1, 1000 Berlin 33, FRG

Mutations in Man, ed. by G. Obe
© Springer-Verlag Berlin Heidelberg 1984

The PL are primed for being stimulated in that they have all cellular components which are important for a quick resumption of full biochemical and physiological activities of a cycling cell. A morphological correlate of this is a pair of centrioles (Bessis 1973; Petrzilka and Schroeder 1979; Trepel 1976). Only a small fraction of the ribosomes in the cytoplasm of the $G_0$ PL are active in protein synthesis and during the first 12 h of stimulation the cells show an enhanced protein synthesis without the formation of new ribosomes (Ahern and Kay 1975; Kay et al. 1971). There seems to be a rate-limiting step in the protein synthesis of unstimulated lymphocytes which is most probably a non-messenger RNA. The amount of this RNA seems to increase upon stimulation and this leads to the observed early increase of the rate of protein synthesis (Cooper and Braverman 1980). In the nuclei of unstimulated PL, 30S proteins can be found which become associated with the newly synthesized RNA (Martin and Okamura 1981).

In T-cell cultures, Petrzilka and Schroeder (1979) found the following: Some 1% activated cells were present in the absence of PHA, and more than 98% 12 h after addition of PHA. Twenty four hours after addition of PHA there were less than 0.1% dividing cells, but 0.9% cells were dividing after 48 h. Forty eight hours after PHA addition the cell volume was 5 times higher as compared to the unstimulated cells (some 500 $\mu m^3$ as compared to 100 $\mu m^3$), and the cytoplasmic volume was some 6 times higher (348 $\mu m^3$ as compared to 56 $\mu m^3$). The nuclear volume is likewise rising (about 48 $\mu m^3$ before stimulation and 170 $\mu m^3$ 48 h after addition of PHA), the nuclear density is falling upon stimulation. The chromatin disperses and there is a considerable euchromatization of the extended heterochromatin, which occupies nearly 70% of the nuclear volume in unstimulated PL but only 13% 48 h after addition of PHA. Surface and volume of the rough endoplasmic reticulum are rising during the stimulation and so is the number of the ribosomes.

During the stimulation with PHA there is a considerable increase in the RNA synthesis. In the $G_0$ lymphocytes there is a very low activity of RNA polymerases I (rRNA), II (tRNA, 5S RNA), and III (hn RNA) which is rising dramatically during the stimulation. Exposure of PL to PHA for 4 d, leads to a 17-fold increase of the activity of RNA polymerase I and III and an 8-fold increase of the activity of RNA polymerase II (Jaehning et al. 1975). After a 3-d stimulation the total RNA polymerase activity increased about 9-fold and the rate of RNA synthesis about 100 fold (Jaehning et al. 1975). The considerable increase of RNA polymerase activity has its morphological correlate in an impressive increase of the nucleolar material. $G_0$ PL usually have one ring shaped nucleolus. During cultivation in the presence of PHA the nucleoli increase considerably in size and number, and show the morphological characteristics of the nucleoli in physiologically active cells (Arrighi et al. 1980; Obe and Beek 1982a; Petrzilka and Schroeder 1979; Schwarzacher and Wachtler 1983; Schwarzacher et al. 1978; Wachtler et al. 1980). The increase in the number of nucleoli and the dispersion of the chromatin is reflected by a decrease of the satellite associations during the stimulation with PHA (Mattevi and Salzano 1975; Nankin 1970; Sigmund et al. 1979) and there is a tendency for the frequencies of satellite associations to be higher in first and lower in second and third in vitro metaphases after the same culture time (Beek 1981a). In prematurely condensed chromosomes (PCC) from PL, Schmiady et al. (1979) found that there is at best one silver positive chromosome per metaphase in PCC from $G_0$ PL, but there are about 6 per metaphase in PCC from PL stimulated for 72 h.

Nucleoids of stimulated PL sediment faster as nucleoids of unstimulated PL (Yew and Johnson 1979b; Johnstone and Williams 1982). This seems to be associated with a rapid rejoining of DNA single strand breaks in lymphocytes after stimulation with PHA, which was maximal 8 h after stimulation and started already 1 h after stimulation (Johnstone and Williams 1982). Already a few minutes after stimulation with PHA there is an increase of acridine orange uptake indicating a weakening of DNA-protein binding in the chromatin (Killander and Rigler 1965, 1969; Rigler and Killander 1969). In concanavalin A (Con A)-stimulated human PL, Polet and Spieker-Polet (1980) found a gradual increase in the uptake of ($^3$H)-actinomycin D upon stimulation, and while the histone level remained constant, there was a gradual increase of the amount of nonhistone proteins in the nuclei, reflecting again the dispersion of the chromatin upon stimulation. During the first $G_1$ phase of PHA stimulated PL, the PCC becomes progressively longer (Hittelman and Rao 1976; Schmiady 1979). PCC of chromosome 1 is about 7 $\mu$ long in $G_0$, 9 $\mu$ after 1 h in culture, and 12 $\mu$ after 20 h (Schmiady 1979).

There are only very low activities of DNA polymerases in unstimulated PL but with the onset of DNA synthesis upon stimulation with PHA the activities of these enzymes are increasing considerably (Agarwal and Leob 1972; Bertazzoni et al. 1976; Coleman et al. 1974; Loeb et al. 1968, 1970; Mayer et al. 1975; Pedrali Noy et al. 1974; Tyrsted et al. 1973).

Mayer et al. (1975) found that all three types of DNA polymerases ($a$, $\beta$, $\gamma$) are induced during stimulation with PHA.

Pedrali Noy et al. (1974) found two waves of DNA polymerase activity in PHA-stimulated human PL. The first wave occurs in correlation with the DNA synthesis with a maximum after 3 d, the second one has a maximum after 5 d and is not related with the DNA synthesis, but rather with a maximal expression of the activity of DNA ligase which is likewise stimulated under the influence of PHA. Other enzymes which are induced with a delay with respect to the maximum of the DNA synthesis are DNAse which is acting as an exonuclease on single stranded DNA, and an endonuclease (Pedrali Noy et al. 1974). In human PL stimulated with PHA, Bertazzoni et al. (1976) found that there are detectable DNA polymerase $a$ and $\beta$ activities in unstimulated cells which are elevated considerably during stimulation (24 times between the 4th and 6th day after stimulation for DNA polymerase $a$ and 7 times for DNA polymerase $\beta$ at the 8th day of stimulation). After the 6th day in culture the activity of polymerase $a$ is falling considerably but the level of polymerase $\beta$ remains high. The activity of polymerase $a$ rises in parallel with the DNA synthesis rate and decreases with the decrease of the DNA synthesis rate. Polymerase $\beta$ rises at the same time as the DNA synthesis (but not as intensively as the polymerase $a$ activity) but is reaching a second peak at 9 d when the activity of polymerase $a$ is already low again. At days 10 to 12 when the polymerase $a$ activity is nearly 0, the activity of polymerase $\beta$ is still high. The pattern of polymerase $\beta$ is concordant with the course of DNA repair synthesis after UV irradiation in these cells and this led the authors to speculate that the polymerase $a$ is associated with the DNA synthesis and the polymerase $\beta$ with the capacity of the cells to repair damage in their DNA (Bertazzoni et al. 1976).

Zöllner et al. (1975) found four groups of endonucleolytic DNAses with different pH optima (A–D) in human PL, one of these being localized in the cytoplasm (A) the others in the nucleus. DNAse D showed a preference for denatured and UV irradiated DNA. One acid and two neutral DNAses have been found to be stimulated when PL are stimulated with PHA (Zöllner et al. 1979).

Otto (1977) found ATPases II and III in unstimulated bovine lymphocytes from retropharyngeal lymph nodes. The activities of these enzymes increased by a factor of about 5 in lymphocytes which have been stimulated with Con A for 50 h, when the cells were active in DNA synthesis. ATPases I was only present in stimulated lymphocytes (Otto 1977).

$G_0$ PL are able to repair DNA damage induced by ultraviolet light, ionizing radiations and chemicals and this repair capacity increases up to 20 times upon stimulation (Hamlet et al. 1982; Forell et al. 1982; Klimov et al. 1982; see Obe et al. 1982 for further references). The ability of lymphocyte extracts to remove $O^6$-methylguanine from exogenous DNA is generally higher when extracts from PHA stimulated PL are used for the assay. The number of methyl-acceptor molecules in unstimulated PL is estimated to be 14,000 to 110,000 and in stimulated PL, 40,000 to 140,000 (Waldstein et al. 1982).

## 3 Inter-Individual Variability in Induced Chromosomal Alterations and in Repair Capacities

There is considerable variation in the yield of induced chromosomal alterations (structural chromosomal aberrations and SCE's) when PL from different donors are used. Irradiation of $G_0$ PL and analysis of first in vitro metaphases results in aberration frequencies which vary with a factor of 4 and more when the results from PL of different donors were compared (Bianchi et al. 1982). Irradiation of PL from 21 donors in the $G_2$ phase of the cell cycle (3 h before fixation) with 100 R gave considerable variation in the aberration frequencies (18 to 127 aberrations in 100 metaphases). Addition of caffeine (final concentration 1 mM) prior to irradiation in $G_2$, resulted in a clear elevation of the aberration frequencies of various extent, or remained without any effect. The caffein effect was not correlated with the aberration rates obtained without caffeine (Natarajan et al. 1982). These findings indicate an interindividual variability with respect to repair capacities of PL.

The spontaneous (Lambert et al. 1982) and the in vivo induced SCE frequencies (see Chap. Gebhart, p. 198ff.) show interindividual variability which may be interpreted as being an outcome of different repair capacities. The same holds true for in vivo induced structural chromosomal aberrations (see Chap. Gebhart, p. 198ff.).

Waldstein et al. (1982) found a large interindividual variability in the capacities of extracts from PL to remove $O^6$-methylguanine from exogenous DNA, which was independent from age and sex of the blood donors. The repair capacities of PL from different donors measured as UV light induced unscheduled DNA synthesis show strong interindividual differencies and are negatively correlated with age (Lambert et al. 1976, 1977, 1979). Using the nucleoid sedimentation method, Yew and Johnson

(1979b) were able to show that the capacity of UV induced excision repair diminishes with age in T and in B lymphocytes. These findings are in line with the observation that there is a tendency of older individuals having more structural aberrations than younger ones (Hedner et al. 1982). There is no clear effect of age on the spontaneous SCE frequencies (Hedner et al. 1982; Evans 1982), but there are indications that PL from cord blood have lower spontaneous SCE frequencies than PL from adults (see Crossen 1982; Obe et al. in preparation). In experimental systems it has been shown that the induced SCE frequencies are lower in aging as compared to young cells (see Schneider 1982).

PL contain the enzyme aryl hydrocarbon hydroxylase (AHH) which is metabolizing polycyclic aromatic hydrocarbons (PAH) and which can be stimulated by PAH (see Natarajan and Obe 1982 for literature). The basal and the induced activity of AHH exhibits considerable inter-individual variability, and in addition to this it has a seasonal variation of 1 year showing a maximum activity in late summer and early fall and a minimum activity in winter (Paigen et al. 1981).

An inter-individual variability has been found in the extent of carcinogen induced unscheduled DNA synthesis, carcinogen binding to DNA, and carcinogen induced chromosomal aberrations (Pero and Mitelman 1979; Pero et al. 1976, 1978). The well-known inter-individual variation in the frequencies of T and B lymphocytes which is dependent on factors such as age, season, day time, bodily exercise, is another source of variability in the outcome of mutation experiments with PL (see Natarajan and Obe 1982 for references).

All these analyses concerning an inter-individual variability do not really allow a decision to be made on whether these variabilities are not rather intra-individual variabilities.

## 4 Intra-Individual Variabilities – the Problem of Subpopulations

One of the major problems concerning the use of human PL cultures in mutation research, in terms of reproducibility and consistency of data and their interpretation, is the existence of subgroups, subsets or subpopulations of cells differing in immunological characteristics, proliferation response (cell cycle length) after mitogenic stimulation, and in mutagen sensitivity. Since Bender and Brewen (1969) published the first results indicating this problem, a considerable body of literature has been accumulated consisting of a variety of experimental approaches and several different genetic endpoints investigated. Publications of this topic up to about 1979 have been reviewed recently (Obe and Beek 1982a), but in the meantime several important papers on the problem of subpopulations in human PL cultures have appeared. So it seems to be justified to try an updating of the current state of knowledge in the present review. In Tables 1–3 "old" and recent publications on this problem are summarized mainly in chronological order. Concerning a more detailed discussion and presentation of the literature up to about 1979 the reader is refered to the review mentioned above (Obe and Beek 1982a), while important recent papers will be discussed in more detail in the following.

**Table 1.** Mutagen sensitivity of human T and B lymphocytes: cell survival studies

| Mutagen | Comments | Ref. |
|---|---|---|
| UV-rays | Higher sensitivity of T as compared to B lymphocytes. B lymphocytes have a less efficient excision repair | Yew and Johnson (1978) |
| γ-rays | Higher sensitivity of B as compared to T lymphocytes | Santos-Mello et al. (1974) |
| γ-rays | Higher sensitivity of B as compared to T lymphocytes. Several subgroups of T lymphocytes with different sensitivities | Kwan and Norman (1977) |
| X-rays | Higher sensitivity of B as compared to T lymphocytes (2–5 times) | Prosser (1976) |
| γ-rays | Higher sensitivity of B as compared to T lymphocytes | Schwartz and Gaulden (1980) |
| X-rays | Indications for the presence of at least two subpopulations of T lymphocytes differing in their radiosensitivity | Rigas et al. (1980) |
| X-rays | Extreme differences in the radio-sensitivity of different subpopulations of both T and B lymphocytes | Szczylik and Wiktor-Jedrzejczak (1981) |
| γ-rays | Different sensitivities of subpopulations of T lymphocytes | Schwartz et al. (1983) |

**Table 2.** Mutagen sensitivity of human early and late replicating lymphocytes: chromosomal aberration studies

| Mutagen | Comments | Ref. |
|---|---|---|
| X-rays | Decreasing aberration frequencies with increasing culture time due to mitotic selection | Heddle et al. (1967) |
| X-rays | Differencies in aberration frequencies at different fixation times not explainable only by mitotic selection | Bender and Brewen (1969) |
| X-rays | Significantly different aberration frequencies at different culture times within the first mitotic wave | Steffen and Michalowski (1973) |
| X-rays | Higher sensitivity of early as compared to late replicating lymphocytes | Beek and Obe (1976) |
| X-rays | After irradiation of cancer patients, higher aberration frequencies in lymphocytes proliferating early in vitro | Steffen et al. (1978) |
| γ-rays | Higher sensitivity of T cells in culture as compared to whole lymphocyte cultures and higher sensitivity of early as compared to late proliferating T lymphocytes | Steffen et al. (1978) |
| X-rays | Equal aberration frequencies at different fixation times within the first mitotic wave | Scott and Lyons (1979) |

**Table 2** (continued)

| Mutagen | Comments | Ref. |
|---------|----------|------|
| X-rays | Different aberration frequencies at different fixation times within the first mitotic wave | Bianchi et al. (1979) |
| X-rays | Equal aberration frequencies at different fixation times within the first mitotic wave | Léonard and Decat (1979) |
| γ-rays | Decrease in aberration frequency with increasing fixation times within the first mitotic wave | Wyszynska and Liniecki (1980) |
| X-rays | Heterogeneous results suggestive of a donor variation in sensitivity as a predominant factor influencing aberration yields | van Buul and Natarajan (1980) |
| X-rays | Lower aberration frequencies in earlier as compared to later dividing cells within the first mitotic wave | Bianchi et al. (1982) |
| A 139, difunctional alkylating agent (2,5-bis-(methoxy-ethoxy)-3,6-bis-ethylene-imino p-benzo-quinone) | Higher sensitivity (about 3 times) of early replicating as compared to late replicating lymphocytes | Beek and Obe (1974a) |
| No treatment | No significant differences in the "spontaneous" chromosomal aberration frequencies in T as compared to B lymphocytes in patients with Fanconi's anemia | Bushkell et al. (1976) |

**Table 3.** Sensitivity of human early and late replicating lymphocytes: sister chromatid exchanges (SCE)

| Mutagen | Comments | Ref. |
|---------|----------|------|
| No treatment | Significant differences in SCE frequencies at different fixation times | Lezana et al. (1977) |
| No treatment | Suggestion of a rapidly growing subpopulation with a lower SCE-frequency and a slower growing one with a higher SCE frequency | Snope and Rary (1979) |
| No treatment | Higher frequency in T as compared to B lymphocytes | Santesson et al. (1979) |
| No treatment | No differences in the SCE frequencies in second divisions at different culture times | Giulotto et al. (1980) |
| No treatment | No differences in the SCE frequencies in second divisions at different culture times | Carrano et al. (1980) |

**Table 3.** (continued)

| Mutagen | Comments | Ref. |
|---|---|---|
| No treatment | Increase of SCE frequencies in second divisions at later culture times | Ockey (1980) |
| No treatment | No differences in the SCE frequencies in second divisions at different culture times | Morimoto and Wolff (1980) |
| No treatment | No correlation between the percentage of B or T lymphocytes with SCE frequencies or with cell proliferation in culture. Higher SCE frequencies in slowly proliferating cultures | Lindblad and Lambert (1981) |
| Two different 9-amino--acridine derivatives and Mitomycin C | Higher SCE frequencies in slower proliferating cells after induction with all three different chemical mutagens | Gibas and Limon (1979) |
| Mitomycin C | Increase of induced SCE frequencies in second divisions at later culture times. Slight increase of spontaneous frequencies with increasing culture time | Evans and Vijayalaxmi (1980) |
| Mitomycin C | No differences in the SCE frequencies, spontaneous or induced, in second division metaphases from early and late proliferating cells | Littlefield et al. (1983) |
| Trenimon, trifunctional alkylating agent (2,3,5-tris-ethylene-imino-p-benzoquinone) | Equal SCE frequencies, spontaneous or induced, in second division metaphases from early and late proliferating cells | Beek and Obe (1979) |
| Trenimon | Decrease of induced SCE frequencies in second division metaphases at later culture times with a higher Trenimon dose as that used by Beek and Obe (1979) | Riedel and Obe (1980) |

Publications on mutagen sensitivity of subpopulations of human PL can be roughly divided into three groups:

4.1  Repair and cell survival studies (Table 1).

4.2  Chromosome aberration studies in early- and late-proliferating human PL after mitogenic stimulation (Table 2).

4.3  Studies on spontaneous and induced sister chromatid exchange frequencies (SCE) in early- and late-proliferating human PL after mitogenic stimulation (Table 3).

## 4.1  Repair and Cell Survival Studies  (Table 1)

A few studies indicate that T and B lymphocytes are different with respect to their repair capacities. Using the nucleoid sedimentation method, Yew and Johnson (1978, 1979a) found that T lymphocytes are slower than B lymphocytes with respect to the repair of UV induced lesions, resulting in a higher sensitivity of T lymphocytes to UV irradiation. With ethidium bromide intercalation as a measure for the superhelicity, Yew and Johnson (1978, 1979b) showed that the DNA of T and B cells has similar superhelical density. Extracts from T lymphocytes are more active to remove $O^6$-methylguanine from DNA than extracts from B lymphocytes (Waldstein et al. 1982).

Survival studies have shown that after X- or $\gamma$-irradiation, human B lymphocytes are up to 5 times more sensitive than T lymphocytes and that there seem to exist several different T cell subpopulations, differing in their radiosensitivity to a high extent as well (Table 1). The general picture that emerges from all survival studies up to now seems to indicate that the "sensitivity" of human PL to mutagens is by no means unitary, or determined by only one single event, but might depend on the type of primary lesion induced, the type of DNA-repair involved, and may be even on some other, non-DNA related, target in the different cell types. Clearly, investigations with chemical mutagens with different modes of action (e.g., cross-linking, alkylating, intercalating, etc.) could add some interesting information to the problem of cell survival and mutagen sensitivity of human lymphocyte subpopulations.

Szczylik and Wiktor-Jedrzejczak (1981) recently investigated the effects of X-irradiation in vitro on cell survival of human lymphocyte subpopulations. The following subpopulations of B and T lymphocytes, as characterized by surface properties, were examined for their radiosensitivity, by this confirming and extending earlier results from Prosser (1976) and Kwan and Norman (1977):

E:    Cells spontaneously forming rosettes with sheep red blood cells (T lymphocytes).
AE:   Cells forming active rosettes with sheep red blood cells (subset of T lymphocytes).
EAC: Cells with the receptor of activated complement (B lymphocytes).
ME:   Cells spontaneously forming rosettes with mouse red blood cells (subset of B lymphocytes).

The authors found that, irrespective of different culture periods employed, the radiosensitivity of these subpopulations increased in the following sequence: E < EAC = ME < AE (T < B = SubB < SubT). This means that most of the T lymphocytes are the most resistant cells to the killing effect of X-irradiation, but a subpopulation of T lymphocytes is the most sensitive one, while B lymphocytes and the subpopulation of B lymphocytes indentified with about the same amount of sensitivity are situated in between them. The authors suggest that these differences may be an outcome of different metabolic activities of the subpopulations, in the sense that higher metabolic activity leads to higher radiosensitivity.

In an investigation of T and B cell sensitivity to $\gamma$-irradiation Schwartz and Gaulden (1980) have confirmed that B lymphocytes are more radiosensitive than T lymphocytes. Moreover, B cells as well as T cells were stimulated up to cell division by PHA. The B cells were comprising up to 15% of the mitoses found in unirradiated cultures at 48 h culture time, and the proportion of B cells in mitosis decreased with increasing dose, probably due to higher cell killing and/or cell cycle delaying effect in B cells as compared with T cells. No differences, however, in chromosome aberration frequencies were found between mitoses from T and B cells.

Recently Schwartz et al. (1983) extended their analysis on the study of the radiosensitivity of three different T lymphocyte subpopulations, identified according to their ability to bind immunoglobuline (Ig)G (T-G), IgM (T-M), or neither of both (T-null). Survival curves revealed that even each subset had radioresistant and radiosensitive portions, T-G cells were the most sensitive subpopulation. The mitotic response varied greatly among unirradiated subpopulations (highest incidence: T-M; lowest T-G cells). Furthermore, the effect of T-G cells to suppress the PHA-induced proliferation of the other T lymphocytes has been shown to be radiosensitive and to be abolished completely by doses of γ-irradiation over 1.0 Gy. In summary, this means that in cultures exposed to more than 1.0 Gy of γ-rays only few mitoses will be found derived from B lymphocytes or T-G cells in 48 h cultures. Additionally, the loss of T-G cell suppressor activity on the other T lymphocyte subpopulations will lead to a shift toward the radioresistant T-M and T-null cells.

So investigators of radiation-induced chromosomal aberrations in 48 h cultures of human PL may examine metaphases derived from a quite complex mixture of cells differing greatly in their radiosensitivity in terms of cell killing and/or delaying effects, the composition of the mixture depending on and changing with culture time and dose. As mentioned above, this already complex picture must by no means necessarily be representative for all different types of mutagens.

It may be interesting in this context that studies of growth kinetics of lymphocytes from BALB/c mice (Rahmsdorf et al. 1981) have shown, in accordance with results from human PL studies described above, a higher sensitivity of B as compared to T lymphocytes after γ-irradiation. Radiation sensitive cells were delayed for a longer period in S- and $G_2$-phase of the cell cycle than resistant cells. No differences were detected in repair capacities of the different cell types (single strand break repair, double strand break repair, unscheduled DNA synthesis).

## 4.2 Chromosome Aberration Studies in Early- and Late-Proliferating Human PL After Mitogenic Stimulation (Table 2)

At least since the investigations of Buckton and Pike (1964) it has been known that the frequencies of chromosomal aberrations in human PL cultures decrease with increasing culture time, after having a maximum aberration frequency of around the peak time of first mitoses after mitogenic stimulation (normally around 48 h culture time). With the occurrence of second or further divisions, a mechanism of "mitotic selection" is leading to lower frequencies of aberrations in the population of mitotic cells. Through a sequence of events, involving the formation of micronuclei and the occurrence of a particular type of premature chromosome condensation (PCC) from the chromatin included in the micronuclei (see Obe and Beek 1982b, for review), cells bearing unstable chromosomal damage are eliminated from the population of dividing cells. After irradiation of human PL in the $G_0$ state of the cell cycle with X-rays and the investigation of dicentric chromosomes and micronucleus-derived PCC's in mitoses definitely determined as first, second and third divisions after the irradiation, with a BUdR-Giemsa labeling technique, it has been shown that the frequency of dicentric chromosomes is reduced by about 50% for each division step, either from first to second or from second to third mitosis (Beek 1981b). Additionally it has been shown that PCC patches do not occur before the second division after induction of chromosomal aberrations (Beek 1981b; Beek et al. 1980), an important argument in favor of their being derived from micronuclei that had to be formed during the preceding division. So it seems to be clear that, to obtain a realistic picture of the amount of damage originally induced in human PL it is necessary to ensure analysis of only first divisions after aberration induction. With human PL cultures this can be achieved in most, but not all cases (see Obe and Beek 1982a, for review) by the use of fixation times not later than 48 h.

Unfortunately, it seems to be that even after restriction of scoring to metaphases from the first mitotic wave in human PL cultures, aberration frequencies may differ sometimes considerably within the population of first mitoses at different fixation times, between cultures set up with PL from different blood donors, and even between different cultures from the same blood donor (Bianchi et al. 1982; Natarajan et al. 1982).

As shown in Table 2, frequencies of dicentric chromosomes, either induced by X- or $\gamma$-rays, have been found to decrease, increase or keep constant within the first mitotic wave in different experiments. The reason for this is still obscure, but several authors assume that the existence of different subpopulations of human PL either T or B cells, or different T subsets, possibly differing in cell cycle kinetics and chromosomal radiosensitivity and occurring in variable amounts, may be responsible for this. Indeed, Steffen et al. (1978) have shown higher aberration frequencies in early- as compared to late-proliferating human T lymphocytes. Wyszynska and Liniecki (1980) found a significant decrease in the yields of dicentric chromosomes not only in human, but also in rabbit and pig PL cultures, while Bianchi et al. (1981) found no difference with pig lymphocytes (all first divisions) so that the problems may be similar with other mammalian leukocyte cultures. Van Buul and Natarajan (1980) found in the same experimental series with human PL increasing, decreasing, heterogeneous, and steady frequencies of dicentric chromosomes in first post-irradiation mitoses. According to these authors "it depended on the donor, and even on the condition of the donor, what picture emerged, because repeats of the same donor did not always produce the same pattern with respect to variation in aberration frequency with sampling time."

Possibly, some kind of not very well-defined "metabolic status" of the irradiated individual human PL may have an influence on chromosomal radiosensitivity (see Beek and Obe 1977; Beek 1982). This "metabolic status", as reflected for example by the presence of different length-classes of prematurely condensed $G_0$-chromosomes (see Obe and Beek 1982a, for review), may be quite different from cell to cell and from blood sample to blood sample, probably depending on a variety of physiological factors.

In conclusion, there may be no serious doubt anymore that human PL are behaving as a heterogeneous group of cells with respect to their chromosomal radiosensitivity, and with resepct to cell killing and delaying effects. This does not mean, however, that human PL cultures cannot be used as a reliable biological indicator for radiation-exposure. In an IAEA-coordinated programme on the evaluation of radiation-induced chromosomal aberrations in human PL in vitro (Bianchi et al. 1982), though showing a quite heterogeneous outcome of the results, it is concluded that if fixations are made in the maximal part of the first mitotic wave (which should be determined and controlled by appropriate BUdR-Giemsa labeling techniques), reliable data should be obtained in most cases.

Apart from radiation studies, there is still only one report on the differential sensitivity of early-, and late-replicating human PL after treatment with a difunctional alkylating chemical mutagen (Beek and Obe 1974a, see Table 2), indicating that early replicating cells may show a higher sensitivity in terms of chromatid damage than late replicating cells.

In B and T lymphocytes from patients with Fanconi's anemia Bushkell et al. (1976) found no difference in "spontaneously" occurring aberration frequencies, but in their investigations with extremely long culture times, mitotic selection should have influenced the outcome of their results in one or the other way.

### 4.3 Studies on Spontaneous and Induced Sister Chromatid Exchange (SCE) Frequencies in Early- and Late-Proliferating Human PL After Mitogenic Stimulation (Table 3)

These studies have revealed constant, increasing and decreasing SCE-frequencies, both "spontaneous" (or BUdR-induced) and induced with several different chemical mutagens, with increasing culture time in second division cells. Even for the same mutagen (Trenimon), at a lower dose, constant SCE frequencies (Beek and Obe 1979) and at a higher dose, decreasing SCE frequencies have been found (Riedel and Obe 1980). Santesson et al. (1979) have described that T lymphocytes exhibit a higher spontaneous SCE frequency than B lymphocytes and that B lymphocytes proliferate more slowly than T cells. Lindblad and Lambert (1981) have recently shown that human B and T lymphocytes, cultured separately, may differ in the rate of cell proliferation and in the frequencies of spontaneous SCE's. There was, however, no correlation found between the SCE frequencies and the total proportion of B and T cells, indicating that differences in SCE frequencies are not the result of differences in the proportions of B and T lymphocytes at different culture times. On the other hand, a significantly higher SCE frequency was found in slowly proliferating cultures than in cultures with a high proliferation rate.

In conclusion, more comparative investigations on spontaneous and induced SCE frequencies in human PL seem to be necessary to clarify the question of homogeneous or heterogeneous sensitivity with respect to the induction of SCE.

## 5 Fragile Sites in Human Chromosomes

As a new variant of structural changes in human chromosomes about 15 different "fragile sites" have been described as an inherited condition up to now (see Table 4

**Table 4.** Location of different fragile sites in human chromosomes known up to now and their classification according to Sutherland (1983)

| Group | Total number of fragile sites | Location of the fragile sites | Comments |
|---|---|---|---|
| 1 | 13 | 2q11; 6p23; 7p11; 8q23; 9p21; 9q32; 10q23; 11q13; 11q23; 12q13; 16p12; 20p11; Xq27 | The folate sensitive fragile sites, including the fra-X |
| 2 | 1 | 16q22 | Expression independent of culture conditions |
| 3 | 1 | 10q25 | Expression dependent on the presence of bromo-deoxyuridine (BUdR) |

and Sutherland 1983, for review). They are inherited in a co-dominant manner, like polymorphisms in C-banding in homologous chromosomes, located at exactly the same position, seemingly representing a constitutive variant of either qualitative or quantitative nature possibly in DNA composition of the respective chromosomal region. In their common expression they look like an isochromatid gap (see Fig. 1). For their definition as real "fragile sites" it is further necessary, however, that at the same location, fragments, tri- and quadriradial configurations as well as deleted chromosomes are to be found.

As may be seen from Table 4, 13 from the total of 15 fragile sites described require particular culture conditions for their expression, while the fragile site at 16q22 is

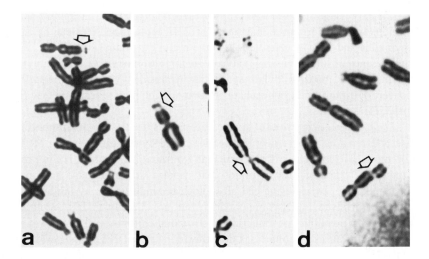

**Fig. 1a–d.** Different types of fragile sites (pointers). **a** Xq; **b** 6p 23; **c** 2q 11; **d** 8q 23. (Photographs were kindly provided by Drs. P.B. Jacky and G.R. Sutherland, Dept. Histopathology, Cytogenetics Unit, Adelaide Children's Hospital, Adelaide, Australia)

expressed independently from the culture conditions. The fragile site at 10q25 requires the presence of the thymidine-analogue bromodeoxyuridine (BUdR) for the expression, by this resembling slightly the phenomenon of centromeric decondensation after BUdR-incorporation (Kim 1974). The particular cellular condition that is required for the expression of the majority of fragile sites is the lack of thymidine and folic acid (the "folate sensitive" fragile sites). In commercial culture media which are poor in the content of these compounds (like TC medium 199) fragile sites are expressed, while they are not to be seen when culture media rich in folic acid and/or thymidine are used (like Ham's F10). The folate sensitive fragile sites have attracted much attention recently because of their clinical relevance: The fragile X-chromosome, related to a common form of mental retardation at least in males belongs to this group (see Sutherland 1983, for review). Even under conditions of folate- and thymidine-deprivation these fragile sites are not expressed in a substantial amount at culture times of about 2 days, where there is no reason to suspect the majority of divisions belonging to the first mitotic wave after PHA-stimulation (see Obe and Beek 1982a). The expression of the folate sensitive fragile sites (as percent of metaphases) may increase thereafter considerably (but never showing a 100% expression), having a maximum usually at day 3 or 4 of culture, and decreasing with increasing further culture time (see Fig. 2). One might speculate that even when cultivated in media lacking folic acid and thymidine, lymphocytes with folate sensitive fragile sites need a sequence of some cell cycles after PHA-stimulation for the expression of the fragile sites, and this may be due to the need of exhaustion of the cellular DNA-precursor pool by DNA replication. At later culture times there seems to be a higher chance for finding breakage manifestations, like triradials, at the fragile site. The expression of the folate sensitive fragile sites can be completely suppressed by the addition of folic acid, thymidine, or its analogue BUdR (see Sutherland 1979a b; Sutherland and Hinton 1981; Sutherland et al. 1983).

The expression of fragile sites probably depends upon reduced amount of thymidine monophosphate during the late stages of DNA replication. This can be achieved by thymidine and folic acid reduction in the culture medium (Sutherland 1979a), or in the presence of folic acid, enzymatically by inhibition of either the folate metabolism (methotrexate) or thymidylate synthetase (fluorodeoxyuridine), the latter enzyme converting uridine monophosphate into thymidine monophosphate (see Sutherland 1983, for review).

As discussed, expression of fragile sites means occurrence of chromosomal breakage and reunion. So the new perspective raised by the phenomenon of fragile sites for mutation research, and in particular for mutations in man, is the possibility of "induction" of structural chromosomal damage in considerable amounts without any "mutagen" by changing cell physiological conditions or by inhibiting particular enzymes involved in DNA-metabolism, but not directly in DNA-replication. This opens a new possibility of induction of structural chromosomal damage in eukaryotic cells by substances or conditions not interfering with DNA as a target, but with proteins like the respective enzymes mentioned. Instead of "induction" one could also speak about "increase of spontaneously occurring" breakage. Although the phenomenon of fragile sites has already been known for several years, this interesting connection with mutation research does not seem to have been realized until recently

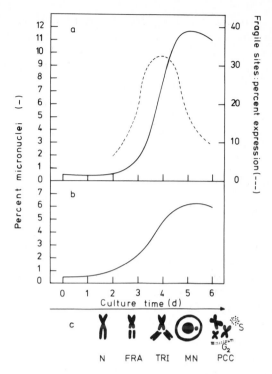

**Fig. 2a–c.** An idealized graphic representation showing the expression of fragile sites (a *right ordinate* percent of metaphases showing the fragile site) and the frequencies of interphases with micronuclei (MN) (**a b** *left ordinate* percent of interphases having a MN), in human lymphocyte cultures set up with the blood of probands with different folate-sensitive fragile sites (**a**) and of normal blood donors (**b**) at days 1 to 6 in culture with the use of thymidine and folic acid deficient media. In **c** the sequence of events, starting with non-expression (*N*), expression of the fragile sites either as a double gap (*FRA*), or as a triradial configuration (*TRI*), micronucleus formation (*MN*) and the occurrence of MN derived premature chromosome condensations (*PCC*) of either the S-phase or the $G_2$-phase type is shown in relation to culture time. [The graphs **a** and **b** are calculated from data on fragile sites expression and micronucleus formation from several different series published by Beek et al. (1983a) and the graph (c) is based on results from Noël et al. (1977) and Beek et al. (1983a)]

(Beek et al. 1983a), probably because of the more applied clinical aspects of at least the fragile X-chromosome.

In a series of experiments it has now been shown that the occurrence of DNA-precursor deprivation induced chromosomal damage is not refined to fragile site probands or to the actual fragile site itself, but may be obtained in lymphocyte cultures from normal probands as well, following obviously the same rules (Beek et al. 1983a,b; Jacky et al. 1983). Investigations of the frequencies of micronuclei (MN) in interphase cells of human PL cultures, incubated for up to 6 days with media lacking folic acid and thymidine (see Fig. 2) show that the maximum of fragile sites expression (about day 4 of culture) is followed by a sharp, considerable increase in MN frequencies about 24 h later (day 5 of culture). In a similar manner, but not to the same extent, MN frequencies increase in human PL cultures from normal blood donors. Addition of thymidine, folic acid and BUdR to PL cultures of either fragile sites or normal probands may suppress MN formation virtually to the control level in a dose dependent manner (Beek et al. 1983a; Jacky et al. 1983).

It has been known for a long time that medium composition in a more general sense may have an influence on chromosome breakage, both "spontaneous" (e.g., Obe et al. 1976) and induced (e.g., Beek and Obe 1974b), but the role of thymidine metabolism has not been identified. As already mentioned in the previous section, MN formation is followed by the occurrence of MN derived PCC, probably as part of a selection mechanism against unstable chromosomal damage. As expected, MN derived PCC has been found also in human PL cultures set up with lymphocytes from fragile sites probands (Beek et al. 1983a). So one might speculate that the apparent decrease in fragile sites expression at later culture times may be due at least in part to mitotic selection against cells with fragile sites expression and following chromosomal breakage, while the portion of non-expressing cells is proliferating and propagating itself undisturbed.

Further studies on fragile sites expression and DNA-precursor deprivation induced chromosomal damage may lead to a better understanding of spontaneous chromosome damage in man.

## References

Agarwal SS, Loeb SS (1972) Studies on the induction of DNA polymerase during the transformation of human lymphocytes. Cancer Res 32:107–113

Ahern T, Kay JE (1975) Protein synthesis and ribosome activation during the early stages of phytohemagglutinin lymphocyte stimulation. Exp Cell Res 92:513–515

Arrighi FE, Lau Y-F, Spallone A (1980) Nucleolar activity in differentiated cells after stimulation. Cytogenet Cell Genet 26:244–250

Beek B (1981a) BUdR-Giemsa labelling and satellite association in human leukocytes. Hum Genet 59:240–244

Beek B (1981b) Cell proliferation and chromosomal damage in human leukocytes: dicentrics and premature chromosome condensations in first, second, and third mitoses after X-irradiation. Hum Genet 57:75–77

Beek B (1982) X-ray-induced cell cycle delay in human leukocytes: differential response within the first poststimulation G1-phase. Int J Radiat Biol 41:227–230

Beek B, Obe G (1974a) The human leukocyte test system. II. Different sensitivities of sub-populations to a chemical mutagen. Mutat Res 24:395–398

Beek B, Obe G (1974b) The effect of lead acetate on human leukocytes in vitro. Experientia 30: 1006–1007

Beek B, Obe G (1976) The human leukocyte test system. X. Higher sensitivity to X-irradiation in the G0-stage of the cell cycle of early as compared to late replicating cells. Hum Genet 35: 57–70

Beek B, Obe G (1977) Differential chromosomal radiosensitivity within the first G1-phase of the cell cycle of early-dividing human leukocytes in vitro after stimulation with PHA. Hum Genet 35:209–218

Beek B, Obe G (1979) Sister chromatid exchanges in human leukocyte chromosomes: spontaneous and induced frequencies in early- and late-proliferating cells in vitro. Hum Genet 49:51 to 61

Beek B, Jacky PB, Sutherland GR (1983a) Heritable fragile sites and micronucleus formation. Ann Genet 26:5–9

Beek B, Jacky PB, Sutherland GR (1983b) DNA precursor deprivation-induced chromosomal damage. Mutat Res 113:331 (Abstr.)

Beek B, Klein G, Obe G (1980) The fate of chromosomal aberrations in a proliferating cell system. Biol Zbl 99:73–84

Bender MA, Brewen JG (1969) Factors influencing chromosome aberration yields in the human peripheral leukocyte system. Mutat Res 8:383–399

Bertazzoni U, Stefanini M, Dedrali Noy G, Giulotto E, Nuzzo F, Falaschi A, Spadari S (1976) Variations of DNA polymerase-$\alpha$ and $\beta$ during prolonged stimulation of human lymphocytes. Proc Natl Acad Sci USA 73:785–789

Bessis M (1973) Living blood cells and their ultrastructure. Springer, Berlin Heidelberg New York

Bianchi NO, Bianchi MS, Larramendy M (1979) Kinetics of human lymphocyte division and chromosomal radiosensitivity. Mutat Res 63:317–324

Bianchi MS, Bianchi NO, Larramendy M, Garcia-Heras J (1981) Chromosomal radioselectivity of pig leukocytes in relation to sampling time. Mutat Res 80:313–320

Bianchi M, Bianchi NO, Brewen JG, Buckton KE, Fabry L, Fischer P, Gooch PC, Kucerova M, Leonard A, Mukherjee RN, Mukherjee U, Nakai S, Natarajan AT, Obe G, Palitti F, Pohl-Rüling J, Schwarzacher HG, Scott D, Sharma T, Takahashi E, Tanzarella C, van Buul PPW (1982) Evaluation of radiation-induced chromosomal aberrations in human peripheral blood lymphocytes in vitro. Result of an IAEA-coordinated programme. Mutat Res 96:233–242

Buckton KE, Evans HJ (1982) Human peripheral blood lymphocyte cultures: an in vitro assay for the cytogenetic effects of environmental mutagens. In: Hsu TC (ed) Cytogenetic assays of environmental mutagens. Allanheld Osmun, Totowa New York, pp 183–202

Buckton KE, Pike MC (1964) Chromosome investigations of lymphocytes from irradiated patients: effect of time in culture. Nature 202:714–715

Bushkell LL, Kersey JA, Cervenka J (1976) Chromosomal breaks in T and B lymphocytes in Fanconi's anemia. Clin Genet 9:583–587

Buul PPW van, Natarajan AT (1980) Chromosomal radiosensitivity of human leukocytes in relation to sampling time. Mutat Res 70:61–69

Cantor H (1981) Regulation of immune responses: analysis with lymphocyte clones. Cell 25:7–8

Carrano AV, Minkler JL, Stetka DG, Moore DH (1980) Variation in the baseline sister chromatid exchange frequency in human lymphocytes. Environ Mutagen 2:325–337

Coleman MS, Hutton JJ, Bollum FJ (1974) Terminal riboadenylate transferase in human lymphocytes. Nature 248:407–409

Cooper HL, Braverman R (1980) Protein synthesis in resting and growth-stimulated human peripheral lymphocytes. Evidence for regulation by a non-messenger RNA. Exp Cell Res 127: 351–359

Crossen PE (1982) SCE in lymphocytes. In: Sandberg AA (ed) Sister chromatid exchange. Liss, New York, pp 175–193

Evans HJ (1982) Sister chromatid exchanges and disease states in man. In: Wolff S (ed) Sister chromatid exchange. Wiley, New York, pp 183–228

Evans HJ, Vijayalaxmi (1980) Storage enhances chromosome damage after exposure of human leukocytes to mitomycin C. Nature 284:370–372

Forell B, Meyers LS Jr, Norman A (1982) DNA repair synthesis in minimally stressed human lymphocytes. Int J Radiat Biol 41:535–545

Gebhart E (1982) The epidemiological approach: chromosome aberrations in persons exposed to chemical mutagens. In: Hsu TC (ed) Cytogenetic assays of environmental mutagens. Allanheld Osmun, Totowa New York, pp 385–408

Gibas Z, Limon J (1979) The induction of sister-chromatid exchanges by 9-aminoacridine derivatives. I. The relation between the yield of SCE induction and cell kinetics in cultured human lymphocytes. Mutat Res 67:93–96

Giulotto E, Mottura A, Giorgi R, de Carli L (1980) Frequencies of sister-chromatid exchanges in relation to cell kinetics in lymphocyte cultures. Mutat Res 70:343–350

Hamlet SM, Lavin MF, Jennings PA (1982) Increased rate of repair of ultraviolet-induced DNA strand breaks in mitogen stimulated lymphocytes. Int J Radiat Biol 41:483–491

Harris H (1968) Nucleus and cytoplasm. Clarendon, Oxford

Harris H (1970) Cell Fusion. Clarendon, Oxford

Heddle JA, Evans HJ, Scott D (1967) Sampling time and the complexity of the human leukocyte culture system. In: Evans HJ, Court-Brown WM, McLean AS (eds) Human radiation cytogenetics. Elsevier/North-Holland, Amsterdam, pp 6–19

Hedner K, Högstedt B, Kolnig A-M, Mark-Vendel E, Strömbeck B, Mitelman F (1982) Sister chromatid exchanges and structural chromosome aberrations in relation to age and sex. Hum Genet 62:305–309

Hittelman WN, Rao PN (1976) Premature chromosome condensation: Conformational changes of chromatin associated with phytohemagglutinin stimulation of peripheral lymphocytes. Exp Cell Res 100:219–222

Jacky PB, Beek B, Sutherland GR (1983) Fragile sites in chromosomes: possible model for the study of spontaneous chromosome breakage. Science 220:69–70

Jaehning JA, Stewart CC, Roeder RG (1975) DNA-dependent RNA polymerase levels during the response of human peripheral lymphocytes to phytohemagglutinin. Cell 4:51–57

Johnstone AP, Williams GT (1982) Role of DNA breaks and ADP-ribosyl transferase activity in eukaryotic differentiation demonstrated in human lymphocytes. Nature 300:368–370

Kay JE, Ahern T, Atkins M (1971) Control of protein synthesis during the activation of lymphocytes by phytohemagglutinin. Biochem Biophys Acta 247:322–334

Killander D, Rigler R (1965) Initial changes of deoxyribonucleoprotein and synthesis of nucleic acid in phytohemagglutinin-stimulated human leukocytes in vitro. Exp Cell Res 39:701–704

Killander D, Rigler R (1969) Activation of deoxyribonucleoprotein in human leukocytes stimulated by phytohemagglutinin. I. Kinetics of the binding of acridine orange to deoxyribonucleoprotein. Exp Cell Res 54:163–170

Kim MA (1974) Chromatidaustausch und Heterochromatinveränderungen menschlicher Chromosomen nach BUdR-Markierung. Hum Genet 25:179–188

Klimov NA, Vashchenko VI, Kolyubaeva SN, Komar VE (1982) Changes in the supercoiled structure of nuclear DNA in rat and human peripheral blood lymphocytes after γ-irradiation. Int J Radiat Biol 41:221–225

Kwan DK, Norman A (1977) Radiosensitivity of human lymphocytes and thymocytes. Radiat Res 69:143–151

Lambert B, Ringborg U, Swanbeck G (1976) Ultraviolet-induced DNA repair synthesis in lymphocytes from patients with actinic keratosis. J Invest Dermatol 67:594–598

Lambert B, Ringborg U, Swanbeck G (1977) Repair of UV-induced DNA lesions in peripheral lymphocytes from healthy subjects of various ages, individuals with Down's syndrome and patients with actinic keratosis. Mutat Res 46:133–134

Lambert B, Ringborg U, Skoog L (1979) Age-related decrease of ultraviolet light-induced DNA repair synthesis in human peripheral leukocytes. Cancer Res 39:2792–2795

Lambert B, Lindblad A, Holmberg K, Francesconi D (1982) The use of sister chromatid exchange to monitor human populations for exposure to toxicologically harmful agents. In: Wolff S (ed) Sister chromatid exchange. Wiley, New York, pp 149–182

Léonard A, Decat G (1979) Relation between cell cycle and yield of aberrations observed in irradiated human lymphocytes. Can J Genet Cytol 21:473–478

Lezana EA, Bianchi NO, Zabala-Suarez JE (1977) Sister chromatid exchanges in Down syndromes and normal human beings. Mutat Res 45:85–90

Lindblad A, Lambert B (1981) Relation between sister chromatid exchange, cell proliferation and proportion of B and T cells in human lymphocyte cultures. Hum Genet 57:31–34

Ling NR, Kay JE (1975) Lymphocyte stimulation. North-Holland, Amsterdam

Littlefield LG, Colyer SP, DuFrain RJ (1983) SCE evaluations in human lymphocytes after G0 exposure to mitomycin C. Lack of expression of MMC-induced SCEs in cells that have undergone greater than two in vitro divisions. Mutat Res 107:119–130

Loeb LA, Agarwal SS, Woodside AM (1968) Induction of DNA polymerase in human lymphocytes by phytohemagglutinin. Proc Natl Acad Sci USA 61:827–834

Loeb LA, Ewald JL, Agarwal SS (1970) DNA polymerase and DNA replication during lymphocyte transformation. Cancer Res 30:2514–2520

Martin TE, Okamura CS (1981) hn RNP protein distribution in various differentiated vertebrate cells. In: Schweiger HG (ed) International cell biology 1980–1981. Springer, Berlin Heidelberg New York, pp 77–84

Mattevi MS, Salzano FM (1975) Effect of sex, age, and cultivation time on number of satellite and acrocentric associations in man. Humangenetik 29:265–270

Mayer RJ, Smith RG, Gallo RC (1975) DNA-metabolizing enzymes in normal human lymphoid cells. VI. Induction of DNA polymerases $\alpha$, $\beta$, and $\gamma$ following stimulation with phytohemagglutinin. Blood 46:509–518

Morimoto K, Wolff S (1980) Increase of sister chromatid exchanges and pertubations of cell division kinetics in human lymphocytes by benzene metabolites. Cancer Res 40:1189–1193

Nankin HR (1970) In vitro alterations of satellite association and nucleolar persistence in mitotic human lymphocytes. Cytogenetics 9:42–51

Natarajan AT, Obe G (1980) Screening of human populations for mutations induced by environmental pollutants: use of human lymphocyte system. Ecotoxicol Environm Safety 4:468–481

Natarajan AT, Obe G (1982) Mutagenicity testing with cultured mammalian cells: cytogenetic assays. In: Heddle JA (ed) Mutagenicity. New Horizons in genetic toxicology. Academic, New York, pp 171–213

Natarajan AT, Meijers M, van Rijn JLS (1982) Individual variability of human cells in induction of chromosomal aberrations by mutagens. In: Sorsa M, Vainio H (eds) Mutagens in our environment. Liss, New York, pp 75–88

Noël B, Quack B, Mottet J, Nantois Y, Dutrillaux B (1977) Selective endoreduplication or branched chromosome? Exp Cell Res 104:423–426

Nowell PC (1960) Phytohemagglutinin: an initiator of mitosis in cultures of normal human leukocytes. Cancer Res 20:462–466

Obe G, Beek B (1982a) The human leukocyte test system. In: deSerres FJ, Hollaender A (eds) Chemical mutagens. Principles and methods for their detection. Plenum, New York, vol 7, pp 337–400

Obe G, Beek B (1982b) Premature chromosome condensation in micronuclei. In: Rao PN, Johnson RT, Sperling K (eds) Premature chromosome condensation: application in basic, clinical, and mutation research. Academic, New York, pp 113–130

Obe G, Madle S (1981) Prüfung der Mutagenität von Medikamenten beim Menschen. Prax Pneumol 35:1027–1033

Obe G, Beek B, Slacik-Erben R (1976) The use of the human leukocyte test system for the evaluation of potential mutagens. Excerpta Med 376:118–126. Elsevier/North-Holland, Amsterdam

Obe G, Kalweit S, Nowak C, Ali-Osman F (1982) Liquid holding experiments with human peripheral lymphocytes. I. Effects of liquid holding on sister chromatid exchanges induced by trenimon, diepoxybutane, bleomycin, and X-rays. Biol Zbl 101:97–113

Ockey CH (1980) Differences between "spontaneous" and induced sister-chromatid exchanges with fixation time and their chromosome localization. Cytogenet Cell Genet 26:223–235

Oppenheim JJ, Rosenstreich DL (eds) (1976) Mitogens in immunobiology. Academic, New York

Otto B (1977) DNA-dependent ATPases in concanavalin A stimulated lymphocytes. FEBS Lett 97:175–178

Paigen B, Ward E, Reilly A, Houten L, Gurtoo HL, Minowada J, Steenland K, Havens MB, Sartori P (1981) Seasonal variation of aryl hydrocarbon hydroxylase activity in human lymphocytes. Cancer Res 41:2757–2761

Pedrali Noy GCF, Dalpra L, Pedrini AM, Ciarrocchi G, Giulotto E, Nuzzo F, Falaschi A (1974) Evidence for two waves of induction of DNA enzymes in stimulated human lymphocytes. Nucleic Acids Res 1:1183–1189

Pero RW, Mitelman F (1979) Another approach to in vivo estimation of genetic damage in humans. Proc Natl Acad Sci USA:462–463

Pero RW, Bryngelsson C, Mitelman F, Thulin T, Nordén Å (1976) High blood pressure related to carcinogen-induced unscheduled DNA synthesis, DNA carcinogen binding, and chromosomal aberrations in human lymphocytes. Proc Natl Acad Sci USA 73:2496–2500

Pero RW, Bryngelsson C, Mitelman F, Kornfält R, Thulin T, Nordén Å (1978) Interindividual variation in the responses of cultured human lymphocytes to exposure from DNA damaging chemical agents. Interindividual variation to carcinogen exposure. Mutat Res 53:327–341

Petrzilka GE, Schroeder HE (1979) Activation of human T-lymphocytes. A kinetic and stereo-
logical study. Cell Tissue Res 201:101–127

Polet H, Spieker-Polet H (1980) Role of nuclear proteins on ($^3$H)actinomycin D binding during
lymphocyte mitogenesis. Exp Cell Res 128:419–429

Prosser JS (1976) Survival of human T and B lymphocytes after X-irradiation. Int J Radiat Biol
30:459–465

Rahmsdorf HJ, Ponta H, Bächle M, Mallick U, Weibezahn K-F, Herrlich P (1981) Differentiated
cells from BALB/c mice differ in their radiosensitivity. Exp Cell Res 136:111–117

Riedel L, Obe G (1980) Trenimon-induced SCEs and structural chromosomal aberrations in early-
and late-dividing lymphocytes. Mutat Res 73:125–131

Rigas D, Eginitis-Rigas C, Bigley RH, Stankova L, Head C (1980) Biphasic radiosensitization of
human lymphocytes by diethyldithiocarbamate: possible involvement of superoxide dismutase.
Int J Radiat Biol 38:257–266

Rigler R, Killander D (1969) Activation of deoxyribonucleoprotein in human leukocytes stimulated
by phytohemagglutinin. II. Structural changes of deoxyribonucleoprotein and synthesis of
RNA. Exp Cell Res 54:171–180

Santesson B, Lindahl-Kiessling K, Mattson A (1979) SCE in B and T lymphocytes: possible
implications for Bloom's syndrome. Clin Genet 16:133–135

Santos-Mello R, Kwan D, Norman A (1974) Chromosome aberrations and T-cell survival in
human lymphocytes. Radiat Res 60:482–488

Schmiady H (1979) Die Länge vorzeitig kondensierter Chromosomen (PCC) und Chromosomen-
abschnitte menschlicher Zellen in Abhängigkeit vom Zellzyklus. Thesis, Freie Universität
Berlin

Schmiady H, Münke M, Sperling K (1979) Ag-staining of nucleolus organizer regions on human
prematurely condensed chromosomes from cells with different ribosomal RNA gene activity.
Exp Cell Res 121:425–428

Schneider EL (1982) Aging and sister chromatid exchange. In: Sandberg AA (ed) Sister chromatid
exchange. Liss, New York, pp 195–203

Schwartz JL, Gaulden ME (1980) The relation contributions of B and T lymphocytes in the
human peripheral blood mutagen test system as determined by cell survival, mitogenic stimulation, and induction of chromosome aberrations by radiation. Environ Mutagen 2:473–485

Schwartz JL, Darr JC, Gaulden ME (1983) Survival and PHA-stimulation of γ-irradiated human
peripheral blood T lymphocyte subpopulations. Mutat Res 107:413–425

Schwarzacher HG, Wachtler F (1983) Nucleolus organizer regions and nucleoli. Hum Genet 63:
89–99

Schwarzacher HG, Mikelsaar A-V, Schnedl W (1978) The nature of Ag-staining of nucleolus
organizer regions. Electron- and light-microscopic studies on human cells in interphase, mitosis,
and meiosis. Cytogenet Cell Genet 20:24–39

Scott D, Lyons CY (1979) Homogeneous sensitivity of human peripheral blood lymphocytes to
radiation-induced chromosome damage. Nature 278:756–758

Sigmund J, Schwarzacher HG, Mikelsaar A-V (1979) Satellite association frequency and number
of mucleoli depend on cell cycle duration and NOR-acitivty. Studies on first, second, and
third mitoses of lymphocyte cultures. Hum Genet 50:81–91

Snope AJ, Rary JM (1979) Cell-cycle duration and sister chromatid exchange frequency in cultured human lymphocytes. Mutat Res 63:345–349

Steffen JA, Michalowski A (1973) Heterogeneous chromosomal radiosensitivity of phytohemagglutinin-stimulated human blood lymphocytes in culture. Mutat Res 17:367–376

Steffen JA, Swierkowska K, Michalowski A, Kling E, Nowakowska A (1978) In vitro kinetics of
human lymphocytes activated by mitogens. In: Evans HJ, Lloyd DC (eds) Mutagen-induced
chromosome damage in man. Edinburgh University Press, Edinburgh, pp 89–107

Sutherland GR (1979a) Heritable fragile sites on human chromosomes. I. Factors affecting expression in lymphocyte culture. Am J Hum Genet 31:125–135

Sutherland GR (1979b) Heritable fragile sites on human chromosomes. II. Distribution, phenotypic effects, and cytogenetics. Am J Hum Genet 31:136–148

Sutherland GR (1983) The fragile X chromosome. Int Rev Cytol 81:107–143

Sutherland GR, Hinton L (1981) Heritable fragile sites on human chromosomes. VI. Characterisation of the fragile site at 12q13. Human Genet 57:217–219

Sutherland GR, Jacky PB, Baker E, Manuel A (1983) Heritable fragile sites on human chromosomes. X. New folate sensitive fragile sites: 6p23, 9p21, 9q32, 11q23. Am J Hum Genet 35: 432–437

Szczylik C, Wiktor-Jedrzejczak W (1981) The effect of X-irradiation in vitro on subpopulations of human lymphocytes. Int J Radiat Biol 39:253–263

Trepel F (1975) Kinetik lymphatischer Zellen. In: Theml H, Begeman H (eds) Lymphozyt und klinische Immunologie. Springer, Berlin Heidelberg New York, pp 16–26

Trepel F (1976) Das lymphatische Zellsystem: Struktur, allgemeine Physiologie und allgemeine Pathophysiologie. In: Begemann H (ed) Handb Inn Medizin, Bd 2, Teil 3. Springer, Berlin Heidelberg New York, pp 1–191

Tyrsted G, Munch-Petersen B, Cloos L (1973) DNA polymerase activity in phytohemagglutinin-stimulated and non-stimulated human lymphocytes. Exp Cell Res 77:415–427

Wachtler F, Ellinger A, Schwarzacher HG (1980) Nucleolar changes in human phytohemagglutinin-stimulated lymphocytes. Cell Tissue Res 213:351–360

Waldstein EA, Cao E-H, Bender MA, Setlow RB (1982) Abilities of extracts of human lymphocytes to remove $O^6$-methylguanine from DNA. Mutat Res 95:405–416

Wyszynska K, Liniecki J (1980) The yield of radiation-induced chromosomal aberrations in lymphocytes as related to the time of arrival at first, poststimulation mitosis. Mutat Res 73: 101–114

Yew F-H, Johnson RT (1978) Human B and T lymphocytes differ in UV-induced repair capacity. Exp Cell Res 113:227–231

Yew F-H, Johnson RT (1979a) Ultraviolet-induced DNA excision repair in human B and T lymphocytes. II. Effects of inhibitors and DNA precursors. Biochem Biophys Acta 562:240–251

Yew F-H, Johnson RT (1979b) Ultraviolet-induced DNA excision repair in human B and T lymphocytes. III. Repair in lymphocytes from chronic lymphocytic leukemia. J Cell Sci 39:329–337

Zöllner EJ, Reitz M, Zahn RK, Slor H (1979) Deoxyribonucleases in phytohemagglutinin-stimulated lymphocytes. Exp Cell Res 123:365–369

Zöllner EJ, Störger H, Breter H-J, Zahn RK (1975) Characterization of different deoxyribonucleases in human lymphocytes. Z Naturforsch 30c:781–784

# Chromosomal Aberrations in Lymphocytes of Patients Under Chemotherapy

E. GEBHART[1]

## 1 Introduction

All results from analyses on the mutagenicity of drugs and/or environmental chemicals should eventually refer to man. Facing the problems of drawing conclusions from model test systems, it is most desirable to include data from studies performed directly on humans whenever possible.

As ethical reasons preclude any experimental work on humans, the possibilities of such studies are very limited. Mutagenicity testing in man, therefore, has to be focussed on those specific cases when individuals or groups of the population have been exposed to mutagenic influences as a therapeutic necessity, occupational inevitability, or accidental reasons.

Besides several theoretically applicable methods, the only suitable method at present for a short-term practical application on limited groups of exposed individuals is to study chromosomes from their somatic cells. This allows a fast and reliable demonstration of chromosome and genome mutations with an acceptable expenditure. The types of mutation analyzed by this method are of great practical importance for man, as has been documented by the findings of clinical and tumor cytogenetics. A supplementation of the scale of indicators of mutagenic action is possible, for example, by the analysis of sister chromatid exchange or the estimation of mutagen-induced micronuclei. Since the first papers on the chromosome damaging activity of chemotherapy in man (Conen and Lansky 1961; Arrighi et al. 1962), the number of reports has grown steadily over many years, thus documenting the practicability as well as the popularity of this method. For a long time most concern was focussed on the assessment of the mutagenic activity of administered therapeutic agents, more recently also studies on individuals occupationally loaded with industrial chemicals have been performed (see Gebhart 1982, for references).

The present paper will consider some methodological aspects and deal with the data on chromosome damaging activity of therapeutics in man given in the literature as well as with the results of more recent studies on specific problems in this field.

---

1 Institut für Humangenetik und Anthropologie, Bismarckstr. 10, 8520 Erlangen, FRG

Mutations in Man, ed. by G. Obe
© Springer-Verlag Berlin Heidelberg 1984

## 2 Methodological Aspects

As the advantages and limitations of this method have been subject of several reports (Kilian and Picciano 1976; Gebhart 1982; Obe and Beek 1982), the considerations presented here will be restricted to some basic problems.

More than 90% of all cytogenetic studies on individuals exposed to chemicals have been performed with cultures of lymphocytes from the peripheral blood of exposed individuals. Only about 8% were based on bone marrow cells and the rest on other cell material. Therefore most of the problems and facts discussed below will refer to studies on lymphocytes (Table 1).

**Table 1.** Frequencies of indications of certain test criteria in cytogenetic studies on individuals exposed to chemicals in vivo (as compiled from the literature)

| Number of analyzed publications | Criteria | Frequencies of papers indicating | | | |
|---|---|---|---|---|---|
| 350 | Target cells | Lymphocytes 91% | | Bone marrow 8.5% | Other 0.5% |
| 322 | Culture period | 40–52 h 25% | 65–72 h 61% | Other 1.5% | No indic. 12.5% |
| 224 | No. of methaphases analyzed | Less 50 10% | 50–99 15% | 100 59% | More 16% |
| 350 | No. of individ. in test group | Less 10 27% | 10–30 45% | More 25% | No indic. 3% |
| 350 | Quality of control group | No contr. 17% | | Average popul. 41% | Matched 20% |
| | | Ident. indiv. before expos. 14% | | No indication 8% | |

From the slides obtained by a standardized chromosome preparation technique (Obe and Beek 1982, for details) the analysis of structural and numerical chromosome aberrations is possible, as is the analysis of sister chromatid exchanges (SCE), if bromo-deoxyuridine (BrDU)-labeling has previously been applied to the cultures. In the qualitative criteria of aberration analysis there should no longer be fundamental differences between different groups working in this field, as clear definitions were given previously (Gebhart 1970; Buckton and Evans 1973; Savage 1975). There are, however, differences as to the number of test individuals, the composition of suitable control groups, and the interpretation of the results. Nevertheless, during the past years there has been an increasing tendency towards a standardization of test conditions (see Gebhart 1982, for references).

For instance, there is now common agreement on the necessity of statistical considerations preceding any study on the size of test and control groups, the criteria of their selection, and on the necessary number of metaphases to be analyzed (Whorton et al. 1979).

As a lower limit of the number of metaphases to be analyzed 100 apparently is generally accepted, while for test group and control group a minimum of 20 individuals are recommended. All these figures, however, must be subject to enlargement as necessary for statistical reasons. The ideal control group for studies on the chromosome-damaging activity of therapeutics would comprise the test individuals before their exposure. If this is not possible, matching control individuals must be selected carefully as to age, sex, and life style. As several studies yielded evidence of a weakly mutagenic or at least comutagenic activity of smoking heavily (Obe and Herha 1978; Hopkins and Evans 1980; Husgafvel-Pursiainen et al. 1980), also the smoking habits of test and control persons should be considered.

Another problem under discussion for a long time is the culture time for the peripheral lymphocytes. Many authors still favor the culture period of 72 h over that of 48 h because of the considerably higher number of metaphases which can be obtained with the former (see Gebhart 1982, for detailed discussion). It could be shown, however, by BrDU-labeling that in most untreated individuals only about 25% to 30% of the lymphocyte metaphases observed in 72-h cultures actually represent "first" mitoses after blood sampling (Crossen and Morgan 1977a; Beek and Obe 1979; Gebhart et al. 1980b). This amount, admittedly, may increase up to more than 60% under the influence of cytostatically acting mutagens, as will be shown below. Though it appears necessary either to limit such studies uniformly to 48 h of culture period, or, as a compromise, to restrict the analysis of chromosome damage to "first" metaphases in BrDU-labeled 72-h cultures easily distinguished by their labelling pattern from "second" or "third" metaphases. Thus SCE analysis may be performed from the second metaphases exhibiting sister chromatid differentiation in the same slides used for aberration analysis.

Some further methodological details will be discussed in connection with results presented below.

## 3 Results

### 3.1 General Review of Data from the Literature

Summarizing the results of chromosome studies on individuals exposed to therapeutic chemicals will yield further evidence on the reliability and the predictive value of this method. As shown in Table 2, by far the most of the substances shown to induce chromosome damage in man in vivo have also been reported to be mutagenic in other test systems. Detailed references on the substances in question are compiled in Table 3.

Since alkylating chemicals were mutagenic in nearly all test systems, it was not surprising to find that even therapeutic doses of these cytostatic agents were sufficient to induce chromosome damage in patients treated with them. In contradiction to most reports, however, Schinzel and Schmid (1976) concluded from their study on cytostatically treated individuals that an increased incidence of chromatid breaks and

**Table 2.** Number of substances tested for their chromosome damaging activity in man in vivo and qualitative comparison of the outcome with data from other test systems

|  | Positive results | Conflicting results | Negative results | Sum |
|---|---|---|---|---|
| No. of substances tested in man | 55 | 13 | 42 | 110 |
| Ident. results with other test systems | 38 | 8 | 20 | 66 |
| Divergent results with other systems | 2 | 3 | 13 | 18 |
| Not yet tested with other systems | 15 | 2 | 9 | 26 |

**Table 3.** Data from chromosome studies in individuals under chemotherapy (or abuse)

| Substance | Number of test individuals | Cell system | Results CA | SCE | Ref. |
|---|---|---|---|---|---|
| Actinomycin D | 12 | L | – | – | [1] |
| Adriamycin | 11 | L | (+) |  | [2] |
| Alcohol abuse | > 200 | L | + | + | [3] |
| Allopurinol | 19 | L | – |  | [4] |
| Amethopterin | > 90 | L, BM, Go | – (+) | (+) – | [2,5] |
| Amphetamine | 12 | L | – |  | [6] |
| Antiepilept. drug combin. | > 90 | L | (+) – | – | [7] |
| Arsenic | 20 | L | ± | + | [8] |
| Azathioprin | > 100 | L, BM | – (+) | – (+) | [9] |
| Bactrim® (Eusaprim®) | 46 | L, BM | – (+) |  | [10] |
| Bleomycin | > 10 | L | + | + | [11] |
| Busulphan (Myleran) | > 40 | L, BM | + | + | [12] |
| Cannabis | > 150 | L | – (+) |  | [13] |
| Carbamazepine | 10 | L | (+) |  | [14] |
| CCNU (incl. Methyl-CCNU) | > 50 | L | + | + | [15,33] |
| Chlorambucil (Leukeran) | > 30 | L, BM | ± |  | [16] |
| Chlordiazepoxide | 10 | L | – |  | [17] |
| Chlorpromazin | 35 | L | – |  | [18] |
| Clozapine | 20 | BM | (+) |  | [19] |
| Colchicine | > 40 | L | – |  | [20] |
| "Contrast material" | 26 | L (Mn) | (+) |  | [21] |
| Cyclamate | 35 | L | (+) – |  | [22] |
| Cyclophosphamide | > 190 | L, BM | + (–) | + | [23,2,25] |
| Cyproheptadine | 40 | L | – |  | [24] |
| Cytembena | 20 | BM | + |  | [25] |
| Cytosine arabinoside | > 20 | BM, L | + | ± | [26] |
| Daunomycin | 20 | L | ± | + | [27,2] |
| Diazepam | 50 | L | – |  | [28] |
| Diphenylhydantoine | 100 | L, BM | – (+) |  | [14,29] |
| Dipropylacetate | 10 | L | – |  | [30] |
| Drug (halucin.) combin. | > 40 | L, BM | ± |  | [31] |
| DTIC | 5 | L |  | – | [32] |
| 5-Fluorouracil | > 60 | L | + | + | [33] |
| Heroin | > 35 | L | ± |  | [34] |
| HMG + HCG | 48 | Ab | (+) |  | [35] |

**Table 3** (continued)

| Substance | Number of test individuals | Cell system | Results CA | SCE | Ref. |
|---|---|---|---|---|---|
| Hycanthone | 13 | L | – | | [36] |
| Hydroxyurea | 11 | L | ± | | [37] |
| Isoniazide | 70 | L | – | | [38] |
| Isoniazide + Thiacetazone | 10 | L | + | | [39] |
| Isoniazide + PAS | 10 | L | + | | [40] |
| Laevamisole | n.s. | L | (+) | (+) | [65] |
| Lincomycin | 15 | L | – | | [41] |
| Lithium salts | > 125 | L | – (+) | – | [42] |
| LSD | > 300 | L | – (+) | | [43] |
| Melphalan (Alkeran) | 51 | L | + | + | [44] |
| 6-Mercaptopurine | 25 | L, F | + | | [45] |
| Mescalin | 50 | L | – | | [46] |
| Methadone | 100 | L | – (+) | | [47, 34] |
| Metronidazol | > 40 | L, BM | – | | [48] |
| Mitomycin C | > 10 | L | | + | [49] |
| Nalidixic acid | 12 | L | | + | [50] |
| Nitrosomethylurea | 32 | L | – | | [51] |
| Opiates | 16 | L | (+) | | [52] |
| Oral contraceptives | >1,000 | L, Ab | – | – | [53] |
| Oxamniquine | 24 | L | – | | [54] |
| Penicillamine | 11 | BM | – | | [55] |
| Penicillines | 11 | L | – | | [41] |
| Perazine | 56 | L | (+) | | [56] |
| Perphenazine | > 50 | L | – (+) | | [57] |
| Phenylbutazon | > 125 | L, BM | – (+) | | [58] |
| Primidon | 50 | L | – (+) | + | [59] |
| Psoralene + UVA | > 20 | L | – | – | [60] |
| Sulfonylureas | 23 | L | (+) | | [61] |
| Thioridazine | 9 | L | (+) | | [62] |
| Thiotepa | 63 | BM | + | | [63] |
| Vincristine | > 50 | L | – (+) | – | [33, 64] |

+ = positive results; – = negative results; (+) = weakly positive results; ± = differing (contradictory?) results

Abbreviations: CCNU: 1-(2-chloroethyl)-3-cyclohexyl-1-nitrosourea; DTIC: 5-(3,3-dimethyl-1-triazeno)imidazole-4-carboxamide; HMG: human menopausal gonadotropine; HCG: human chorionic gonadotropine; LSD: lysergic acid diethylamide; L = lymphocytes; BM = bone marrow; F = fibroblasts; Ab = abortus material; Go = gonadal cells; CA = Chromosomal aberrations; SCE = Sister chromatid exchanges; MN = micronuclei

*References to Table 3:* [1] Cohen et al. (1971), Lambert et al. (1979); [2] Schinzel and Schmid (1976); [3] Obe and Herha (1975), Obe et al. (1980), Butler et al. (1981); [4] Stevenson et al. (1976); [5] Jensen (1967, Locher and Fränz (1967), Jensen and Nyfors (1979), Lawler and Walden (1979), Düker (1981); [6] Fu et al. (1978); [7] Große et al. (1972), Obe and Beek (1982); [8] Burgdorf et al. (1977), Nordenson et al. (1979); [9] Jensen (1967, 1970), Eberle et al. (1968), Friedrich and Zeuthen (1970), Ganner et al. (1973), Rossi et al. (1973), Apelt et al. (1981), Kucerova et al. (1982); [10] Gebhart (1973, 1975), Stevenson et al. (1973), Sörensen and Jensen (1981); [11] Bornstein et al. (1971), Dresp et al. (1978); [12] Gebhart et al. (1974), Misawa et al. (1980); [13] Dorrance et al. (1970), Matsuyama et al. (1973, 1977), Herha and Obe (1974), Martin et al. (1974), Nichols et al. (1974), Stenchever (1974), Matsuyama and Jarvik (1975), Obe and

exchanges was not a typical finding in lymphocyte cultures of persons exposed to chromosome-breaking agents. It should, however, be noted that in their study the cytogenetic examination was conducted one to several weeks (in some cases even months) after the cessation of the therapy.

Antimetabolites also were shown to be mutagenic in most test systems. When administered for cytostatic therapy, with a few exceptions, they also proved clastogenic (= chromosome-damaging) in man in vivo, while contradictory results were obtained from studies on individuals after immunosuppressive therapy with these substances. Thus the data on antimetabolites do not appear as conclusive as those on alkylating agents. This may be due to the fact reported from experimental studies on mammals or mammalian cell cultures that the clastogenic activity of antimetabolites requires higher final concentrations than that of alkylants. Therefore, therapeutic doses in some cases may not have reached that level necessary for the induction of chromosome damage. Furthermore most of these substances are known to act preferentially on cells during the S-phase, which is absent in the circulating lymphocyte. Thus, bone marrow in vivo under certain circumstances may be a better target for chromosome studies than are peripheral lymphocytes (e.g. Jensen 1970; Jensen and Nyfors 1979). Only a rather limited amount of information as yet is available on the clastogenic action of cytostatic antibiotics in man in vivo. In both experimental animals and human lymphocyte cultures these antibiotics were found to be clasto-

*References to Table 3* (continued)

Beek (1982); [14] Herha and Obe (1976); [15] Lambert et al. (1978); [16] Stevenson and Patel (1973), Reeves et al. (1975); [17] Cohen et al. (1969), Matsuyama and Jarvik (1975); [18] Nielsen et al. (1969), Roman (1969), Gabilondo et al. (1971), Cohen et al. (1972), Matsuyama and Jarvik (1975); [19] Knuutila et al. (1977); [20] Cohen et al. (1977); [21] Cochran et al. (1980); [22] Bauchinger et al. (1970), Dick et al. (1974); [23] Bauchinger and Schmid (1969), Dobos et al. (1974), Tolchin et al. (1974), Fischer et al. (1978), Schuler et al. (1979), Düker (1981); [24] Gebhart (1971), Murken (1971); [25] Goetz et al. (1973); [26] Bell et al. (1966), Raposa (1978, 1982); [27] Whang-Peng et al. (1969); [28] Stenchever et al. (1970), White et al. (1974), Matsuyama and Jarvik (1975); [29] Marquez-Monter et al. (1970), Alving et al. (1977), Knuutila et al. (1977), Torigoe et al. (1978), Eßer et al. (1981), Obe and Beek (1982); [30] Kotlarek and Faust (1978); [31] Amarose and Schuster (1971), Gilmour et al. (1971); [32] Lambert et al. (1979); [33] Gebhart et al. (1980a,b); [34] Amarose and Norusis (1976); [35] Boue and Boue (1973); [36] Frota-Pessoa et al. (1975); [37] Rossi et al. (1973); [38] Bauchinger et al. (1978); [39] Ahuja et al. (1981); [40] Jaju et al. (1981); [41] Gebhart (unpublished data); [42] Genest and Villeneuve (1971), Jarvik et al. (1971), Bille et al. (1975), De La Torre and Krompotic (1976), Matsuyama and Jarvik (1975), Banduhn et al. (1980), Garson (1981); [43] Cohen et al. (1967), Robinson et al. (1974), Matsuyama and Jarvik (1975), Obe and Beek (1982); [44] Einhorn et al. (1982); [45] Fedortzeva et al. (1973); [46] Dorrance et al. (1975); [47] Matsuyama et al. (1978); [48] Mitelman et al. (1976, 1980, 1982), Hartley-Asp (1979), Salamanca-Gomez et al. (1980); [49] Abe et al. (1979), Ohtsuru et al. (1980); [50] Kowalczyk (1980); [51] Selezneva and Korman (1973); [52] Falek et al. (1972), Matsuyama and Jarvik (1975); [53] McQuarrie et al. (1970), De Gutierrez and Lisker (1973), Fitzgerald et al. (1973), Bishun et al. (1975), Littelfield et al. (1975), Müller and Ritter (1978), Murthy and Prema (1979), Husum et al. (1982); [54] Monsalva et al. (1976); [55] Jensen et al. (1979); [56] Mdale et al. (1980); [57] Nielsen et al. (1969), Gabilondo et al. (1971), Cohen et al. (1972), Matsuyama and Jarvik (1975); [58] Stevenson et al. (1971), Walker et al. (1975), Crippa et al. (1976), Vormittag and Kolarz (1979); [59] Galle (1978), Eßer et al. (1981); [60] Brögger et al. (1978), Mourelatos et al. (1977), Wolff-Schreiner et al. (1977); [61] Watson et al. (1976); [62] Saxena and Ahuja (1982); [63] Blank (1979); [64] Gehbart et al. (1969), Raposa (1982); [65] Berger et al. (1980)

genic. On the other hand, however, most antibacterial and antiparasitic antibiotics have failed to produce distinct clastogenic activities on peripheral lymphocytes when administered therapeutically. Another group of therapeutics, the psychotropic drugs, has gained importance in many industrialized countries. But only a few of these agents have been shown to be weakly clastogenic in man. By studies on combinations of antiepileptic drugs, however, Grosse et al. (1972) discovered an increased incidence of chromosomal aberrations not only in patients directly treated with these drugs but also even in newborn infants who had contact with these agents in utero. An increased frequency of chromatid exchanges was found by Herha and Obe (1976) in peripheral blood lymphocytes from patients treated with Carbamazepine as compared with untreated control individuals.

Although such halucinogens as lysergic acid diethylamide (LSD) were first alleged to be clastogenic, subsequent studies with pure LSD yielded negative results. Probably the drug addicts in most cases are exposed to several drugs in combination or in sequence, and these drugs contain far more impurities than substances administered in a controlled study. While a higher level of chromosome changes was found in heroin and opiate addicts than in controls, marijuana and mescalin, but also methadone thus far have yielded predominantly negative results.

There are also some other data on chromosome studies in patients loaded with several other therapeutics included in Table 3. Most of them thus far have yielded negative results. Such negative findings obtained from peripheral lymphocytes, however, must be interpreted with some caution: The substance in vivo displays its action on differentiated non-dividing lymphocytes of the peripheral blood, being at that time in a "resting stage" ($G_0$) of interphase.

Therefore, chemicals must either cause primary molecular lesions in this $G_0$-phase or, if only active during S-phase (e.g. some antimetabolites), be stored within the lymphocyte in sufficient amount until reaching S-phase in culture, to manifest their detrimental action on chromosomes. For this reason agents exclusively active on S-phase as well as weak mutagens may yield false negative results in this system.

## 3.2 Contributions to Specific Problems

As shown in Table 3, seemingly contradictory results have been obtained on several substances by chromosome and SCE studies on peripheral lymphocytes from exposed individuals. This fact reflects technical as well as fundamental problems involved in these studies. Some of them are considered here.

### 3.2.1 The Question of the Most Suitable Culture Time

Many investigators favor the 48-h harvest time of lymphocyte cultures because most of the transformed lymphocytes should have entered their first mitosis at this point unless the substances the test individuals had been exposed to strongly inhibit the lymphocyte proliferation in culture. In most cases the data on 48-h samples should give a highly representative picture of the chromosome damage and/or the primary lesions induced in vivo.

At 72 h of culture, second or even third cell cycles may have been completed; therefore, selection and differential multiplication rates between damaged and un-damaged lymphocytes may obscure the true picture. On the other hand, a considerably higher number of metaphases can be obtained from 72-h samples than from 48-h samples, particularly after treatment with cytostatic agents.

Table 4 gives recent data from studies using BrDU-labeling for differentiation of lymphocytes with different proliferative activity pointing to a strong shift in favor of first metaphases in 72-h cultures if the test substances are inhibitors of cell prolifera-tion. The proportion of lymphocytes completing one, two, or three cell cycles dur-ing 72 h in the presence of BrDU, in addition, appears to be dependent on the culture medium used (Obe et al. 1975). Even in 48 h cultures, some authors reported a con-siderable number of metaphases in the second division cycle: from 5% to as much as 40% in some cases (e.g. Crossen and Morgan 1977a,b; Beek and Obe 1979).

**Table 4.** Percentage of "first", "second", and "third" metaphases (as differentiated by BrDU-labeling) in PHA-stimulated human lymphocyte cultures after 72 h culture time from normal subjects and subjects undergoing chemotherapy

| Indication | Number of metaphases analyzed | Percentage of | | | Culture medium | Ref. |
|---|---|---|---|---|---|---|
| | | M1 | M2 | M3 | | |
| *Normal subjects* | 1,000 (10) | 33,1 | 44,6 | 22,3 | F 10 | [1] |
| | 3,000 (30) | 47,6 | 45,2 | 7,2 | F 10 | [2] |
| | 800 (4) | 25,5 | 37,8 | 36,8 | McCoy | [3] |
| | 1,000 (1) | 19 | 36 | 45 | n.s. | [4] |
| | 600 (1) | 5 | 18,7 | 76,3 | RPMI | [5] |
| | 469 (6) | 53,1 | 40,5 | 6,4 | MEM (Eagle) | [6] |
| | 600 (1) | 22,7 | 49,2 | 28,1 | TC 199 | [7] |
| | 1,200 (12) | 8 | 33,1 | 58,9 | MEM | [8] |
| | 2,000 (20) | 28,5 | 49,9 | 21,6 | n.s. | [9] |
| | 2,000 (2) | 12,2 | 31 | 56,8 | RPMI | [10] |
| | 11,000 (56) | 36,2 | 38,1 | 25,7 | F 10 | [11] |
| | 900 (7) | 36,3 | 36,3 | 27,3 | F 10 | [12] |
| *Cancer patients* | | | | | | |
| Before therapy | 2,200 (11) | 30,6 | 35 | 34,4 | F 10 | [13] |
| i.a.t.c.[a] | 4,400 (22) | 51,2 | 32,3 | 16,5 | F 10 | [13] |
| 4 w.a.t.c.[a] | 10,400 (52) | 39 | 34,6 | 26,4 | F 10 | [13] |
| After 1st t.[a] | 2,200 (11) | 35,9 | 36 | 28,1 | F 10 | [13] |
| After 5th t.[a] | 2,400 (12) | 46 | 36,4 | 17,6 | F 10 | [13] |
| After 9th t.[a] | 1,600 (8) | 58,1 | 32,4 | 9,5 | F 10 | [13] |

Numbers in parantheses indicate the number of subjects

[a] i.a.t.c.: immediately after all therapy courses;
   4 w.a.t.c.: four weeks after all therapy courses;
   t.: therapy course

*References:* [1] Crossen and Morgan (1977a); [2] Crossen and Morgan (1977b); [3] Alvarez et al. (1980); [4] Purrott et al. (1980); [5] Morimoto and Wolff (1980); [6] Giulotto et al. (1980); [7] Madle (1981); [8] Abdel Fadil et al. (1982); [9] Crossen (1982); [10] Hill and Wolff (1982); [11] Gebhart (1982); [12] Gebhart (unpublished); [13] Gebhart et al. (1980b)

A number of investigators (see Table 5, for references) claimed that the chromosome aberration rates between the 48 h and the 72 h samples showed no significant difference. Our own study on individuals exposed to a cytostatic interval therapy (Gebhart et al. 1980a,b) also documented no grave differences of the aberration rates in lymphocytes cultured for these two respective times (Table 5). One of the possible main reasons for this observation will be discussed in Sect. 3.2.4. Other investigators, however, did observe some minor differences. In some cases higher aberration rates were observed in 48-h samples (Table 5), but in others also the reverse was reported.

**Table 5.** Examples of average aberration rates in peripheral lymphocytes of chemically loaded persons after 48 h and 72 h in culture (Gebhart 1982)

| No. of individ. | No. of metaphases analyzed | Percentage of aberrant metaphases | | Breakage rate | | Ref. |
|---|---|---|---|---|---|---|
| | | 48 h | 72 h | 48 h | 72 h | |
| 5 | 850 | 20.8 | 15.9 | 0.33 | 0.32 | Bauchinger and Schmid (1969) |
| 16 | 2,034 | 0.23 | 2,56 | 0.002 | 0.02 | Falek et al. (1972) |
| 3 | 600 | 43.0 | 48.6 | 0.35 | 0.42 | Gebhart et al. (1974) |
| 10 | 1,850 | 9.6 | 12.6 | 0.05 | 0.06 | Gebhart (1975) |
| 30[a] | 3,200 | 4.6 | 4.9 | 0.035 | 0.028 | Gebhart et al. (1980a) |
| 75[b] | 7,530 | 29.7 | 22.6 | 0.26 | 0.27 | Gebhart et al. (1980a) |
| 123[c] | 12,222 | 19.9 | 18.3 | 0.21 | 0.19 | Gebhart et al. (1980a) |

[a]  Cancer patients before therapy
[b]  Cancer patients immediately after courses of cytostatic interval therapy
[c]  Cancer patients four weeks after courses of cytostatic interval therapy

### 3.2.2 Dose-Effect Correlations

Although dose-effect relations are helpful and important data to verify and to support the conclusions on mutagenic and clastogenic effects, most in vivo cytogenetic studies in man have not stressed this point. Various technical difficulties and limitations have hampered progress in this direction. However, cytogenetic studies on patients receiving chemotherapy should be able to shed some light on this problem.

We have tried to find out, in patients treated with busulphan for chronic myelocytic leukemia, whether a dose-effect relation existed for chromosome damage (Gebhart et al. 1974). In these patients, a clear correlation was established between the frequency of chromosome aberrations and the total additive dose administered over a long period of time. Another study on this problem was performed on lymphocyte cultures from the peripheral blood of 43 patients undergoing a cytostatic interval therapy with a standardized regimen of methyl-CCNU, 5 fluorouracil, and vincristine, yielding a total of 229 individual chromosome examinations (Gebhart et al. 1980a). Although there was a striking variability of individual data, nevertheless, a distinct increase of the average breakage rate was observed depending on the number of rounds of drug administration (Fig. 1).

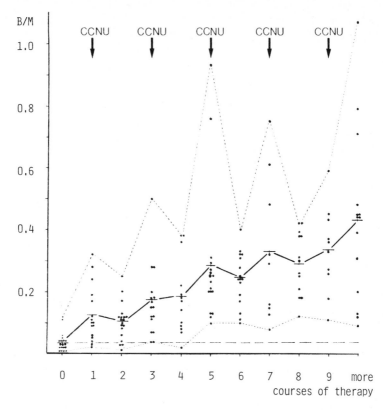

**Fig. 1.** Breakage rates in individual blood samples obtained after the indicated courses of cytostatic interval therapy with methyl-CCNU, 5-fluorouracil, and vincristine (Gebhart et al. 1980a). *Upper dotted line* connects maximal, *lower line* minimal breakage rates; the mean values are connected by *solid line*; therapy courses with methyl-CCNU are indicated by *vertical arrows*

It seems noteworthy that not only the highest individual breakage rates were observed after therapy courses with methyl-CCNU-administration but also the maxima of average exchange rates (courses 1, 3, 5, 7 in Figs. 1 and 2). Together with respective data of other authors (e.g. Bauchinger and Schmid 1969; Nevstad 1978; Ohtsuru et al. 1980; Düker 1981; Obe et al. 1981), these results well demonstrate the reliability and sensitivity of this method for control of mutagenic (carcinogenic) damage by certain types of chemotherapy.

## 3.2.3 Interindividual Variation of Cytogenetic Damage Induced by Chemotherapy

A good many of contradictory results of cytogenetic studies on subjects exposed to chemical mutagens may be ascribed to the influence of the variability of the individual response to the mutagen, if too few individual samples were studied. In contrast to most highly homogenous test systems used for studies on the mutagenicity of chemicals, man is a "highly inhomogenous test system".

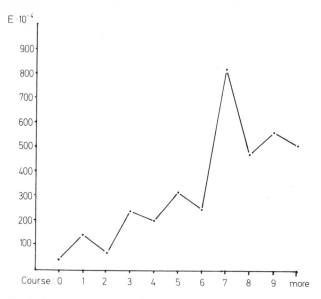

**Fig. 2.** Interchange rate per $10^4$ metaphases from lymphocytes sampled after the indicated courses of cytostatic interval therapy (see Fig. 1)

The chromosome study on patients undergoing a cytostatic interval therapy which has just been discussed above, rendered it possible to examine this inter-individual variation of mutagen sensitivity in more detail. All individuals included in the test group underwent exactly the same schedule of interval therapy not only with identical cytostatics (see above) but also with comparable dosage corresponding with the body surface.

As reported previously (Gebhart et al. 1982), a high inter-individual variability of the parameters studied (i.e. rates of breakage and SCE) was found before, as well as during the different courses of therapy (Fig. 1). This was particularly evident if cyto-genetic data obtained before and immediately after single courses of therapy were compared from different individuals (Fig. 3).

The authors concluded from these studies that the observed inter-individual varia-tion of response of human lymphocyte chromosomes to identical courses of cytostatic therapy may reflect a general feature of human cells as well as human populations (see also Schwanitz et al. 1975; Gebhart 1981; Bora et al. 1982, for references). The extent of this variation apparently depends on the kind of exposure and is strongly influenced by a few individuals' high sensitivity or resistance to the mutagenic influ-ences. This latter fact justifies any effort to figure out "high-risk" individuals by those cytogenetic studies. It should be pointed out that inter-individual variations in the proportion of abnormal cells determined in untreated individuals, as performed under statistical aspects by Whorton et al. (1979), must not necessarily reflect the variability observed after exposure to a clastogen. Reasons for the observed inter-individual variability may be found in individual peculiarities of enzymatic make up (see Motulsky 1982, for references) and other internal causes as well as in additional external factors (see Berg 1979, for references).

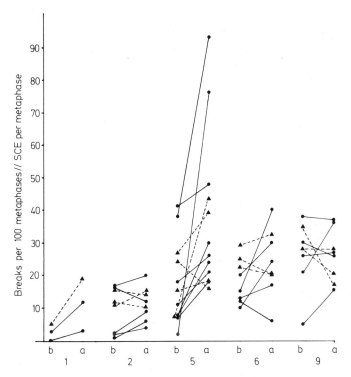

**Fig. 3.** Breakage (*circles*) and SCE (*triangles*) rates of individual patients before and after the indicated courses of cytostatic interval therapy; *solid lines* connect the individual rates of breakage, *dotted lines* those of SCE of identical individuals. (Gebhart et al. 1982; with kind permission of the copyright-holder)

Thus, for instance, data of other authors (Lambert et al. 1978; Obe and Herha 1978; Husgafvel-Pursiainen et al. 1980; Mäki-Paakkanen et al. 1981) strongly suggest an influence of smoking habits on the rate and yield of chromosome damage induced by environmental clastogens. Although in our study cited above, data on individual smoking habits were not available, it may not be too misleading to assume that smoking in general is incompatible with cytostatic therapy, and therefore during the therapy it may be excluded that a considerable number of the patients under study actually did smoke. There were, however, no other evident external factors to be accused of influencing the chromosomal response in individual cases which led us to the suggestion that internal factors discussed above might mainly contribute to the observed variability.

### 3.2.4  Studies Using Bromodeoxyuridine (BrDU)-Labeling

The method of BrDU-labeling of cells for two or more rounds of DNA replication in culture has yielded a substantial enlargement of the spectrum of possibilities for cytogenetic studies on individuals exposed in vivo to chemotherapeutics.

*SCE-Analyses.* The evaluation of sister chromatid exchanges (SCE) from BrDU-differentiated chromosomes has become an important additional cytogenetic method for recording mutagenic/carcinogenic influences on somatic cells. It is applicable for in vitro and in vivo studies (see Sandberg 1982, for references). For a large number of mutagens a good correlation between the chromosome breaking and the SCE-inducing activity could be demonstrated, the frequency of SCE exceeding that of breaks about a hundred-fold in most cases (see Abe and Sasaki 1982; Littlefield 1982, for references). Although there is a series of observations on clearcut differences between the induction of breakage and SCE, some strong clastogens having been shown not to induce SCE (see Gebhart 1981, for references), it still appears imperative to compare the evidence and the reliability of the SCE test and the classical cytogenetic methods by further detailed analyses in man in vivo.

First examinations of a rather limited number of patients treated with cytostatics indicated the use of the SCE technique for cytogenetic studies in man (review: Raposa 1982). From a total of 92 individual SCE analyses on the group of cytostatically treated patients reported above, the following observations could be made (Gebhart et al. 1980b):

A distinct increase of SCE frequency over the control level (i.e. lymphocyte cultures of patients before the start of the therapy) was observed after all courses of interval therapy. It was clearly correlated with the number of courses of therapy up to course 7, later the SCE rate remained more or less at the level reached. The influence of the composition of each drug regimen was less pronounced than it was on the breakage rate. Moreover, although a clear correlation existed between the individual rates of breakage and SCE (Fig. 4), the formation of the latter was interpreted as reflecting a long-term effect of the therapy rather than did the formation of break aberrations: It seems to be of particular interest that the frequency of SCE observed 4 weeks after the preceding course of therapy in most individual cases, where a direct comparison was possible, was higher (or at least only slightly decreased) than immediately after the respective drug administration. This observation is in contrast to the data obtained from the analysis of breakage induced by the same therapy in the same patients, but it is in agreement with the observations of other authors on the SCE induction by cytostatic therapy (e.g. Musilova et al. 1979; Aronson et al. 1982). These authors reported an increase of the SCE rate even 4 months after discontinuation of a cytostatic therapy with comparatively low doses of cyclophosphamide, busulphan, and adriamycin. After busulphan therapy these authors found the highest SCE rates only 5 weeks after the last administration of the drug. Thus SCE analysis apparently gives evidence on cytogenetic long-term effects which break analysis alone would not. In addition, as the intercellular variability of the number of SCE was much higher than that of breaks, the inter-individual variability (= variation of the average SCE frequency of each patient) was small compared to the respective variability of breakage rates. Thus, as in studies of other authors reviewed by Raposa (1982) SCE analysis proved to be a valuable tool for additional information on the action of chemotherapeutics in man. Certainly an increasing number of observations will contribute to a better understanding of the meaning of increased SCE rates in individuals exposed to potential mutagens in vivo, as well as of their relation to breakage (Hedner et al. 1982).

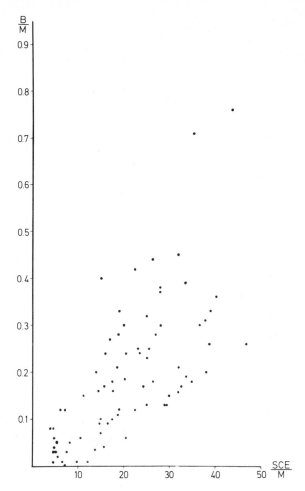

**Fig. 4.** Correlation of individual breakage rates from conventionally stained metaphases and SCE rates from BrDU-labeled second metaphases obtained from the same blood samples after cytostatic interval therapy. (Gebhart et al. 1980b)

*Proliferation Kinetics and Aberration Yields.* Cytogenetic studies with BrDU-labeling, however, beyond SCE-analysis, offer additional opportunities: Thus, for instance, the proliferation kinetics of cultured lymphocytes can be studied by comparing the proportions of "first", "second", and "third" metaphases in cultures grown in the presence of BrDU, which can be discerned by their characteristic labeling pattern (Crossen and Morgan 1977a,b). The influence of the cytostatic therapy on the proliferation activity and, in this case, the consequences of this influence on the aberration yield may be determined.

As shown previously (Gebhart et al. 1980b) the average portions of metaphases with the respective labeling pattern vary in a wide range strictly depending on the respective therapeutic influences. Immediately after courses of cytostatic interval therapy, in 72-h lymphocyte cultures a percentage of first metaphases up to over 60% could be observed on average. This fact must be taken into consideration when the influences of selective factors on the results of cytogenetic studies on exposed individuals are discussed (Table 4).

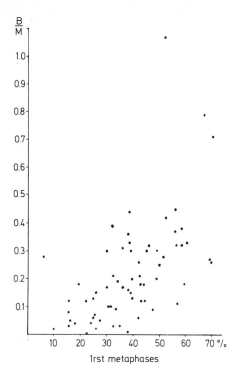

**Fig. 5.** Correlation of breakage rates observed in 72-h lymphocyte cultures without BrDU-labeling and the portion of "first" metaphases in BrDU-labeled cultures from the same blood sample. (Gebhart et al. 1980b)

There was some correlation between the frequency of break aberrations and the respective amount of "first" metaphases in individual samples, that is, between the inhibition of proliferation and chromosome damage (Fig. 5).

*Aberration Yields in Lymphocytes of Different Proliferation Activity.* It should be pointed out that the first, second, and third metaphases coexisting in 72-h cultures of peripheral blood lymphocytes do not represent successive divisions of the same lymphocyte populations, but simultaneously performed divisions of lymphocyte populations proliferating with different speed. Therefore, BrDU-labeling also permits aberration analysis on these differently proliferating lymphocytes.

In a study on the distribution of structural chromosome aberrations on BrDU-labeled first, second, and third metaphases of lymphocytes from patients undergoing cytostatic interval therapy (Gebhart and Mueller 1982), most aberrations were observed in first division metaphases in every case. But also second and third metaphases exhibited increased aberration rates, which clearly were dependent on the number of therapy courses (Fig. 6). Chromatid breaks were predominating in all metaphases irrespective of the number of performed replication cycles. This distribution of aberrations in M1-, M2-, and M3-metaphases was not as markedly influenced by peculiarities of the therapy or by individual sensitivities, as was the absolute number of aberrations. There was also no distinct correlation between the frequency of breaks in the different categories of labeled metaphases and delay of proliferation. Differences of the breakage rates between M1-metaphases observed in 48-h cultures and those observed in 72-h cultures were inconsistent (Table 6).

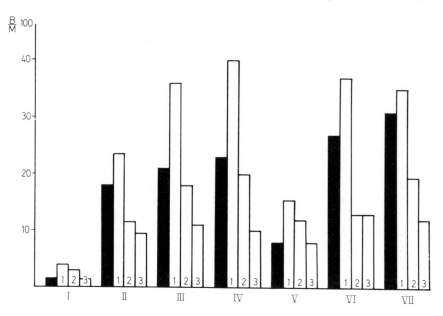

**Fig. 6.** Comparison of breakage rates in conventionally stained (= undifferentiated) metaphases (*black columns*) and in BrDU-differentiated M1, M2, and M3 at different stages of the cytostatic interval therapy (*I* before treatment; *II* immediately after courses 1 and 2; *III* immediately after courses 3 to 6; *IV* immediately after courses 7 or more; *V* 4 weeks after courses 1 and 2; *VI* 4 weeks after courses 3 to 6; *VII* 4 weeks after courses 7 or more)

**Table 6.** Frequency of break aberrations in M1-metaphases of 48 h and 72 h lymphocyte cultures identified by their BrDU-labeling pattern and in conventionally stained metaphases of 72-h cultures from patients undergoing cytostatic interval therapy. (Data from Gebhart and Mueller 1982)

| Patient code | Time of sampling[a] | 48-h period | | 72-h period | | | period |
|---|---|---|---|---|---|---|---|
| | | M 1[a] | B/M[c] | M 1[b] | B/M[c] | Mc[b] | B/M[c] |
| 389 | i.a.1 | 50 | 0.28 | 50 | 0.20 | 100 | 0.28 |
| 66 | i.a.2 | 50 | 0.14 | 50 | 0.24 | 100 | 0.04 |
| 92 | i.a.2 | 50 | 0.04 | 50 | 0.06 | 100 | 0.06 |
| 133 | i.a.4 | 50 | 0.06 | 50 | 0.38 | 100 | 0.17 |
| 134 | i.a.5 | 50 | 0.30 | 50 | 0.22 | 100 | 0.24 |
| 144 | 4 w.a.1 | 50 | 0.12 | 50 | 0.18 | 100 | 0.16 |
| 407 | 4 w.a.3 | 50 | 0.32 | 50 | 0.40 | 100 | 0.28 |
| 431 | 4 w.a.4 | 50 | 0.32 | 50 | 0.38 | 100 | 0.18 |
| 391 | 4 w.a.6 | 50 | 0.46 | 50 | 0.62 | 100 | 0.42 |
| 432 | 4 w.a.6 | 50 | 0.20 | 50 | 0.14 | 100 | 0.19 |
| Average of B/M[c] | | | 0.204 | | 0.268 | | 0.202 |

[a] i.a. = immediately after, 4 w.a. = four weeks after the indicated course number of interval therapy

[b] Number of first (M1) or conventionally stained (Mc) metaphases from which the data were obtained

[c] Number of break events per metaphase (breakage rate)

To interpret these results we have to realize that most chromosome damage was found in slowly proliferating lymphocytes which have reached only their first mitosis after 72 h in culture. The frequency of breaks observed in lymphocytes having reached the second or even third division in culture at the same time may be reduced by superimposed factors of selection.

Including BrDU-labeling into all cytogenetic studies on subjects exposed to potential mutagens seems imperative. The expenditure on preparation may be reduced if SCE-analyses can be performed from differentiated second metaphases and if break analysis is performed from undifferentiated first metaphases of the same slide. This, however, will only be possible if the amount of available metaphases is not reduced too much by the synergistic effect of BrDU and the environmental chemical under study, and unless other aspects of the respective study rule out this procedure. Some authors (e.g. Bauchinger et al. 1982) apply BrDU-labeling techniques to their studies on individuals exposed in vivo to chemicals to ensure that analysis of chromosome aberrations is restricted to first mitoses only.

### 3.2.5 Clinical Aspects of Cytogenetic Studies After Chemotherapy

As shown above, chromosome damage induced by chemotherapy is readily recorded by reliable standardized techniques. Thus data on cytogenetic consequences of the application of therapeutics yield a useful, informative, and appropriate tool for estimating the mutagenicity of those chemicals in man under realistic in vivo conditions. Positive results give us evidence of the genetic and cancer risk for the exposed individuals of chemical mutagens, which can help to develop safer therapies and/or to offer a more proper genetic counseling. Some authors have also pointed out the opportunity offered by cytogenetic studies of monitoring late side effects of chemotherapy and of comparing the obtained cytogenetic data with cancer risks (e.g. Einhorn et al. 1982; Fekete et al. 1981).

These cytogenetic methods, however, are not restricted to obtaining evidence of mutagenicity/carcinogenicity only. They are also suitable to figure out individuals with an outstanding specific sensitivity or resistancy to the respective therapeutics. Although pertinent data are lacking so far, cytogenetic discovery of these individual differences of reactivity to chemical influences, particularly to chemotherapy, may prove useful for selecting an appropriate schedule of therapy in individual cases. It may also be valuable for studies on genetically determined differences of metabolic potency. In consequence, measures of health protection and of genetic counseling can be focussed on this "high-risk" groups within the population.

## 4 Conclusions

Monitoring mutagenic influences on human genetic material is an imperative task of modern mutagenicity testing. As pointed out in the introduction, the possibilities of studies in man are rather limited. The most usual approach has been discussed in detail and the data obtained by it so far have been summarized.

If criteria of standardization, as shown above, are seriously kept in mind the in vivo approach presented here of monitoring cytogenetic damage in individuals undergoing any form of chemotherapy allows a fast and reliable qualitative and in some cases even quantitative recording of highly actual types of mutation with justifiable expenditure. The activity of metabolites and ultimate forms of the chemotherapeutic of interest actually active in the human organism is analyzed. Long-term studies yield data on the detrimental cytogenetic activity of drugs in man over a long period. Therefore a routine application of cytogenetic studies in the framework of clinical trials should be considered. This is also emphasized by the fact that mutations in somatic cells, as considered by this approach, apparently are of very high importance in the process of malignant transformation.

There are of course evident disadvantages of cytogenetic studies in subjects exposed to therapeutic chemicals: There is no opportunity for systematic experimental mutagen screening so that the pertinent studies have to be restricted on special occasions. Additionally, negative results obtained from those studies are not reliable, for reasons discussed above.

Nevertheless the cytogenetic approach of recording mutagenic action in man in vivo may be, and always has been considered a valuable completion fo the spectrum of systems available for mutagenicity testing. As positive results from these studies give us direct evidence of genetic and cancer risks of chemical mutagens in man, cytogenetic monitoring of exposed individuals as well as populations should be performed whenever possible. This is important not only for subjects exposed to chemicals for therapeutic reasons but also for individuals or populations exposed by their occupation.

# References

Abdel-Fadil MR, Palmer CG, Heerema N (1982) Effect of temperature variation on sister-chromatid exchange and cell-cycle duration in cultured human lymphocytes. Mutat Res 104:267–273

Abe T, Morita M, Matsuno K, Ogawa H, Misawa S (1979) Sister chromatid exchanges of human cultured lymphocytes induced by chemotherapeutic drugs in vivo and in vitro (Abstr.). Nippon Eiseigaku Zasshi 34:302

Abe S, Sasaki M (1982) SCE as an index of mutagenesis and/or carcinogenesis. In: Sandberg AA (ed) Sister chromatid exchange. Liss, New York, pp 461–514

Ahuja YR, Jaju M, Jaju M (1981) Chromosome-damaging action of isoniazid and thiacetazone on human lymphocyte cultures in vivo. Hum Genet 57:321–322

Alvarez MR, Cimino LE, Cory MJ, Gordon RE (1980) Ethanol induction of sister chromatid exchanges in human cells in vitro. Cytogenet Cell Genet 27:66–69

Alving J, Jensen MK, Meyer H (1977) Chromosome studies of bone marrow cells and peripheral blood lymphocytes from diphenyldantoine-treated patients. Mutat Res 48:361–366

Amarose AP, Schuster CR (1971) Chromosomal analyses of bone marrow and peripheral blood in subjects with a history of illicit drug use. Arch Gen Psychiatry 25:181–185

Amarose AP, Norusis MJ (1976) Cytogenesis of methadone-managed and heroin-addicted pregnant women and their newborn infants. Am J Obstet Gynecol 124:635–640

Apelt F, Kolin-Gerresheim J, Bauchinger M (1981) Azathioprine, a clastogen in human somatic cells? Mutat Res 88:61–72

Aronson MM, Miller RC, Hill RB, Nichols WW, Meadows AT (1982) Acute and long-term cytogenetic effects of treatment in childhood cancer: sister-chromatid exchanges and chromosome aberrations. Mutat Res 92:291–307

Arrighi FE, Hsu TC, Bergsagel DE (1962) Chromosome damage in murine and human cells following cytoxan therapy. Tex Rep Biol Med 20:545–549

Banduhn N, Obe G, Müller-Oerlinghausen B (1980) Is lithium mutagenic in man? Pharmakopsychiatr Neuropsychopharmakol 13:218–227

Bauchinger M, Schmid E (1969) Cytogenetische Veränderungen in weißen Blutzellen nach Cyclophosphamidtherapie. Z Krebsforsch 72:77–87

Bauchinger M, Schmid E, Pieper M, Zöllner N (1970) Zytogenetische Wirkung von Cyclamat in menschlichen peripheren Lymphocyten in vivo. Dtsch Med Wochenschr 95:2220–2223

Bauchinger M, Gebhart E, Fonatsch C, Schmid E, Müller W, Obe G, Beek B, Göbel D, Radenbach KL (1978) Chromosome analyses in man in the course of chemoprophylaxis against tuberculosis and of antituberculosis chemotherapy with isoniazid. Hum Genet 42:31–43

Bauchinger M, Schmid E, Dresp J, Kolin-Gerresheim J, Hauf R, Suhr E (1982) Chromosome changes in lymphocytes after occupational exposure to toluene. Mutat Res 102:439–445

Bauchinger M, Dresp J, Schmid E, Hauf R (1982) Chromosome changes in lymphocytes after occupational exposure to pentachlorophenol (PCP). Mutat Res 102:83–88

Beek B, Obe G (1979) Sister chromatid exchanges in human leukocyte chromosomes: spontaneous and induced frequencies in early- and late-proliferating cells in vitro. Hum Genet 49:51–61

Bell WR, Whang JJ, Carbone PP, Brecher G, Block JB (1966) Cytogenetic and morphologic abnormalities in human bone marrow cells during cytosine arabinoside therapy. Blood 27:771 to 781

Berg K (ed) (1979) Genetic damage in man caused by environmental agents. Academic, New York San Francisco London

Berger R, Bernheim A, Feingold J, Andrieu JM (1980) The effect of levamisole on human chromosomes. Pathol Biol 28:323–324

Bille PE, Jensen MK, Jensen JPK, Poulsen JC (1975) Studies on the haematologic and cytogenetic effect of lithium. Acta Med Scand 198:281–286

Bishun N, Mills J, Parke DV, Williams DC (1975) A cytogenetic study in women who had used oral contraceptives and in their progeny. Mutat Res 33:299–310

Blank MA (1979) Cytogenetic analysis of the damaging effect of thiophosphamide on bone marrow of patients with malignant tumors. Vopr Onkol 25:72–73

Bora KC, Douglas GR, Nestmann ER (eds) (1982) Chemical mutagenesis, human population monitoring and genetic risk assessment. Elsevier, Amsterdam Oxford New York, p 364

Bornstein RS, Hungerford DA, Haller G, Engstrom PF, Yarbro JW (1971) Cytogenetic effects of bleomycin therapy in man. Cancer Res 31:2004–2007

Boue JG, Boue A (1973) Increased frequency of chromosomal anomalies in abortions after induced ovulation. Lancet I:679–680

Brögger A, Waksvik H, Thune P (1978) Psoralen/UVA treatment and chromosomes. II Analysis of psoriasis patients. Arch Dermatol 261:287–294

Buckton KE, Evans HJ (1973) Methods for the analyiss of human chromosome aberrations. WHO, Geneva

Burgdorf W, Kurvink K, Cervenka J (1977) Elevated sister chromatid exchange rate in lymphocytes of subjects treated with arsenic. Hum Genet 36:69–72

Butler MG, Sanger WG (1981) Increased frequency of sister-chromatid exchanges in alcoholics. Mutat Res 85:71–76

Cochran ST, Khodadoust A, Norman A (1980) Cytogenetic effects of contrast material in patients undergoing excretory urography. Radiol 136:43–46

Cohen MM, Hirschhorn K, Frosch WA (1967) In vivo and in vitro chromosomaldamage induced by LSD-25. New Engl J Med 277:1043–1049

Cohen MM, Hirschhorn K, Frosch WA (1969) Cytogenetic effects of tranquilizing drugs in vivo and in vitro. J Am Med Ass 207:2425–2426

Cohen MM, Gerbie AB, Nadler HL (1971) Chromosomal investigations in pregnancies following chemotherapy for choriocarcinoma. Lancet II:219

Cohen MM, Lieber E, Schwartz HN (1972) In-vivo cytogenetic effect of perphenazine and chlorpromazine: a negative study. Br Med J 3:21–23

Cohen MM, Levy M, Ehakim M (1977) A cytogenetic evaluation of long-term colchicine therapy in the treatment of familial Mediterranean fever. Am J Med Sci 274:147–152

Conen PE, Lansky GS (1961) Chromosome damage during nitrogen mustard therapy. Br Med J 2:1055–1057

Crippa L, Klein D, Linder A (1976) Etude cytogenetique et statistique doun groupe de patients traites a la butazolidine (phenylbutazone). J Genet Hum 24:1–13

Crossen PE, Morgan WF (1977a) Analysis of human lymphocyte cell cycle time in culture measured by sister chromatid differential staining. Exp Cell Res 104:453–457

Crossen PE, Morgan WF (1977b) Proliferation of PHA- and PWM-stimulated lymphocytes measured by sister chromatid differential staining. Cell Immunol 32:432–438

Crossen PE (1982) SCE in lymphocytes. In: Sandberg AA (ed) Sister chromatid exchange. Liss, New York, pp 175–193

De Gutierrez AC, Lisker R (1973) Longitudinal study of the effects of oral contraceptives on human chromosomes. Ann Genet 16:259–262

De la Torre R, Krompotic E (1976) The in vivo and in vitro effects of lithium on human chromosomes and cell replication. Teratology 13:131–138

Dick CE, Schniepp ML, Sonders RC, Wiegand RG (1974) Cyclamate and cyclohexylamine: lack of effects on the chromosomes of man and rats in vivo. Mutat Res 26:199–203

Dobos M, Schuler D, Fekete G (1974) Cyclophosphamide-induced chromosomal aberrations in nontumorous patients. Human Genet 22:221–227

Dorrance D, Janiger O, Teplitz RL (1970) In vivo effects of illicit halucinogens on human lymphocyte chromosomes. J Am Med Ass 212:1488–1491

Dorrance D, Janiger O, Teplitz RL (1975) Effect of peyote on human chromosomes. Cytogenetic study of the Huichol Indians of Northern Mexico. J Am Med Ass 234:299–302

Dresp J, Schmid E, Bauchinger M (1978) The cytogenetic effect of bleomycin on human peripheral lymphocytes in vitro and in vivo. Mutat Res 56:341–353

Düker D (1981) Investigations into sister chromatid exchange in patients under cytostatic therapy. Hum Genet 58:198–203

Eberle P, Hunstein W, Perings E (1968) Chromosomes in patients treated with imuran. Human Genet 6:69–73

Einhorn N, Eklund G, Franzen S, Lambert B, Lindsten J, Söderhäll S (1982) Late side effects of chemotherapy in ovarian carcinoma. Cancer 49:2234–2241

Eßler KJ, Kotlarek F, Habedank M, Mühler U, Mühler E (1981) Chromosomal investigations in epileptic children during long-term therapy with phenytoin or primidone. Hum Genet 56: 345–348

Falek A, Jordan RB, King BJ, Arnold PJ, Skelton WD (1972) Human chromosomes and opiates. Arch Gen Psychiatry 27:511–515

Fedortzeva RF, Dygin VP, Mamaeva SE, Goroshchenko YL (1973) Cytogenetical analysis of the effect of 6-mercaptopurine on human chromosomes. I Effects on blood cells of acute leukemia patients. Tsitologiia 15:1172–1173

Fekete G, Szollar J, Dobos M (1981) Die klinische Bedeutung der chemischen Mutagenese. Genetpädiat Arb Tgg, Budapest

Fischer P, Nacheva E, Pohl-Rühling J, Krepler P (1978) Cytogenetic effects of chemotherapy and cranial irradiation on the peripheral blood lymphocytes of children with leukemia. In: Evans HJ, Lloyd DC (eds) Mutagen-induced chromosome damage in man. University Press, Edinburgh, pp 247–257

Fitzgerald PH, Pickering HF, Ferguson DN, Hamer JW (1973) Longterm use of oral contraceptives: a study of chromosomes and lymphocyte transformation. Aust NZ J Med 3:572–575

Friedrich U, Zeuthen E (1970) Chromosomenabnormitäten und Behandlung mit Imuran (Azathioprin) nach Nierentransplantationen. Humangenet 8:289–294

Frota-Pessoa O, Ferraira NR, Pedroso MB, Moro AM, Otto PA, Chamone DAF, Da Silva LC (1975) A study of chromosomes of lymphocytes from patients treated with hycanthone. J Toxicol Environ Health 1:305–307

Fu TK, Jarvik LF, Matsuyama SS (1978) Amphetamine and human chromosomes. Mutat Res 53:127–128

Gabilondo F, Cobo A, Lisker R (1971) Anormalidades cromosomicas en patientes tratados con chlorpromazina y perfenazina: Rev Invest Clin 23:177–180

Galle G (1978) Cytogenetische und biochemische Untersuchungen bei Kindern mit Primidon-Monotherapie. MD Thesis, Erlangen

Ganner E, Osment J, Dittrich P, Huber H (1973) Chromosomes in patients treated with azathioprine. Humangenet 18:231–236

Garson OM (1981) Chromosome studies of patients on long-term lithium therapy for psychiatric disorders. Med J Aust 2:37–39

Gebhart E, Schwanitz G, Hartwich G (1969) Zytogenetische Wirkung von Vincristin auf menschliche Leukozyten in vivo und in vitro. Med Klin 64:2366–2371

Gebhart E (1970) The treatment of human chromosomes in vitro: Results. In: Vogel F, Röhrborn G (eds) Chemical mutagenesis in mammals and man. Springer, Berlin Heidelberg New York, pp 367–382

Gebhart E (1971) Chromosomenuntersuchungen bei Nuran®-Therapie. Z Kinderheilk 111:109 to 117

Gebhart E (1973) Chromosomenuntersuchungen bei Bactrim®-Therapie. Med Klin 68:878–881

Gebhart E, Schwanitz G, Hartwich G (1974) Chromosomenaberrationen bei Busulfan-Behandlung. Dtsch Med Wochenschr 99:52–56

Gebhart E (1975) Chromosomenuntersuchungen bei Bactrim®-behandelten Kindern. Z Kinderheilk 119:47–52

Gebhart E, Lösing J, Wopfner F (1980a) Chromosome studies on lymphocytes of patients under cytostatic therapy. I Conventional chromosome studies in cytostatic interval therapy. Hum Genet 55:53–63

Gebhart E, Windolph B, Wopfner F (1980b) Chromosome studies on lymphocytes of patients under cytostatic therapy. II. Studies using BUDR-labelling technique in cytostatic interval therapy. Hum Genet 56:157–167

Gebhart E (1981) Sister chromatid exchange (SCE) and structural chromosome aberration in mutagenicity testing. Hum Genet 58:235–254

Gebhart E (1982) The epidemiological approach: chromosome aberrations in persons exposed to chemical mutagens. In: Hsu TC (ed) Cytogenetic assays of environmental mutagens. Allanheld Osmun, Totowa, pp 385–408

Gebhart E, Mueller RL (1982) Cell kinetics and chromosome damage in BrDU-Labelled lymphocyte culture from patients undergoing cytostatic interval therapy. Biol Zbl 101:513–526

Gebhart E, Lösing J, Mueller RL, Windolph B (1982) Interindividual variation of cytogenetic damage by cytostatic therapy. In: Sorsa M, Vainio H (eds) Mutagens in our environment. Liss, New York, pp 89–98

Genest P, Villeneuve A (1971) Lithium, chromosomes, and mitotic index. Lancet I:1132

Gilmour DG, Bloom AD, Lele KP, Robbins ES, Maximilian C (1971) Chromosomal aberrations in useres of psychoactive drugs. Arch Gen Psychiatry 24:268–272

Giulotto E, Mottura A, Giorgi R, De Carli L, Nuzzo F (1980) Frequencies of sister-chromatid exchanges in relation to cell kinetics in lymphocyte cultures. Mutat Res 70:343–350

Goetz P, Sram RJ, Dohnalova L (1975) Relationship between experimental results in mammals and man. I Cytogenetic analysis of bone marrow injury induced by a single dose of cyclophosphamide. Mutat Res 31:247–254

Große KP, Schwanitz G, Rott HD, Wißmüller HF (1972) Chromosomenuntersuchungen bei Behandlung mit Antikonvulsiva. Humangenet 16:209–216

Hartley-Asp B (1979) Absence of chromosomal damage in the lymphocytes of patients treated with metronidazole for Trichomoniasis vulgaris. Toxicol Lett 4:15–19

Hedner K, Högstedt B, Kolnig AM, Mark-Vendel E, Strömbeck B, Mitelman F (1982) Relationship between sister chromatid exchanges and structural chromosome aberrations in lymphocytes of 100 individuals. Hereditas 97:237–245

Herha J, Obe G (1974) Chromosomal damage in chronic users of Cannabis: in vivo investigation with two-day leukocyte cultures. Pharmakopsychiatr. Neuro-Psychopharmakol 7:328–337

Herha J, Obe G (1976) Chromosomal damage in epileptics on monotherapy with carbamazepine and diphenylhydanoin. Hum Genet 34:255–263

Hill A, Wolff S (1982) Increased induction of sister chromatid exchange by diethylstilbestrol in lymphocytes from pregnant and premenopausal women. Cancer Res 42:893–896

Hopkin JM, Evans HJ (1980) Cigarette smoke-induced DNA damage and lung cancer risks. Nature 283:388–390

Husgafvel-Pursiainen K, Mäki-Paakkanen J, Norppa H, Sorsa M (1980) Smoking and sister chromatid exchange. Hereditas 92:247–250

Husum B, Wulf HC, Niebuhr E (1982) Normal sister-chromatid exchanges in oral contraceptive users. Mutat Res 103:161–164

Jaju M, Jaju M, Ahuja YR (1981) Combined action of isoniazid and para-aminosalicylic acid in vivo on human chromosomes in lymphocyte cultures. Hum Genet 56:375–377

Jarvik LF, Bishun NP, Bleiweiss H, Kato T, Moralishvili E (1971) Chromosome examinations in patients on lithium carbonate. Arch Gen Psychiatry 24:166–168

Jensen MK (1967) Chromosome studies in patients treated with azathioprine and amethopterin. Acta Med Scand 182:445–455

Jensen MK (1970) Effect of azathioprine on the chromosome complement of human bone marrow cells. Int J Cancer 5:147–151

Jensen MK, Nyfors A (1979) Cytogenetic effect of methotrexate on human cells in vivo. Comparison between results obtained by chromosome studies on bone-marrow cells and blood lymphocytes and by the micronucleus test. Mutat Res 64:339–343

Jensen MK, Rasmussen GG, Ingeberg S (1979) Cytogenetic studies in patients treated with penicillamine. Mutat Res 67:357–359

Kilian DJ, Picciano DJ (1976) Cytogenetic surveillance of industrial population. In: Hollaender A (ed) Chemical mutagens, vol 4. Plenum, New York, pp 321–339

Knuutila S, Helminen E, Knuutila L, Leist S, Siimes M, Tammisto P, Westermarck T (1977) Role of clozapine in the occurrence of chromosomal abnormalities in human bone-marrow cells in vivo and in cultured lymphocytes in vitro. Hum Genet 38:77–89

Kotlarek F, Faust J (1978) Chromosomal investigations in children with pyknolepsy on dipropylacetate monotherapy. Hum Genet 43:329–331

Kowalczyk J (1980) Sister-chromatid exchanges in children treated with nalidixic acid. Mutat Res 77:371–375

Kucerova M, Kocandrle V, Matousek V, Polivkova Z, Reneltova I (1982) Chromosomal aberrations and sister-chromatid exchanges (SCEs) patients undergoing long-term Imuran therapy. Mutat Res 94:501–509

Lambert B, Ringborg U, Harper E, Lindblad A (1978) Sister chromatid exchanges in lymphocyte cultures of patients receiving chemotherapy for malignant disorders. Cancer Treat Rep 62: 1413–1419

Lambert B, Ringborg U, Lindblad A, Sten M (1979) The effects of DTIC, Melphalan, Actinomycin D, and CCNU on the frequency of sister chromatid exchanges in peripheral blood lymphocytes of melanoma patients. In: Jones SE, Salmon SE (eds) Adjuvant therapy of cancer II. Grune, Stratton, pp 55–62

Lawler SD, Walden PAM (1978) Chromosome studies in patients treated with chemotherapy for trophoblastic tumours. In: Evans HJ, Lloyd DC (eds) Mutagen-induced chromosome damage in man. University Press, Edinburgh

Littlefield LG, Lever WE, Miller FL, Goh K (1975) Chromosome breakage studies in lymphocytes from normal women, pregnant women, and women taking oral contraceptives. Am J Obstet Gynecol 121:976–979

Littlefield LG (1982) Effects of DNA-damaging agents on SCE. In: Sandberg AA (ed) Sister chromatid exchange. Liss, New York

Locher H, Fränz J (1967) Chromosomenveränderungen bei Methotrexatbehandlungen. Med Welt 34:1965–1967

Madle S, Obe G, Schroeter H, Herha J, Pietzcker A (1980) Possible mutagenicity of the psychoactive phenothiazine derivative perazine in vivo and in vitro. Hum Genet 53:357–361

Madle S (1981) Evaluation of experimental parameters in a S9/human leukocyte SCE test with cyclophosphamide. Mutat Res 85:347–356

Mäki-Paakkanen J, Sorsa M, Vainio H (1981) Chromosome aberrations and sister chromatid exchanges in lead-exposed workers. Hereditas 94:269–275

Marquez-Monter H, Ruiz-Fargoso E, Velasco M (1970) Anticonvulsant drugs and chromosomes. Lancet II:426–427

Martin PA, Thorburn MJ, Bryant SA (1974) In vivo and in vitro studies of the cytogenetic effects of Cannabis sativa in rats and men. Teratology 9:81–86

Matsuyama S, Yen FS, Jarvik LF, Fu TK (1973) Marijuana and human chromosomes (Abstr.). Genetics 74:175

Matsuyama SS, Jarvik LF (1975) Cytogenetic effects of psychoactive drugs. In: Mendlewicz J (ed) Genetics and psychopharmacology. Karger, Basel, pp 99–132

Matsuyama SS, Yen FS, Jarvik LF, Sparkes RS, Fu TK, Fisher H, Reccius N, Frank IM (1977) Marijuana exposure in vivo and human lymphocyte chromosomes. Mutat Res 48:255–266

Matsuyama SS, Charuvastra VC, Jarvik LF, Fu TK, Sanders K, Yen FS (1978) Chromosomes in patients receiving methadone and methadyl acetate. Arch Gen Psychiatry 35:989–991

McQuarrie HG, Scott CD, Ellsworth HS, Harris JW, Stone RA (1970) Cytogenetic studies in women using oral contraceptives and their progeny. Am J Obstet Gynecol 108:659–665

Misawa S, Abe T, Takino T, Kawai K (1980) Increased frequencies of sister chromatid exchanges in lymphocytes of patients on busulfan therapy. Acta Haematol Jap 43:1–6

Mitelman F, Hartley-Asp B, Ursing B (1976) Chromosome aberrations and metronidazole. Lancet II:802

Mitelman F, Strömbeck B, Ursing B (1980) No cytogenetic effect of metronidazole. Lancet II: 1249–1250

Mitelman F, Strömbeck B, Ursing B, Nordle Ö, Hartley Asp B (1982) Metronidazole exhibits no clastogenic activity in a double-blind cross-over study on Crohn's patients. Hereditas 96:279 to 286

Monsalva MV, Frota-Pessoa O, Garcia Campos AM, Sette H (1976) A study of chromosomes of Schistosomiasis patients under oxamniquine (UK 4271) treatment. J Toxicol Environ Health 1:1023–1026

Morimoto K, Wolff S (1980) Increase of sister chromatid exchanges and perturbations of cell division kinetics in human lymphocytes by benzene metabolites. Cancer Res 40:1189–1193

Motulsky AG (1982) Interspecies and human genetic variation, problems of risk assessment in chemical mutagenesis and carcinogenesis. In: Bora KC, Douglas GR, Nestmann ER (1982) Chemical mutagenesis, human population monitoring and genetic risk assessment. Elsevier, Amsterdam Oxford New York, pp 75–83

Mourelatos D, Faed MJW, Gould PW, Johnson BE, Frain-Bell W (1977) Sister chromatid exchanges in lymphocytes of psoriatics after treatment with 8-methoxypsoralen and long wave ultraviolet radiation. Br J Dermatol 97:649–654

Müller R, Ritter C (1978) Zytogenetische Untersuchungen von Frauen während und nach der Einnahme hormonaler Kontrazeptiva. Zentralbl Gynäkol 100:347–354

Murken JD (1971) Chromosomenschäden durch Appetitanreger? Dtsch Med Wochenschr 96: 1696–1697

Murthy PBK, Prema K (1979) Sister-chromatid exchanges in oral contraceptive users. Mutat Res 68:149–152

Musilova J, Michalova K, Urban J (1979) Sister-chromatid exchanges and chromosomal breakage in patients treated with cytostatics. Mutat Res 67:289–294

Nevstad NP (1978) Sister chromatid exchanges and chromosomal aberrations induced in human lymphocytes by the cytostatic drug adriamycin in vivo and in vitro. Mutat Res 57:253–258

Nichols WW, Miller RC, Heneen W, Bradt C, Hollister L, Kanter S (1974) Cytogenetic studies on human subjects receiving marihuana and 9-tetrahydrocannabinol. Mutat Res 26:413–417

Nielsen J, Friedrich U, Tsuboi T (1969) Chromosome abnormalities in patients treated with chlorpromazine, perphenazine, and lysergide. Br Med J 3:634–637

Nordenson I, Salmonsson S, Brun E, Beckman G (1979) Chromosome aberrations in psoriatic patients treated with arsenic. Hum Genet 48:1–6

Obe G, Beek B, Dudin G (1975) The human leukocyte test system. V) DNA synthesis and mitoses in PHA-stimulated 3-day cultures. Humangenet 28:295–302

Obe G, Herha J (1975) Chromosome damage in chronic alcohol users. Humangenet 29:191–200

Obe G, Herha J (1978) Chromosomal aberrations in heavy smokers. Hum Genet 41:259–263

Obe G, Göbel D, Engeln H, Herha J, Natarajan AT (1980) Chromosomal aberrations in peripheral lymphocytes of alcoholics. Mutat Res 73:377–386

Obe G, Matthiessen W, Göbel D (1981) Chromosomal aberrations in the peripheral lymphocytes of cancer patients treated with high-energy electrons and bleomycin. Mutat Res 81:133–141

Obe G, Riedel L, Herha J (1981) Mutagenicity of antiepileptic drugs. In: Grob-Selbeck G, Doose H (eds) Epilepsy-problems of marriage, pregnancy, genetic counseling. Thieme, Stuttgart New York, pp 54–60

Obe G, Beek B (1982) The human leukocyte test system. In: De Serres F, Hollaender A (eds) Chemical mutagens, vol 7. Plenum, New York London, pp 337–400

Ohtsuru M, Ishii Y, Takai S, Higashi H, Kosaki G (1980) Sister chromatid exchanges in lymphocytes of cancer patients receiving mitomycin C treatment. Cancer Res 40:477–480

Purrott RJ, Vulpis N, Lloyd DC (1980) The use of harlequin staining to measure delay in the human lymphocyte cell cycle induced by in vitro irradiation. Mutat Res 69:275–282

Raposa T (1978) Sister chromatid exchange studies for monitoring DNA damage and repair capacity after cytostatics in vitro and in lymphocytes of leukaemic patients under cytostatic therapy. Mutat Res 57:241–251

Raposa T (1982) SCE and chemotherapy of non-cancerous and cancerous conditions. In: Sandberg AA (ed) Sister chromatid exchange. Liss, New York, pp 579–617

Reeves BR, Pickup VL, Lawler SD, Dinning WJ, Perkins ES (1975) A chromosome study of patients with uveitis treated with chlorambucil. Br Med J 4:22–23

Robinson JT, Chitham RG, Greenwood RM, Taylor JW (1974) Chromosome aberrations and LSD. A controlled study in 50 psychiatric patients. Br J Psychiatry 125:238–244

Roman IC (1969) Rat and human chromosome studies after promazine medication. Br Med J 4:172

Rossi A, Sebastio L, Ventruto V (1973) Studio cromosomico in 14 soggetti psoriasici trattati con idrossiurea o azathioprino. Minerva Med 64:1728–1732

Salamanca-Gomez F, Castaneda G, Farfan J, Del Castro Santillan M, Munoz O, Armendares S (1980) Chromosome studies of bone marrow cells from metronidazole-treated patients. Ann Genet 23:63–64

Sandberg AA (ed) (1982) Sister chromatid exchange. Liss, New York

Savage JRK (1975) Classification and relationship of induced chromosomal structural changes. J Med Genet 12:103–122

Saxena R, Ahuja YR (1982) Clastogenic effect of the psychotropic drug thioridazine on human chromosomes in vivo. Hum Genet 62:198–200

Schinzel A, Schmid W (1976) Lymphocyte chromosome studies in humans exposed to chemical mutagens. Validity of the method in 67 patients under cytostatic therapy. Mutat Res 40:139 to 166

Schuler D, Dobos M, Fekete G, Miltenyi M, Kalmar L (1979) Chromosome mutations and chromosome stability in children treated with different regimes of immunosuppressive drugs. Hum Hered 29:100–105

Schwanitz G, Gebhart E, Rott HD, Schaller KH, Essing HG, Lauer O, Prestele H (1975) Chromosomenuntersuchungen bei Personen mit beruflicher Bleiexposition. Dtsch Med Wochenschr 100:1007–1011

Selezneva T, Korman NP (1973) Analysis of chromosomes of somatic cells in patients treated with antitumor drugs. Genetika 9/12:112–118 (russ)

Sörensen PJ, Jensen MK (1981) Cytogenetic studies in patients treated with trimethoprim-sulfamethoxazole. Mutat Res 89:91–94

Stenchever MA, Frankel RS, Jarvis JA (1970) Effect of diazepam on chromosomes of human leukocytes in vivo. Am J Obstet Gynecol 107:456–460

Stenchever MA, Kunysz TJ, Allen MA (1974) Chromosome breakage in users of marihuana. Am J Obstet Gynecol 118:106–113

Stevenson AC, Bedford J, Hill AGS, Hill HFH (1971) Chromosomal studies in patients taking phenylbutazone. Ann Rheum Dis 30:487–500

Stevenson AC, Patel C (1973) Effect of chlorambucil on human chromosomes. Mutat Res 18:333–351

Stevenson AC, Clarke G, Patel CR, Hughes DTD (1973) Chromosomal studies in vivo and in vitro of trimethoprim and sulphamethoxazole. Mutat Res 17:255–260

Stevenson AC, Silcock SR, Scott JT (1976) Absence of chromosome damage in human lymphocytes exposed to allopurinol and oxipurinol. In vivo and in vitro studies. Ann Rheum Dis 35:143–147

Tolchin SF, Winkelstein A, Rodnan GP, Pan SF, Nankin HR (1974) Chromosome abnormalities from cyclophosphamide therapy in rheumatoid arthritis and progressive systemic sclerosis (scleroderma). Arthritis Rheum 17:375–382

Torigoe K, Oota U, Uchiya S, Sakai K (1978) Sister chromatid exchanges in children treated with anticonvulsants (Abstr.). Teratology 18:162

Vormittag W, Kolarz G (1979) Chromosomenuntersuchungen vor und nach Infusionstherapie mit Phenylbutazon. Arzneimittel-Forsch (Drug Res) 29:1163–1168

Walker S, Price Evans DA, Benn PA, Littler TR, Halliday LDC (1975) Phenylbutazone and chromosomal damage. Ann Rheum Dis 34:409–415

Watson WAF, Petrie JC, Galloway DB, Bullock I, Gilbert JC (1976) In vivo cytogenetic activity of sulphonylurea drugs in man. Mutat Res 38:71–80

Whang-Peng J, Leventhal BG, Adamson JW, Perry S (1969) The effect of daunomycin on human cells in vivo and in vitro. Cancer 23:113–121

White BJ, Driscoll EJ, Tjio JH, Smilack ZH (1974) Chromosomal aberration rates and intravenously given diazepam. A negative study. J Am Med Ass 230:414–417

Whorton EB, Bee DE, Kilian DJ (1979) Variations in the proportion of abnormal cells and required sample sizes for human cytogenetic studies. Mutat Res 64:79–86

Wolff-Schreiner EC, Carter DM, Schwarzacher HG, Wolff K (1977) Sister chromatid exchanges in photochemotherapy. J Invest Dermatol 69:387–391

# Mutagenic Activity of Cigarette Smoke

G. OBE[1], W.-D. HELLER[2], and H.-J. VOGT[3]

## 1 Introduction

Smoking is made responsible for a variety of cancers in man and it is estimated that
some 30% of all cancer deaths in the USA are caused by smoking (Doll and Peto 1981;
The Health Consequences of Smoking 1982; Smoking and Health 1979). The finding
that cigarette smoke is mutagenic in a variety of test systems including somatic cells
of man may be helpful to a better understanding of the mechanisms by which smoking leads to cancer. These data also support the mutation theory of the origin of cancer
which was already formulated by Boveri in 1914 and elaborated by Bauer in 1928
(see also Bauer 1963).

### 1.1 Experimental Analyses

Cigarette smoke, its condensates (CSC), and their fractions have been shown to be
mutagenic in a variety of test systems, a topic which has been reviewed recently (de
Marini 1983; Obe 1981).

Figure 1 summarizes data concerning mutagenic (Ames test), cell-transforming
and tumor-inducing activities of CSC and some of its fractions. The mutagenic potential
of CSC is considerable. 1%–5% of the CSC of one cigarette have been shown to induce
mutations in the Ames test Kier et al. (1974) and sister chromatid exchanges SCE's
in human peripheral lymphocytes PL in vitro (Evans 1981). CSC's are complex mixtures of thousands of compounds and it is not possible to say which of these compounds are responsible for the mutagenicity. Different carcinogens/mutagens have
been found in cigarette smoke and in CSC's (Van Duuren 1980; Weisburger et al.
1977; The Health Consequences of Smoking 1982; Smoking and Health 1979)
and they may act together to produce the mutagenic activities (see also Table 13).
The highly complex mixture of chemicals found in CSC's is mainly formed during
burning of the cigarette by pyrolysis. In the last years it has been found that specific compounds are formed when proteins and amino acids are pyrolyzed. Such

1 Institut für Genetik, Freie Universität Berlin, Arnimallee 5–7, 1000 Berlin 33, FRG
2 Institut für Statistik und mathematische Wirtschaftstheorie, Universität Karlsruhe, Postf. 6380,
7500 Karlsruhe 1, FRG
3 Dermatologische Klinik und Poliklinik der Technischen Universität München, Biedersteiner
Straße 29, 8000 München 40, FRG

Mutations in Man, ed. by G. Obe
© Springer-Verlag Berlin Heidelberg 1984

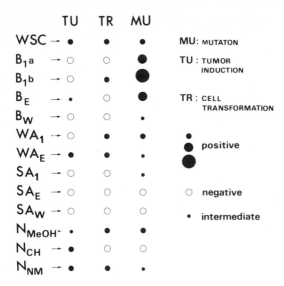

**Fig. 1.** Tumor induction (*TU*) as measured with the skin-painting test with ICR mice (Bock et al. 1969), transformation in vitro (*TR*) of different cell types (Benedict et al. 1975; Rhim and Huebner 1973) and induction of mutations in *Salmonella typhimurium* TA 1538 (frameshift mutations) in the presence of liver microsomes (Hutton and Hackney 1975; Kier et al. 1974), by different fractions of CSC (Swain et al. 1969). *CSC* Cigarette smoke condensate (unfractionated); $B_1a$ Bases, insoluble before alkaline extraction; $B_1b$ Bases, insoluble after alkaline extraction; $B_E$ Bases, ether soluble; $B_W$ Bases, water soluble; $WA_1$ Weak acids, insoluble; $WA_E$ Weak acids, ether soluble; $SA_1$ Strong acids, insoluble; $SA_E$ Strong acids, ether soluble; $SA_W$ Strong acids, water soluble; $N_{MeOH}$ Neutrals, 80% methanol soluble; $N_{CH}$ Neutrals, cyclohexane soluble; $N_{NH}$ Neutrals, nitromethane soluble. The basic fractions contain aromatic amines for which the tester strain TA 1538 is especially sensitive. The neutral fractions contain polycyclic aromatic hydrocarbons for which TA 1538 is not very sensitive, these fractions are especially active with respect to tumor induction. Even the small mutagenic activity found in fraction $N_{NM}$ cannot result from its content of benzo(a)pyrene alone. (Kier et al. (1974)

mutagens are also formed when fish and meat are broiled and when food is fried, toasted, baked, and heated (Diet, Nutrition and Cancer 1982; Inue et al. 1983; Matsushima and Sugimura 1981; Nagao et al. 1977d; Powrie et al. 1982; Stich et al. 1982; Sugimura 1981, 1982; Sugimura and Nagao 1982; Thompson et al. 1983; Tohda et al. 1980; Fabry et al. 1982, see Chap. Sobels, p. 1ff.). The mutagenic activity of CSC's is much higher than that which can be attributed to its contents of benzo(a)-pyrene [B(a)P]. This holds true for the SCE-inducing capacity in mammalian cells (Hopkin and Perry 1980) and for point mutations in *Salmonella* (Hutton and Hackney 1975; Sato et al. 1977). The mutagenicity of CSC's is up to 20,000 times higher than that expected from its B(a)P contents (Evans 1981; Sugimura et al. 1977). B(a)P is also not responsible for the mutagenic potential of the pyrolysis products from tryptophan and lysine hydrochloride (Matsumoto et al. 1977). Smoke condensates from broiled fish have higher mutagenic activities in *Salmonella* than would be expected from its B(a)P content (Sugimura et al. 1977).

Two compounds which are formed during the pyrolysis of tobacco, but which also occur as natural products (see Madle et al. 1981), are the β-carboline derivatives, harman and norharman. These compounds are especially interesting in that under some experimental conditions they have been shown to be comutagenic in *Salmonella*, without being mutagenic alone (Nagao et al. 1977a,b,c; Sugimura 1979; Sugimura and Nagao 1980; Suzuki et al. 1983; Umezawa et al. 1978; Wakabayashi et al. 1981; Riebe et al. 1982; see also Levitt et al. 1977; White and Rock 1981). With respect to the induction of SCE's such an activity of the β-carbolines was not found (Takehisa and Kanaya 1982).

The finding that pyrolysis of proteins and amino acids produces mutagenic compounds is in accord with the observation that the mutagenic activity of smoke condensates from different tobaccos is correlated with the contents of the tobacco leaves of protein and soluble nitrogen from amino acids and alkaloids (Mizusaki et al. 1977a). Nicotine itself is not mutagenic in *Salmonella* (Hutton and Hackney 1975; Mizusaki et al. 1977a) and in *Saccharomyces cerevisiae* (Gairola 1982), but may form mutagenic compounds when pyrolyzed. High sugar content of the tobacco leaves leads to a lower mutagenic activity of the smoke condensates (Mizusaki et al. 1977b), the same holds true when cigarettes are soaked in sugar solutions before they are smoked (Sato et al. 1979). Findings of this type may offer possibilities to make cigarettes which give rise to less mutagenic smoke.

## 1.2 Analyses in Man

Analyses concerning the possible mutagenic activity of cigarette smoke in man have been performed by cultivating PL from smokers S and nonsmokers NS. The results concerning SCE's are compiled in Table 1a and b. An elevation of the rate of chromosomal aberrations CA in S as compared to NS has been found in different analyses (Evans 1981; Fredga et al. 1982; Hüttner and Schöneich 1981; Obe and Herha 1978; Obe et al. 1980, 1982; Vijayalaxmi and Evans 1982). In an unpublished study we found no elevation of the rate of CA of the exchange type in cord blood of neonates of 30 smoking mothers are compared to neonates of 17 nonsmoking mothers. Hedner et al. (1983) found no elevation of CA in the PL of 47 S as compared to 44 NS, but these authors used an inadequate culture time of 72 h.

Our own analysis concerning the effect of cigarette smoke on human PL is the most extensive one and some of the results have been published recently (Obe et al. 1982). In the following we want to present the full and reevaluated data of this analysis.

## 2 Selection of Subjects and Culture Conditions

The selection of the subjects is described in Chap. Vogt et al., p. 247ff. We analyzed 170 S and 124 NS for CA in cultures set up with Ham's F-10 medium. Some cultures were set up with Minimal Essential Medium (MEM) with Earle's salts and with Tissue Culture Medium 199 (M 199) with Hank's salts. The following genetic end-points were analyzed (see Table 2):

**Table 1a.** SCE's per cell found in peripheral lymphocytes of smokers (S) and nonsmokers (NS): elevation of SCE rates in smokers

| Ref. | SCE's per cell (number of probands) | |
|------|------|------|
| | S | NS |
| Lambert et al. (1978) | 15.2  (9)[a] <br> 17.2  (15)[b] | 13.2   (37) |
| Murthi (1979) | 7.7  (20) | 6.4   (12) |
| Hopkin and Evans (1980) | 8.38 (10) | 7.43  (10) |
| Husgafvel-Pursiainen et al. (1980) | 9.6  (43) | 8.1    (40) |
| Husum et al. (1981) | 10.49 (45) | 9.26  (86) |
| Hüttner and Schöneich (1981) | 15.24 (23) | 12.50  (32) |
| Fredga et al. (1982) | 10.86  (6) <br> 9.94  (6) | 7.56   (6) <br> 7.39   (6) |
| Lambert et al. (1982a) | 16.3  (91) | 13.5  (128) |
| Meiying et al. (1982) | 8.33  (6) | 4.41   (6) |
| Wulf et al. (1982) | 9.60 (44)[c] <br> 9.23 (25)[d] <br> 9.00 (24)[e] <br> 9.37 (22)[f] <br> 10.45 (26)[g] <br> 9.63 (29)[h] | 8.25  (60) |
| Livingston and Fineman (1983) | 10.80 (24) | 8.46   (24) |

[a] Less than 10 cigarettes per day
[b] Ten and more cigarettes per day
[c] High tar cigarettes with filter
[d] High tar cigarettes without filter
[e] Low tar cigarettes with filter
[f] Cannabis and cigarettes
[g] Cheroots
[h] Pipe

**Table 1b.** SCE's per cell found in peripheral lymphocytes of smokers (S) and nonsmokers (NS): no elevation of SCE rates in smokers

| Ref. | SCE's per cell (number of probands) | |
|------|------|------|
| | S | NS |
| Ardito et al. (1980) | 8.13 (10) | 8.14 (10) |
| Crossen and Morgan (1980) | 10.56 (35) | 10.02 (85) |
| Fredga et al. (1982) | 10.46  (6) <br> 8.56  (6) | 7.76  (6) <br> 7.27  (6) |
| Hollander et al. (1978) | 11.70 (69) | 11.60  (6) |
| Seshadri et al. (1982) | 11.88 (23)[a] <br> 9.01 (25)[b] | 11.35 (30)[a] <br> 8.95 (30)[b] |
| Hedner et al. (1983) | 9.70 (47) | 9.10 (44) |

[a] Smoking and nonsmoking mothers
[b] Cord blood from smoking and nonsmoking mothers

**Table 2.** Culture media, culture times in h and genetical end-points analyzed in smokers (S) and nonsmokers (NS)

| Medium | Culture time | Number of probands | | Genetical end-point |
|--------|--------------|----|----|---------------------|
| | | S | NS | |
| Ham's F-10 without BUdR without L-glutamine | 48 | 170 | 124 | Exchange type of aberrations in the chromosomes: Dicentric chromosomes (DIC) Ring chromosomes (RING) Minutes (MIN) |
| MEM without BUdR with L-glutamine | | 47 | 24 | Chromatid exchanges (RB') |
| M 199 without BUdR with L-glutamine | | 44 | 58 | In addition to chromosomal aberrations (CA) mitotic indices (MI) were determined |
| Ham's F-10 with BUdR without L-glutamine | 48 | 141 | 99 | Frequencies of 2nd in vitro metaphases, i.e , metaphases with differentially stained chromosomes (M2) |
| Ham's F-10 without BUdR without L-glutamine | 72 | 95 | 39 | Frequencies of cells containing micronuclei (MN) |
| M 199 with BUdR with L-glutamine | 72 | 24 | 20 | Frequencies of sister chromatid exchanges (SCE) |

1. Exchange type of aberrations: Dicentric chromosomes (DIC), ring chromosomes (RING), minutes (MIN), chromatid translocations (RB').
2. SCE.
3. Micronuclei (MN).
4. 2nd in vitro metaphases in 48-h culture (M2).
5. Mitotic indices (MI) in 48-h cultures.

CA were determined by scoring 200 metaphases in Ham's F-10 and in MEM cultures and 100 metaphases in M 199 cultures, whenever possible. The frequencies were normalized to the basis of 1,000 metaphases and these were used to calculate the total aberration rates (aberrations per 1,000 metaphases). The SCE frequencies were determined by scoring 40 metaphases. The MN were determined by scoring 1,000 cells, the M2 by scoring 100 metaphases, and the MI by scoring 1,000 cells.

## 2.1 Statistical Analyses

SPSS 8.1 (Beutel et al. 1980) as well as our own computer programs were used. Comparisons of different populations (S versus NS) for the CA and MN data were carried out using the Mann-Whitney U test. The comparison of the SCE data was done using the t-test. Goodness of fit tests were used for the distribution of the numbers of CA. Analyzing trends in proportion, Armitage's chi square test for linear trend was used (Armitage 1971).

## 3 Results

The results concerning the frequencies of RB' plus DIC, M2, MN, and SCE's can be seen in Table 3. There is a significantly higher rate of RB' plus DIC in Ham's F-10 cultures in S than in NS. There is no significance when the blood was cultured in MEM or M 199. S have significantly less M2 as NS. There is no significant difference between S and NS with respect to MN (Fig. 2). S have one SCE per metaphase more than NS and this is significant at the 5% level (Fig. 3). In the following we want to discuss some of these results in more detail.

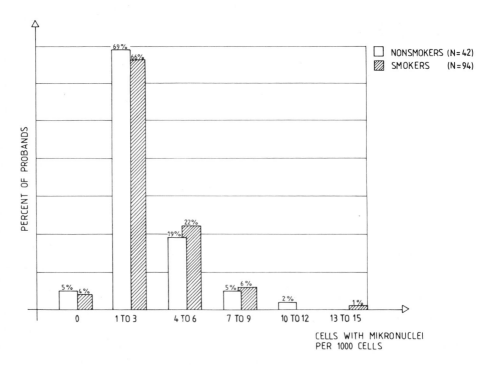

**Fig. 2.** Percent of smokers and nonsmokers with different frequencies of cells with micronuclei (per 1,000 cells) as obtained in 72-h Ham's F-10 cultures

**Table 3.** Effect of smoking on human peripheral lymphocytes in vivo. (Data from Obe et al. 1982 and unpublished)

| Type of culture | Incubation time (h) | Biological end-point analyzed | Number of metaphases (M) or of cells (C) analyzed, number of probands | | Frequencies of biological end-points | | Statistical analyses, differences between smokers and non-smokers |
|---|---|---|---|---|---|---|---|
| | | | Smokers (S) | Nonsmokers (NS) | S | NS | |
| Ham's F-10 without BUdR | 48 | RB' + DIC | 32,118 (M) 170 | 23,831 (M) 124 | 1.418[a] | 0.686[a] | s (1%) |
| MEM without BUdR | 48 | RB' + DIC | 8,056 (M) 47 | 3,764 (M) 24 | 0.787[a] | 0.667[a] | ns |
| TC 199 without BUdR | 48 | RB' + DIC | 4,310 (M) 44 | 5,576 (M) 58 | 0.909[a] | 1.776[a] | ns |
| Ham's F-10 with BUdR | 48 | M2 | 14,100 (M) 141 | 9,850 (M) 99 | 1.738[b] | 2.576[b] | s (5%) |
| Ham's F-10 without BUdR | 72 | MN | 95,000 (C) 95 | 39,000 (C) 39 | 3.120[c] | 2.970[c] | ns |
| M 199 with BUdR | 72 | SCE | 867 (M) 24 | 710 (M) 20 | 5.775[d] | 4.793[d] | s (5%) |

[a] Sum of chromatid exchanges and dicentric chromosomes per 1,000 metaphases
[b] In percent
[c] Per 1,000 cells
[d] Per metaphase
S = Significant
NS = Not significant

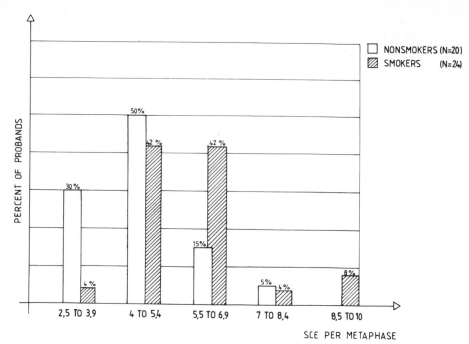

**Fig. 3.** Percent of smokers and nonsmokers with different SCE frequencies as obtained in 72-h M 199 cultures

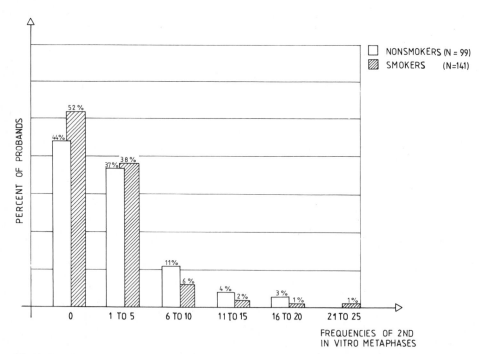

**Fig. 4.** Percent of smokers and nonsmokers with different frequencies of 2nd in vitro metaphases in 48-h Ham's F-10 cultures

## 3.1 M2 and MI

As can be seen in Fig. 4, there are more S in the O class with no M2 at all. In the class 1%–5% there is no difference between S and NS. With the exception of the highest class with 20%–25% M2 which was not represented in the NS, the S had always less M2 than the NS.

MI were determined in 48 h M 199 cultures from 46 S and 61 NS, the results gave no significant differences (73.4 mitoses in 1,000 cells in S and 60.9 in NS, t-test at the level of significance of 5%). A comparison of the MI with the M2 of S and NS can be seen in Fig. 5 (the number of cases in each entry is 86). There is no correlation of these parameters, the coefficient of correlation being 0.05. This analysis shows that the frequencies of M2 in Ham's F-10 cultures are independent of the MI in M 199 cultures.

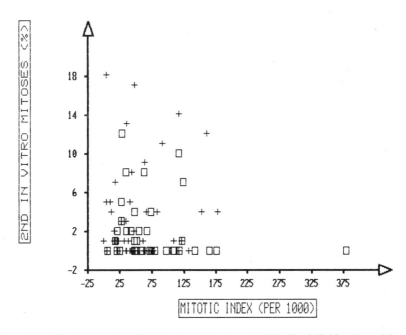

**Fig. 5.** Scatter plot of the frequencies of 2nd in vitro metaphases in 48-h Ham's F-10 cultures (+) and of mitotic indices in 48-h M 199 cultures (mitoses per 1,000 cells) (□) from 37 smokers and 49 nonsmokers

## 3.2 CA

### 3.2.1 Ham's F-10 Cultures

The frequencies of all aberration types analyzed are higher in S as compared to NS (Tables 3 and 4, Fig. 6). The difference is significant when RB′ and DIC are pooled and for all aberrations (Obe et al. 1982).

**Table 4.** Aberrations per 1,000 metaphases (in brackets absolute numbers of aberrations) found in cultured peripheral lymphocytes from smokers (S) and nonsmokers (NS) in dependency on the culture medium used

| Aberration type | Culture media used | | | | | |
|---|---|---|---|---|---|---|
| | Ham's F-10 | | MEM[a] | | TC 199[a] | |
| | S | NS | S | NS | S | NS |
| RB′ | 0.77 (25) | 0.36  (9) | 0.34  (3) | 0.00 (0) | 0.23 (1) | 0.69  (4) |
| DIC | 0.65 (21) | 0.32  (8) | 0.45  (4) | 0.67 (2) | 0.68 (3) | 1.09  (6) |
| RING | 0.52 (17) | 0.16  (4) | 0.34  (3) | 1.50 (3) | 0.45 (2) | 0.52  (3) |
| MIN | 0.13  (4) | 0.08  (2) | 0.26  (2) | 0.00 (0) | 0.23 (1) | 0.00  (0) |
| ALL | 2.07 (67) | 0.93 (25) | 1.38 (12) | 2.17 (5) | 1.59 (7) | 2.29 (13) |

[a]   The blood from different groups of probands was cultivated in these media

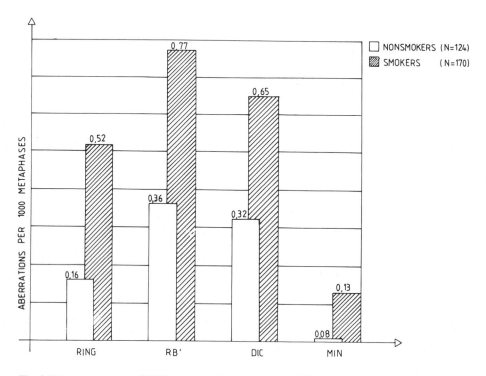

**Fig. 6.** Ring chromosomes (*RING*), chromatid translocations (*RB′*), dicentric chromosomes (*DIC*), and minutes (*MIN*) from 48-h Ham's F-10 cultures of smokers and nonsmokers

### 3.2.2  Time-Effect Relationship

There was no dependency of the CA on the duration of smoking. According to the time intervals used, an upward or downward trend of the CA frequencies was found. In no case did we find a significance (data not shown).

### 3.2.3 Dose-Effect Relationship

No significant relationship was found between CA frequencies and the number of cigarettes smoked per day by current smokers (data not shown). There was, however, a significant relationship between the frequencies of CA and the estimated daily uptake of condensate ($P = 0.02$, using a chi-squared test for trend) (Table 5). The significance persists even when S of hand-rolled cigarettes were excluded (data not shown).

**Table 5.** Dependency of the aberration frequencies in the smokers on the amount of condensate (computed by multiplying the amount of condensate of the type of cigarette smoked with the amount of cigarettes smoked per day). The dose-effect relationship is significant at the 2% level as tested with the chi square test for trend, using the data on the percentage of probands having aberrations in the different condensate groups

| Condensate (mg/day) | Aberrations per 1,000 metaphases (number of smokers) | Percent of probands having aberrations |
|---|---|---|
| 100–200 | 1.03 (39) | 18 |
| 200–300 | 1.77 (47) | 28 |
| 300–400 | 2.67 (36) | 44 |
| > 400[a] | 2.72 (43) | 37 |

[a] Including smokers of hand-rolled cigarettes

### 3.2.4 Inter-individual Distribution of CA

With the data set generated from 200 metaphases each from Ham's F-10 cultures we analyzed the interindividual distribution of the CA. For S and NS the distributions are nearly identical with the theoretical Poisson distributions. The 95% confidence intervals for $\lambda$ S and $\lambda$ NS are separate from each other and by this confirm a significant difference between S and NS (Table 6).

### 3.2.5 CA and Culture Media

Unexpectedly there was no significant differences in the CA from S and NS when cultures were set up with MEM or M 199 (Tables 3 and 4). This could be a chance result because of the smaller number of probands analyzed with these media. A comparison of the CA frequencies obtained with the three media can be seen in Table 7. With respect to Ham's F-10 there is no significant difference between S and NS with the blood also cultured in MEM, but numerically there is a doubling of the aberration rates in S as compared to NS (entry 1 in Table 7). With respect to M 199 the respective Ham's F-10 cultures gave significant differences between S and NS (entry 2 in Table 7). When the data obtained with such Ham's F-10 cultures are pooled which have also been obtained with the other media, there are again significant differences between S and NS (entry 3 in Table 7).

**Table 6.** Interindividual distribution of exchange type of aberrations as compared with the respective Poisson distribution. Data are taken from Ham's F-10 cultures in which 200 metaphases were analyzed and refer to aberrations in 200 metaphases

| Number of aberrations | Nonsmokers (N = 108) | | Smokers (N = 130) | |
|---|---|---|---|---|
| | Incidence observed (percent of probands) | Incidence expected | Incidence observed (percent of probands) | Incidence expected |
| 0 | 89 (82) | 87.3 | 83 (64) | 82.6 |
| 1 | 15 (14) | 18.6 | 39 (30) | 37.5 |
| 2 | 4 (4) | 2.0 | 6 (5) | 8.5 |
| 3 | 0 | 0.1 | 1 (0.7) | 1.3 |
| 4 | 0 | 0 | 0 | 0.2 |
| 5 | 0 | 0 | 1 (0.7) | 0.01 |
| $\lambda$ | 0.213 | | 0.454 | |
| 95% confidence interval | 0.142–0.320 | | 0.352–0.586 | |

**Table 7.** Aberrations per 1,000 metaphases (RB' + DIC + RING + MIN) found in cultures incubated with different media in smokers (S) and nonsmokers (NS). In brackets the numbers of probands are given

| Entry | Media used | S | NS | Statistics |
|---|---|---|---|---|
| 1 | Ham's F-10: blood samples which have also been cultivated in MEM | 1.95 (41) | 0.91 (22) | ns |
| | MEM | 1.46 (41) | 1.86 (22) | ns |
| 2 | Ham's F-10: blood samples which have also been cultivated in M 199 | 2.20 (44) | 0.96 (57) | s (1%) |
| | M 199 | 1.59 (44) | 2.33 (57) | ns |
| 3 | Ham's F-10: blood samples which have also been cultivated in M 199 or in MEM | 2.08 (85) | 0.95 (79) | s (1%) |
| | MEM | 1.46 (41) | 1.86 (22) | ns |
| | M 199 | 1.59 (44) | 2.33 (57) | ns |
| | MEM and M 199 | 1.53 (85) | 2.20 (79) | ns |

## 3.2.6 Confounding Factors

Forty percent of the NS were actually ex-smokers but this had no significant influence on the CA in the group of NS, passive smoking was likewise without influence on the CA in NS. In S and NS the following factors were without influence on the CA rates: Consumption of alcohol, chronic intake of drugs and illicit drug use, unusually high exposure to X-rays, viral infections (hepatitis, measles, herpes, mononucleosis) (see Obe et al. 1982). All the variables which have been tested for a possible influence on CA frequencies have been asked for in the questionnaire which each participant had to answer (see Chap. Vogt et al., p. 247ff.).

*Mumps.* Correlating the CA frequencies with the variable mumps we obtained an unexpected result. The frequencies of CA and M2 were no longer different between S and NS (Obe et al. 1982). One possible explanation of this result may be that some of the probands simply gave wrong answers concerning a previous mumps infection. Since we still had frozen serum probes of our probands, a mumps hemagglutination inhibition test (in the following called serological test) was run (Table 8). The comparison of these results with that from the questionnaires can be seen in Table 9. Only some 50% of the probands gave a correct answer, most of the others incorrectly thought they had no mumps infection. Table 10 shows the frequencies of CA and of

Table 8. Percent of probands having had a mumps infection as revealed by a serological test. The number of probands is given in brackets

| Mumps infection | Smokers | Nonsmokers | All |
|---|---|---|---|
| No | 18 (33) | 18 (24) | 18 (57) |
| Yes | 82 (152) | 82 (109) | 82 (261) |

Table 9. Comparison of the results concerning a mumps infection as obtained from the questionnaire (MQ) or from the serological test (MS). Percent of probands and in brackets number of probands

| Comparisons | Smokers | Nonsmokers | All |
|---|---|---|---|
| MQ same as MS | 48 (88) | 53 (70) | 50 (158) |
| MQ positive, MS negative | 4 (8) | 5 (7) | 5 (15) |
| MQ negative, MS positive | 48 (89) | 42 (56) | 46 (145) |

Table 10. Aberrations per 1,000 metaphases and frequencies of 2nd in vitro metaphases in smokers (S) and nonsmokers (NS) with (+) and without (−) mumps infection as revealed by a serological test. The number of probands is given in brackets

| Genetical end-point tested | + | − | Statistics |
|---|---|---|---|
| **RB′ + DIC + RING + MIN** | | | |
| S | 2.20 (134) | 1.55 (29) | ns |
| NS | 0.79 (95) | 1.73 (23) | s (5%) |
| Statistics | s (1%) | ns | |
| **RB′ + DIC** | | | |
| S | 1.46 (134) | 1.21 (29) | ns |
| NS | 0.58 (95) | 1.30 (23) | ns |
| Statistics | s (1%) | ns | |
| **2nd in vitro metaphases** | | | |
| S | 1.64 (112) | 2.45 (22) | ns |
| NS | 2.74 (78) | 2.27 (15) | ns |
| Statistics | s (2%) | ns | |

M2 in S and NS correlated with the results of the serological tests. There are significant differences between S and NS in the mumps group and no differences in the small group of probands without mumps.


## 4 Discussion

Our results support the findings of others that the rate of CA and SCS's is higher in S than in NS. The differences between S and NS is less impressive as in other studies.

### 4.1 M2

No satisfying explanation can be given for the finding of S having less M2 than NS (Table 3). It is known that in vivo exposure to mutagenic agents leads to a retardation of the cell cycle of the PL in vitro (see Chap. Gebhart, p. 198ff.). Our results may therefore be an indication that cigarette smoke is mutagenic in man. Smoking influences the immune system (Smoking and Health 1979) and this may have some effect on the M2 frequencies. T lymphocytes from S and NS are not different with respect to their stimulation in vitro, but lymphocytes obtained from bronchopulmonary lavage of S have a diminished response to PHA as compared to NS (Daniele et al. 1977). Watersoluble fractions of cigarette smoke and CSC's suppress the stimulation of PL from rabbits (Roszman et al. 1975) and from humans (Hopkin and Steel 1980). We do not know what factors are responsible for the presence of some M2 in 48-h cultures. There are already some "stimulated" lymphocytes in the peripheral blood (Trepel 1976) and it is possible that these cells give rise to M2; these cells may be influenced by smoking. There is no correlation between M2 in Ham's F-10 cultures and MI in M 199 cultures (Fig. 5).

### 4.2 MN

With respect to MN there is no difference between S and NS (Table 3, Fig. 2). This result is not unexpected because of the low aberration frequencies. Only some of the aberrations found can be expected to give rise to MN (see Obe and Beek 1982a,b; Beek et al. 1980). The MN test is clearly inferior to the CA test in cases where the CA frequencies are low.

### 4.3 SCE

S had one SCE per metaphase more than NS (Table 3, Fig. 3). As can be seen from Table 1, such small elevations of the SCE rates in S is not unusual. It seems that the lesions leading to SCE's are repaired in a shorter time than the lesions leading to CA, which would explain the clear elevation of the CA frequencies and the only slight effect with respect to SCE's (Anderson et al. 1981). There are data indicating a positive correlation between the number of cigarettes smoked and SCE frequencies, supporting the SCE-inducing influence of smoking in man (Lambert et al. 1978, 1982b; Hüttner and Schöneich 1981).

## 4.4 CA in Ham's F-10 Cultures

Another study in which moderate smokers have been analyzed is the one by Vijaya-laxmi and Evans (1982). Blood from 41 NS and 55 S was cultivated for 48 h in Ham's F-10 medium supplemented with 15% fetal bovine serum (we used 10%) and with 1% glutamine (we used no glutamine). One hundred metaphases were analyzed in each proband and there were 0.02% RB′ and 0.14% DIC in NS and 0.22% RB′ and 0.42% DIC in S, the differences were significant at the 1% level (chi squared statistics). S had 4 times more exchange type of aberrations than NS, and this is clearly more than in our own study (Table 3, Fig. 6). In view of the fact that the Edinburgh smokers smoked less cigarettes than the Munich smokers ["Few individuals smoked 20 cigaret-tes per day, the majority smoking 10 cigarettes or less ...", Vijayalaxmi and Evans (1982)] (the mean cigarette consumption in our smokers was 20.4), there is a dis-crepancy in the results of the two studies.

In our earlier study with 20 heavy smokers, the mean age of the probands was 43.8 years (mean age of the Edinburgh group 47 years). Of these probands, 19 had smoked 40 and more cigarettes per day for at least 9 years. One smoker reported having smoked 10 cigarettes and 8 cigars per day for more than 40 years. Twelve probands were under treatment either as in- or out-patients for different types of vascular diseases (Obe and Herha 1978). Calculating the CA frequencies in the same way as in this paper, we found 7.23 aberrations per 1,000 metaphases (Table 11). Additionally we analyzed the chromosomes from 12 probands (mean age 50 years, duration of smoking 10 years and more, mean 23.8 years; amount of cigarettes smok-ed more than 20 per day) (Obe and Herha unpublished). Ten of these probands were included in the data given by Obe (1981) and by Obe and Beek (1982a) and were there erroneously described as smoking 40 cigarettes and more per day. Only 5 of these smokers reported having smoked 40 cigarettes, the others were smoking 25-40, more than 30, 20-30, 30, and 25. The two additional smokers included here reported having smoked 25 and more than 20 cigarettes per day for 30 and 40 years, respec-

**Table 11.** Exchange type of aberrations per 1,000 metaphases in 20 heavy smokers (I: Obe and Herha 1978; number of metaphases analyzed = 3,823) and in 12 heavy smokers (II: Obe and Herha, unpublished; number of metaphases analyzed = 2,322) and in both groups of smokers (III). In both analyses Ham's F-10 medium was used, the culture time was 48 h

| Aberration types | Aberrations per 1,000 metaphases (numbers of aberrations found) | | |
| --- | --- | --- | --- |
| | I | II | III |
| RB′ | 3.98 (13)[a] | 1.25  (3) | 2.96 (16) |
| DIC | 2.50 (10) | 2.08  (5) | 2.34 (15) |
| RING + MIN | 0.75  (3) | 1.25  (3) | 0.94  (6) |
| ALL | 7.23 (26) | 4.58 (11) | 6.24 (37) |

[a]  One RB′B″ (triradial) was found

tively. With only one exception the S in this group were all internal medicine in-patients. In these 12 smokers we found 4.58 CA per 1,000 metaphases (Table 11). Pooling the data from both groups of heavy smokers results in 6.24 CA per 1,000 meta-phases, this is 3 times higher than the frequencies reported in this paper.

The mean age of the probands in the Edinburgh study and in the study with heavy smokers from our laboratory is higher than the mean age of the S reported in the present study. In human PL it has been shown that the capacity to repair UV damage is negatively correlated with the age of the blood donors (Lambert et al. 1977, 1979; Yew and Johnson 1979). The finding that chromosomal aberrations tend to increase with age (Hedner et al. 1982) may also be interpreted in the sense of a diminished repair. In our investigation of alcoholics we found no age dependency of the CA fre-quencies (Obe et al. 1980). Apart from age, different confounding factors can be expected to influence the outcome of CA studies with human PL.

Even the results of control studies compiled in Table 12 show a considerable varia-tion. Most of the "controls" have not been controlled for smoking, to mention only one factor which may add to the variability.

## 4.5 Time-Effect Relationship

No time-effect relationship was found for the CA frequencies. The cause of this negative result could be that only relatively wide time spans were analyzed. Only 20 S fell in the time span of 1–3 years, which would be important to subdivide.

Table 12. Control data concerning the frequencies of chromatid exchanges (RB') and dicentric chromosomes (DIC) in human peripheral lymphocytes cultivated for 2 days in vitro

| Entry | Number of probands | Number of metaphases | $RB' \times 10^{-4}$ (Number) | $DIC \times 10^{-4}$ (Number) | Source |
|---|---|---|---|---|---|
| 1 | 83 | 14,652 | 8.19 (12) | 3.41 (5) | Obe and Beek (1982a) |
| 2 | 591 | 24,011 | 2.08 (5) | – | |
| 3 | 20 | 9,000 | 3.33 (3) | 5.60 (5) | This laboratory |
| 4 | 16 | 3,200 | 12.50 (4) | 3.13 (1) | |
| 5 | 124 | 23,831 | 3.78 (9) | 3.36 (8) | This paper |
| 6 | 41 | 4,100 | 2.44 (1) | 14.63 (6) | Vijayalaxmi and Evans (1982) |
| 7 | 105 | 16,076 | 1.24 (2) | –[a] | Ivanov et al. (1978) |
| 8 | 79 | 8,000 | 2.50 (2) | 3.75 (3) | Natarajan (unpublish-ed) |
| 9 | 1,305 | 136,340 | – | 5.57 (76) | Lloyd et al. (1980) |
| 1–9 | | | 3.69 (38) | 5.22 (104) | |

[a] Data for DIC are included in (9)

(5) and (6) are nonsmokers, all the other entries are not controlled for smoking habits
Entry (9) includes data from 54 analyses. The data from Obe and Herha (1978) which are includ-ed in entry (1) and from two investigations resulting in extreme values of $0.91 \times 10^{-4}$ and $21.20 \times 10^{-4}$ DIC are omitted. Inclusion of the latter two data sets results in entry (9) in 1,929 pro-bands and 197,497 metaphases with a frequency of $8.15 \times 10^{-4}$ (161) DIC

### 4.6 Dose-Effect Relationship

We found a significant dose-effect relationship between the CA frequencies and the estimated daily uptake of condensate (Table 5). It should be noted that the highest tar delivery group (400 mg/day) includes 29 S (67%) of hand-rolled cigarettes, the condensate content of which can only be estimated with considerable uncertainty. Nevertheless, the remaining three dose groups still show a dose-effect relationship. This result fits the idea that the CA are induced by mutagenic compounds in the cigarette smoke which occur mainly in the condensate. Our result is also in line with the epidemiological finding that high tar cigarettes are more cancerogenic than low tar cigarettes (Doll and Peto 1981; The Health Consequences of Smoking 1982). A simple interpretation of the dose-effect relationship is difficult. It has been shown that many smokers of "light" cigarettes intensify the puff volume drawn from the cigarette and inhale more deeply (Herning et al. 1981; Hill and Marquardt 1980; Russell et al. 1980), or block the holes in the filter and thus elevate the amount of smoke ingested (Kozlowski et al. 1980). It seems that these behavioral attitudes in the smokers cannot fully compensate for the less tar in low-tar cigarettes (see The Health Consequences of Smoking 1982).

### 4.7 Inter-individual Distribution of CA

The inter-individual distribution of CA follows a Poisson distribution, indicating that there are no hyper- or hyposensitive probands in our groups of S and NS with respect to the induction of CA (Table 6). The induction of CA seems to be a chance result.

### 4.8 CA in Cultures Set Up with MEM or M 199

Culturing PL in these media does not result in a higher rate of CA in S as compared to NS (Tables 3 and 4). Husgafvel-Pursiainen et al. (1980) cultured PL from 31 S and 20 NS for 50 h in M 199 and found no significant difference in the CA frequencies. Probably this is also a result of the medium used. An explanation of the medium effect cannot be given. Among other differences the media used are different in the amounts of components which are important for the cellular DNA synthesis, namely thymine or thymidine (T) and folic acid (F). The amount of these components in mg/l medium is: Ham's F-10 = 0.73 T and 1.32 F; MEM = 0.00 T and 1.00 F; M 199 = 0.30 T and 0.01 F. The expression of heritable fragile sites and "spontaneous" CA measured as frequencies of MN is dependent on the culture medium, in that an increase of these events is found when the medium is deficient in T and F (Beek et al. 1983; Jacky et al. 1983; Sutherland 1983; see Chap. Obe and Beek, p. 177ff.). These findings do not compare with our results directly, because the expression of fragile sites and spontaneous CA is only found after a prolonged culture time. One possibility of explaining our results could be that in the deficient media the cellular repair is disturbed, leading to more open breaks and to less exchange type of aberrations. Since we did not analyze open breaks we cannot decide whether this is true. Most of the chromosome type of aberrations originate already in the $G_0$ PL, the frequencies of these aberrations should not be influenced by the culture medium. The RB' originate during the first S-phase in vitro and may thus be influenced by the medium. Our data do not support this (Table 4). Another possible explanation of the medium effect could be that different subpopulations of lymphocytes are stimulated in the different media (see Chap. Obe and Beek, p. 177ff.).

Recently Gundy and Varga (1983) reported on "spontaneous" CA frequencies in 175 blood donors (17,500 metaphases scored). PL were cultivated in M 199 for 48 h and there were 12 DIC ($6.86 \times 10^{-4}$) and no rings and RB'. The absence of RB' is unexpected and does not fit to the data presented in Table 12. Probably this is also an effect of the medium used.

## 4.9 Confounding Factors

The example mumps shows that confounding factors are a great problem with this type of analysis (Tables 8–10). The highly variable results with "control" populations underline this point (Table 12).

With respect to CA frequencies we found no significant difference between ex-smokers (ES, N = 51) and never smokers (NS, N = 71). Nevertheless there seems to be a tendency of ES having lower CA frequencies than the NS (RB' = 0.392; 0.352; DIC = 0.196; 0.394; RING = 0.0; 0.282; MIN = 0.0; 0.141; all aberrations = 0.588; 1.169). Comparing the CA frequencies in NS with those in current smokers (CS; N = 170; 2.06 aberrations per 1,000 metaphases) we found a significant difference ($P \leqslant 0.05$), but there is no difference between the CA frequencies in NS and smokers of manufactured cigarettes (SM; N = 144; 1.86 aberrations per 1,000 metaphases). The CA frequencies are significantly different between NS and the smokers of hand-rolled cigarettes (SH; N = 26; 3.19 aberrations per 1,000 metaphases, $P \leqslant 0.02$). The low CA frequency in ES as compared to NS, though not significant, is nevertheless striking and this aspect should be analyzed further.

## 5 Is There a Genetic Risk Associated with Cigarette Smoking?

In their 1979 report to the International Commission for Protection against Environmental Mutagens and Carcinogens (ICPEMC) entitled: *Cigarette smoking – does it carry a genetic risk?*, Bridges, Clemmesen, and Sugimura came to the conclusion "... that cigarette smoking may prove a significant genetic hazard for the children of smokers and for subsequent generation. This hypothesis should be subjected to further examination without delay." (Bridges et al. 1979, p 79). Since this report appeared, additional information has been generated adding more weight to the conclusion. Convincing epidemiological evidence for a genetic hazard though cigarette smoking is still missing. The only hint is the report by Mau and Netter (1974) on a connection between paternal smoking and an elevation of perinatal mortality, which was already discussed in the report to ICPEMC. The higher rate of spontaneous abortions in smoking as compared to nonsmoking mothers (see Smoking and Health 1979) may be rather an epigenetic effect. Many of the known cancerogenic compounds of cigarette smoke induce mutations in a variety of test systems in vitro and in vivo. In Table 13 we compiled carcinogens of cigarette smoke which induce sex-linked recessive mutations in the male germ cells of *Drosophila melanogaster*. In this table we also give the amount of these compounds in the sidestream as compared to the mainstream smoke.

**Table 13.** Carcinogens from cigarette smoke which induce sex linked recessive lethal mutations in male germ cells of *Drosophila melanogaster*

| Compound | Amount per cigarette[a] | Ratio in side-stream versus mainstream smoke[a] |
|---|---|---|
| *Nitrosamines* | | |
| Dimethylnitrosamine | 4 −180 ng | 10 −830 |
| Diethylnitrosamine | 0.1− 28 ng | 4 − 25 |
| Nitrosomorpholine[b] | ? | ? |
| Nitrosopyrrolidine | 0 −110 ng | 3 − 76 |
| Nitrosopiperidine | 0 − 9 ng | ? |
| *Polycyclic hydrocarbons* | | |
| Benzo(a)pyrene | 8 − 40 ng | 2.7− 3.4 |
| *Others* | | |
| Hydrazine | 24 − 43 μg | 3 |
| Formaldehyde | 20 − 90 μg | 51 |
| β-Naphthylamine | 1 − 2 μg | 39 |
| Urethane (Ethyl carbamate) | 10 − 35 μg | ? |
| Vinyl chloride | 1 − 16 μg | ? |
| Styrene | 10 μg | ? |

Most of the data are from a review of Vogel et al. (1980), data for formaldehyde are from Auerbach (1976), data for styrene are from Donner et al. (1979)

[a] See The Health Consequences of Smoking (1982) and Smoking and Health (1979)
[b] In chewing tobacco or snuff: 20–700 ppb

# 6 Conclusion

With respect to exchange type of aberrations in the PL of S our results confirm the findings of others, in that S have more CA than NS. This is in line with the epidemiological finding of different types of cancers being induced by smoking and by this support the mutation theory of the origin of cancer. Most of the carcinogenic/mutagenic compounds in tobacco smoke originate during smoking and represent pyrolysis products of tobacco. Pyrolysis products of similar types are also produced when other organic materials are transformed by heating.

# References

Anderson D, Richardson CR, Purchase IFH, Evans HJ, O'Riordan ML (1981) Chromosomal analysis in vinyl chloride workers: Comparison of the standard technique with the sister-chromatid exchange technique. Mutat Res 83:137–144

Ardito G, Lamberti L, Ansaldi E, Ponzetto P (1980) Sister chromatid exchanges in cigarette-smoking human females and their newborns. Mutat Res 78:209–212

Armitage P (1971) Statistical methods in medical research. Blackwell, London

Auerbach C (1976) Mutation research. Problems, results, and perspectives. Chapman and Hall, London

Bauer KH (1928) Mutationstheorie der Geschwulst-Entstehung. Übergang von Körperzellen in Geschwulstzellen durch Gen-Änderung. Springer, Berlin Heidelberg New York

Bauer KH (1963) Das Krebsproblem, 2nd edn. Springer, Berlin Heidelberg New York, pp 521–590

Beek B, Klein G, Obe G (1980) The fate of chromosomal aberrations in a proliferating cell system. Biol Zbl 99:73–84

Beek B, Jacky PB, Sutherland GR (1983) Heritable fragile sites and micronucleus formation. Ann Genet 26:5–9

Benedict WF, Rucker N, Faust J, Kouri RE (1975) Malignant transformation of mouse cells by cigarette smoke condensate. Cancer Res 35:857–860

Beutel P, Küffner H, Schubö B (1980) SPSS 8, 3rd edn. Fischer, Stuttgart

Bock FG, Swain AP, Stedman RL (1969) Bioassay of major fractions of cigarette smoke condensate by an accelerated technique. Cancer Res 29:584–587

Boveri T (1914) Zur Frage der Entstehung maligner Tumoren. Fischer, Jena

Bridges BA, Clemmesen J, Sugimura T (1979) Cigarette smoking – does it carry a genetic risk? Mutat Res 65:71–81

Crossen PE, Morgan WF (1980) Sister chromatid exchanges in cigarette smokers. Hum Genet 53: 425–426

Daniele RP, Daubner JH, Altose MD, Rowlands DT, Gorenberg DJ (1977) Lymphocyte studies in asymptomatic cigarette smokers. A comparison between lung and peripheral blood. Am Rev Respir Dis 116:997–1105

de Marini DM (1983) Genotoxicity of tobacco smoke and tobacco smoke condensate. Mutat Res 114:59–89

Diet, Nutrition, and Cancer (1982) National Academy, Washington, DC

Doll R, Peto R (1981) The causes of cancer. Quantitative estimates of avoidable risks of cancer in the United States today. Oxford University Press, Oxford New York

Donner M, Sorsa M, Vainio H (1979) Recessive lethals induced by styrene oxide in Drosophila melanogaster. Mutat Res 67:373–376

Evans HJ (1981) Cigarette smoke induced DNA damage in man. In: Kappas A (ed) Progress in environmental mutagenesis and carcinogenesis. Elsevier Biomedical, Amsterdam, pp 111–128

Fabry L, Pairon D, Saint-Ruf G, Poncelet F (1982) Cytogenetic study, in vitro, on the pyrolysate mutagens Glu-P-2 and SR 25. Mutat Res 97:241–242 (abstr)

Fredga K, Dävring L, Summer M, Bengtsson BO, Elinder C-G, Sigtrygsson P, Berlin M (1982) Chromosome changes in workers (smokers and nonsmokers) exposed to automobile fuels and exhaust gases. Scand J Work Environ Health 8:209–221

Gairola C (1982) Genetic effects of fresh cigarette smoke in Saccharomyces cerevisiae. Mutat Res 102:123–136

Gundy S, Varga LP (1983) Chromosomal aberrations in healthy persons. Mutat Res 120:187–191

Hedner K, Högstedt B, Kolnig A-M, Mark-Vendel E, Strömbeck B, Mitelman F (1982) Sister chromatid exchanges and structural chromosome aberrations in relation to age and sex. Hum Genet 62:305–309

Hedner K, Högstedt B, Kolnig A-M, Mark-Vendel E, Strömbeck B, Mitelman F (1983) Sister chromatid exchanges and structural chromosome aberrations in relation to smoking in 91 individuals. Hereditas 98:77–81

Herning RI, Jones RT, Bachman J, Mines AH (1981) Puff volume increases when low-nicotine cigarettes are smoked. Br Med J 283:187–189

Hill P, Marquardt H (1980) Plasma and urine changes after smoking different brands of cigarettes. Clin Pharmacol Ther 27:652–658

Hollander DH, Tockman MS, Liang YW, Borgaonkar DS, Frost JK (1978) Sister chromatid exchanges in peripheral blood of cigarette smokers and in lung cancer patients; and the effect of chemotherapy. Hum Genet 44:165–171

Hopkin JM, Evans HJ (1980) Cigarette smoke-induced DNA damage and lung cancer risks. Nature 283:388–390

Hopkin JM, Perry PE (1980) Benzo(a)pyrene does not contribute to the SCEs induced by cigarette smoke condensates. Mutat Res 77:377–381

Hopkin JM, Steel CM (1980) Variation in individual responses to the cytotoxicity of cigarette smoke. Thorax 35:751–753

Husgafvel-Pursiainen K, Mäkipaakkanen J, Norppa H, Sorsa M (1980) Smoking and sister chromatid exchanges. Hereditas 92:247–250

Husum B, Wulf HC, Niebuhr E (1981) Sister-chromatid exchanges in lymphocytes in women with cancer of the breast. Mutat Res 85:357–362

Hüttner E, Schöneich J (1981) Effect of smoking on the frequencies of chromosomal aberrations and SCE in man. Mutat Res 85:255 (abstr)

Hutton JJ, Hackney C (1975) Metabolism of cigarette smoke condensates by human and rat homogenates to form mutagens detectable by Salmonella typhimurium TA 1538. Cancer Res 35:2461–2468

Inoue K, Shibath T, Abe T (1983) Induction of sister-chromatid exchanges in human lymphocytes by indirect carcinogens with and without metabolic activation. Mutat Res 117:301–309

Ivanow B, Praskova L, Mileva M, Bulanova M, Georgieva I (1978) Spontaneous chromosomal aberration levels in human peripheral lymphocytes. Mutat Res 52:421–426

Jacky PB, Beek B, Sutherland GR (1983) Fragile sites in chromosomes: Possible model for the study of spontaneous chromosome breakage. Science 220:69–70

Kier LD, Yamasaki E, Ames BN (1974) Detection of mutagenic activity in cigarette smoke condensates. Proc Natl Acad Sci USA 71:4159–4163

Kozlowski LT, Frecker RC, Khouw V, Pope MA (1980) Hole blocking of ventilated filters. Am J Public Health 70:1202–1203

Lambert B, Ringborg U, Skoog L (1979) Age-related decrease of ultraviolet light-induced DNA repair synthesis in human peripheral leukocytes. Cancer Res 39:2792–2795

Lambert B, Ringborg U, Swanbeck G (1977) Repair of UV-induced DNA lesions in peripheral lymphocytes from healthy subjects of various ages, individuals with Down's syndrome and patients with actinic keratosis. Mutat Res 46:133–134

Lambert B, Lindblad A, Nordenskjöld M, Werelius B (1978) Increased frequency of sister chromatid exchanges in cigarette smokers. Hereditas 88:147–149

Lambert B, Lindblad A, Holmberg K, Francesconi D (1982a) The use of sister chromatid exchange to monitor human populations for exposure to toxicologically harmful agents. In: Wolff S (ed) Sister chromatid exchange. Wiley, New York, pp 149–182

Lambert B, Berndtsson I, Lindsten J, Nordenskjöld M, Söderhäll S, Holmstedt B, Palmer L, Jernström B, Marsk L (1982b) Smoking and sister chromatid exchange. In: Sorsa M, Vainio H (eds) Mutagens in our environment. Liss, New York, pp 401–414

Levitt RC, Legraverend C, Nebert DW, Pelkonen O (1977) Effects of harman and norharman on the mutagenicity and binding to DNA of benzo(a)pyrene metabolites and on aryl hydrocarbon hydroxylase induction in cell culture. Biochem Biophys Res Commun 79:1167–1175

Livingston GK, Fineman RM (1983) Correlation of human lymphocyte SCE frequency with smoking history. Mutat Res 119:59–64

Lloyd DC, Purrott RJ, Reeder EJ (1980) The incidence of unstable chromosome aberrations in peripheral blood lymphocytes from unirradiated and occupationally exposed people. Mutat Res 72:523–532

Madle E, Obe G, Hansen J, Ristow H (1981) Harman and norharman: Induction of sister-chromatid exchanges in human peripheral lymphocytes in vitro and interaction with isolated DNA. Mutat Res 90:433–442

Matsumoto T, Yoshida D, Mizusaki S, Okamoto H (1977) Mutagenic activity of amino acid pyrolysates in Salmonella typhimurium TA 98. Mutat Res 48:279–286

Matsushima T, Sugimura T (1981) Mutagen-carcinogens in amino acid and protein pyrolysates and in cooked food. In: Kappas A (ed) Progress in environmental mutagenesis and carcinogenesis. Elsevier/North-Holland Biomedical, Amsterdam, pp 49–56

Mau G, Netter P (1974) Die Auswirkungen des väterlichen Zigarettenkonsums auf die perinatale Sterblichkeit und die Mißbildungshäufigkeit. Dtsch Med Wochenschr 99:1113–1118

Meiying C, Jiujin X, Xianting Z (1982) Comparative studies on spontaneous and mitomycin C – induced sister-chromatid exchanges in smokers and non-smokers. Mutat Res 105:195–200

Mizusaki S, Takashima T, Tomaru K (1977a) Factors affecting mutagenic activity of cigarette smoke condensate in Salmonella typhimurium TA 1538. Mutat Res 48:29–36

Mizusaki S, Okamoto H, Akiyama A, Fukuhara Y (1977b) Relation between chemical constituents of tobacco and mutagenic activity of cigarette smoke condensate. Mutat Res 48:319 to 326

Murthy PBK (1979) Frequency of sister chromatid exchange in cigarette smokers. Hum Genet 52:343–345

Nagao M, Yahagi T, Honda M, Seino Y, Matsushima T, Sugimura T (1977a) Demonstration of mutagenicity of aniline and o-toluidine by norharman. Proc Jpn Acad 53:34–37

Nagao M, Yahagi T, Kawachi T, Sugimura T, Kosuge T, Tsuji K, Wakabayashi K, Mizusaki S, Matsumoto T (1977b) Comutagenic action of norharman and harman. Proc Jpn Acad 53: 95–98

Nagao M, Yahagi T, Honda M, Seino Y, Kawachi T, Sugimura T, Wakabayashi K, Tsuji K, Kosuge T (1977c) Comutagenic actions of norharman derivatives with 4 dimethylaminoazobenzene and related compounds. Cancer Lett 3:339–346

Nagao M, Yahagi T, Kawachi T, Seino Y, Honda M, Matsukura N, Sugimura T, Wakabayashi K, Tsuji K, Kosuge T (1977d) Mutagens in foods, and especially pyrolysis products of protein. In: Scott D, Bridges BA, Sobels FH (eds) Progress in genetic toxicology. Elsevier/Nort-Holland, Amsterdam, pp 259–264

Obe G (1981) Mutagenicity of alcohol and tobacco smoke. In: Israel Y, Glaser FB, Kalant H, Popham RE, Schmidt W, Smart RG (eds) Research advances in alcohol and drug problems, vol 6. Plenum, New York, pp 281–318

Obe G, Beek B (1982a) The human leukocyte test system. In: de Serres FJ, Hollaender A (eds) Chemical mutagens, principles, and methods for their detection. Plenum, New York, pp 337 to 400

Obe G, Beek B (1982b) Premature chromosome condensation in micronuclei. In: Rao PN, Johnson RT, Sperling K (eds) Premature chromosome condensation. Application in basic, clinical, and mutation research. Academic, New York, pp 113–130

Obe G, Herha J (1978) Chromosomal aberrations in heavy smokers. Hum Genet 41:259–263

Obe G, Göbel D, Engeln H, Herha J, Natarajan AT (1980) Chromosomal aberrations in peripheral lymphocytes of alcoholics. Mutat Res 73:377–386

Obe G, Vogt H-J, Madle S, Fahning A, Heller W-D (1982) Double-blind study on the effect of cigarette smoking on the chromosomes of human peripheral blood lymphocytes in vivo. Mutat Res 92:309–319

Powrie WD, Wu CH, Rosin MP, Stich HF (1982) Mutagens and carcinogens in food. In: Bora KC, Douglas GR, Nestmann RE (eds) Chemical mutagenesis, human population monitoring, and genetic risk assessment. Elsevier Biomedical, Amsterdam, pp 187–199

Rhim JS, Huebner RJ (1973) In vitro transformation assay of major fractions of cigarette smoke condensate (CSC) in mammalian cell lines. Proc Soc Exp Biol Med 142:1003–1007

Riebe M, Westphal K, Keller H (1982) Test-condition-dependent influence of harman and norharman on benzo(a)pyrene mutagenesis in Salmonella. Mutat Res 104:9–15

Roszman TL, Elliott LH, Rogers AS (1975) Suppression of lymphocyte function by products derived from cigarette smoke. Am Rev Respir Dis 111:453–457

Russell MAH, Jarvis M, Iyer R, Feyerabend C (1980) Relation of nicotine yield of cigarettes to blood nicotine concentrations in smokers. Br Med J 280:972–976

Sato S, Seino Y, Ohka T, Yahagi T, Nagao M, Matsushima T, Sugimura T (1977) Mutagenicity of smoke condensates from cigarettes, cigars and pipe tobacco. Cancer Lett 3:1–8

Sato S, Ohka T, Nagao M, Tsuji K, Kosuge T (1979) Reduction in mutagenicity of cigarette smoke condensate by added sugar. Mutat Res 60:155–161

Seshadri R, Baker E, Sutherland GR (1982) Sister-chromatid exchange (SCE) analysis in mothers exposed to DNA-damaging agents and their newborn infants. Mutat Res 97:139–146

Smoking and Health, a report of the Surgeon General (1979) U.S. Department of Health, Education and Welfare, Washington, DC

Stich HF, Rosin MP, Wu CH, Powrie WD (1982) The use of mutagenicity testing to evaluate food products. In: Heddle JA (ed) Mutagenicity: New horizons in genetic toxicology. Academic, New York, pp 117–142

Sugimura T (1979) Naturally occuring genotoxic carcinogens. In: Miller EC, Miller JA, Hirono I, Sugimura T, Takayama S (eds) Naturally occuring carcinogens-mutagens and modulators of carcinogenesis. University Park Press, Baltimore, pp 241–261

Sugimura T (1981) A view of a cancer researcher on environmental mutagens. In: Sugimura T, Kondo S, Takebe H (eds) Environmental mutagens and carcinogens. University Tokyo Press, Tokyo and Liss, New York, pp 3–20

Sugimura T (1982) Mutagens, carcinogens, and tumor promoters in our daily food. Cancer 49: 1970–1984

Sugimura T, Nagao M (1980) Modification of mutagenic activity. In: de Serres FJ, Hollaender A (eds) Chemical mutagens. Principles and methods for their detection, vol 6. Plenum, New York, pp 41–60

Sugimura T, Nagao M (1982) The use of mutagenicity to evaluate carcinogenic hazards in our daily lives. In: Heddle JA (ed) Mutagenicity: new horizons in genetic toxicology. Academic, New York, pp 73–88

Sugimura T, Nagao M, Kawachi T, Honda M, Yahagi T, Seino Y, Sato S, Matsukura N, Matsuchima T, Shirai A, Sawamura M, Matsumoto H (1977) Mutagen-carcinogens in food, with special reference to highly mutagenic pyrolytic products in broiled foods. In: Hiatt HH, Watson JD, Winsten JA (eds) Origins of human cancer, Book C, Human risk assessment. Cold Spring Harbor, pp 1561–1577

Sutherland GR (1983) The fragile X chromosome. Int Rev Cytol 81:107–143

Suzuki J, Koyama T, Suzuki S (1983) Mutagenicities of mononitrobenzene derivatives in the presence of norharman. Mutat Res 120:105–110

Swain AP, Cooper JE, Stedman RL (1969) Large-scale fractionation of cigarette smoke condensate for chemical and biological investigations. Cancer Res 29:579–583

Thompson LH, Carrano AV, Salazar E, Felton JS, Hatch FT (1983) Comparative genotoxic effects of cooked-food-related mutagens Trp P-2 and IQ in bacteria and cultured mammalian cells. Mutat Res 117:243–257

Takehisa S, Kanaya N (1982) Aniline induction of sister chromatid exchanges in human lymphocytes. Mutat Res 101:165–172

The Health Consequences of Smoking. Cancer. A Report of the Surgeon General (1982) U.S. Department of Health and Human Services. Public Health Service, Office of Smoking and Health, Washington, D.C.

Tohda H, Oikawa A, Kawachi T, Sugimura T (1980) Induction of sister-chromatid exchanges by mutagens from amino acid and protein pyrolysates. Mutat Res 77:65–69

Trepel F (1976) Das lymphatische Zellsytem: Struktur, allgemeine Physiologie und allgemeine Pathophysiologie. In: Begemann H (ed) Blut und Blutkrankheiten, Teil 3, Leukocytäres und retikuläres System I. Springer, Berlin Heidelberg New York, pp 1–191

Umezawa K, Shirai A, Matsushima T, Sugimura T (1978) Comutagenic effect of norharman and harman with 2-acetylaminofluorene derivatives. Proc Natl Acad Sci USA 75:928–930

Van Duuren BL (1980) Carcinogens, cocarcinogens, and tumor inhibitors in cigarette smoke condensate. In: Gori GB, Bock FG (eds) A safe cigarette? Banbury Report 3. Cold Spring Harbor, pp 105–112

Vijayalaxmi, Evans HJ (1982) In vivo and in vitro effects of cigarette smoke on chromosomal damage and sister-chromatid exchanges in human peripheral blood lymphocytes. Mutat Res 92:321–332

Vogel E, Blijleven WGH, Klapwijk PM, Zijlstra JA (1980) Some current perspectives of the application of Drosophila in the evaluation of carcinogens. In: Williams GM, Kroes R, Waaijers HW, Van de Poll KW (eds) The predictive value of short-term screening tests in carcinogenicity evaluation. Elsevier/North-Holland Biomedical, Amsterdam, pp 125–147

Wakabayashi K, Nagao M, Kawachi T, Sugimura T (1981) Co-mutagenic effect of norharman with N-nitrosoamine derivatives. Mutat Res 80:1–7

Weisburger JH, Cohen LA, Wynder EL (1977) On the etiology and metabolic epidemiology of the main human cancers. In: Hiatt HH, Watson JD, Winsten JA (eds) Origins of human cancer, Book A. Incidence of cancer in humans. Cold Spring Harbor, pp 567–602

White WE Jr, Rock G (1981) Lack of comutagenicity by norharman and other compounds on revertants induced by photolabelling with 2-azido 9-fluorenone oxime. Mutat Res 84:263 to 271

Wulf HC, Husum B, Niebuhr E (1982) SCE in smokers of cigarettes, cheroots, pipe tobacco, and cannabis. 12th annual meeting EEMS, Dipoli, Espoo, Finland, 20–24 June, 1982. Abstracts volume, p 119

Yew FH, Johnson RT (1979) Ultraviolet-induced DNA excision repair in human B and T lymphocytes. III. Repair in lymphocytes from chronic lymphocytic leukemia. J Cell Sci 39:329–337

# Spermatogenesis in Smokers and Non-Smokers: An Andrological and Genetic Study

H.-J. VOGT[1], W.-D. HELLER[2], and G. OBE[3]

## 1 Introduction

Recent decades have seen significant advances both in the diagnosis and treatment of andrological and gynaecological disorders. There was a time when couples who could not have children had no alternative but to adopt, but with modern methods of treatment many childless couples can now be helped to start a family of their own. The advances in andrology are due not only to improvements in the interpretation of spermatological findings, but also to a fuller understanding of biochemical, immunological, endocrinological, cytological, histomorphological, and psychological factors. Thus we have learned to assess more precisely the way in which fertility is affected by diseases and their therapy, by self-medication, drugs and environmental pollution. Yet despite these advances, not all those who seek medical advice can be helped. Three fundamental causes of infertility have long been recognized: irreparable congenital or traumatic damage to the testes or the regulatory mechanism of the hypothalamic-pituitary-testicular axis, disorders not yet identifiable with present diagnostic methods and irresolvable partner conflicts. A knowledge of these factors obviously influences the therapy chosen. The better our basic understanding of the problems involved, the better are our chances of knowing when a particular case of impaired fertility is likely to respond to treatment.

### 1.1 The Physiology and Pathophysiology of Spermatogenesis

Clinical examination of the genital organs can indicate possible fertility failure. An abnormally small testis (less than 10 ml), for example, or changes in testicular consistency could mean impaired fertility. A hard enlargement of the epididymis suggests that inflammation may have resulted in atrophy of the seminiferous tubules. In the case of testicular variococeles – whether it involves one side of the scrotum or both – there is no correlation between the size of the varicocele and the degree of possible spermatogenic damage.

1 Dermatologische Klinik und Poliklinik der Technischen Universität München, Biedersteiner Straße 29, 8000 München 40, FRG
2 Institut für Statistik und mathematische Wirtschaftstheorie, Universität Karlsruhe, Postf. 6380, 7500 Karlsruhe 1, FRG
3 Institut für Genetik, Freie Universität Berlin, Arnimallee 5–7, 1000 Berlin 33, FRG

Mutations in Man, ed. by G. Obe
© Springer-Verlag Berlin Heidelberg 1984

Spermatogonia possess a diploid set of chromosomes. The process of spermato-genesis begins with a multiplication phase, during which the spermatogonia multiply by mitosis and at the same time enter upon their growth phase. This gives rise to the primary spermatocytes. The first and second meiotic divisions give rise in turn to the secondary spermatocytes (prespermatids) and the spermatids. The number of chromosomes is halved by meiosis. During spermiohistogenesis the haploid spermatids mature into spermatozoa. The latter consist of a head, neck, middle piece, and tail. The head is $3-5$ $\mu$ in length, and contains the nucleus with the haploid number of chromosomes. The chromatin is tightly packed in the nucleus. The acrosome carries enzymes, and forms a caplike covering to the anterior end of the head of the sperma-tozoon. The compression thus exerted on the head of the spermatozoon gives it its characteristic flattened oval shape, so that it looks rather like a pear when viewed from the side.

An indication of the biochemical composition of cells can be obtained by stain-ing with various dyes. This provides both morphological and functional data. For the morphological differentiation of the sperms a preparation is obtained as for a blood smear, and after fixing with methanol is stained using hematoxylin-eosin or other staining methods. With contrast staining the anterior portion of the head stains ambophilically, and the rest of the head basophilically. The middle piece and tail stain eosinophilically.

The focus of all andrological examinations is the ejaculate, which is examined for its physiological, biochemical, and morphological characteristics. A brief period of sexual continence, frequent sexual intercourse or inflammation of the prostate gland or seminal vesicles can reduce the quantity of ejaculate produced. Exposure of the ejaculate to air (oxidation) and inflammation of the prostata, seminal vesicles or epididymis will cause a higher pH value (i.e. more alkaline) in the ejaculate, while chronic prostatitis or anomalies in the Ductus deferentes will produce a more acidic pH value. Any changes in the normal pH value lead to a diminution in fertility. The liquefaction time and consistency of the ejaculate are both factors which affect the ascension of the spermatozoa. The odor and color of the ejaculate are characteristic. Any major deviations from the norm indicate some kind of disorder in the urogenital tract.

The motility of the spermatozoa, as measured in the native ejaculate, is a significant indication of fertility. It is also significant for the morphology of the spermatozoa, as evaluated in the smear preparation after staining. In addition to spermatozoa of normal configuration the ejaculate contains 10%-40% of morphologically atypical sperms, which exhibit various abnormalities of the head, neck, middle piece or tail. Accord-ing to Nebe and Schirren (1980b), an even higher percentage may well be regarded as normal. By far the most common degenerate forms are mis-shapen heads. It is not possible to correlate individual morphological abnormalities with specific damage of the seminal ducts or toxic influences to which the sperms may have been exposed during maturation. Sperms are normally classified into the following categories: normal cells, immature sperms, degraded forms, microcephalous, bicephalous, mis-shapen head, mis-shapen middle piece, bicaudal, and other degenerate forms.

The European Andrology Club (Eliasson et al. 1970) has devised a recommended nomenclature for diagnostic purposes which has been widely accepted (Table 1).

**Table 1.** Spermatozoal characteristics in various andrological diagnoses

| Diagnosis | Sperm count (millions/ml) | Morphology (% normal) | Motility (% normal) |
|---|---|---|---|
| Normozoospermia | 40–250 | Over 60 | Over 60 |
| Polyzoospermia | Over 250 | Over 60 | Over 60 |
| Oligozoospermia | Below 40 | | |
| Cryptozoospermia | Below 1 | | |
| Asthenozoospermia | Over 40 | | Below 60 |
| Teratozoospermia | Over 40 | Below 60 | |
| Necrozoospermia | All spermatozoa dead (stained red in eosin test) | | |
| Azoospermia | No spermatozoa present | – | – |

Further information about the structure of the spermatozoa is revealed by biochemical and histochemical examination. A factor of particular significance for the present study is the nucleic acid content of the spermatozoa. According to studies carried out by Sandritter et al. (1958), the RNA content in the head of the spermatozoon is so small that it is negligible. The DNA content of the head, which was measured for the first time by Miescher in 1869 (findings first published in 1897), is half that of a somatic cell, since spermatozoa only contain the haploid number of chromosomes. The quantity of DNA in polyploid cells is correspondingly larger. Ultraviolet microspectrophotometry has largely confirmed the results obtained using Feulgen's method of nuclear staining (Sandritter 1958; Meyhöfer et al. 1960). The DNA concentration in the heads gives some indication of possible damage of the spermatozoa.

The DNA concentration in spermatozoa has been measured by Leuchtenberger et al. (1956) and Meyhöfer (1963). In the case of fertile males the DNA concentration was found to be $2.18-2.60 \times 10^{-12}$ g, while for patients with clinical infertility the DNA concentration was normally below $2.00 \times 10^{-12}$ g.

DNA concentration correlated with the surface area of the spermatozoal heads. Mis-shapen heads were found to contain either unusually large or unusually small quantities of nucleic acid. Thus the average nucleic acid concentration in amorphous spermatozoa was $1.40 \times 10^{-12}$ g, while that of bicephalous spermatozoa was $3.33 \times 10^{-12}$ g. Meyhöfer (1963) reports substantially lower DNA values in the spermatozoa of men with polyzoospermia. He suggests that this may be one explanation for the higher incidence of miscarriages in the wives of men with polyzoospermia.

A rapid method for determining the DNA concentration in sperm is by means of impulse cytophotometry. After pretreatment of the ejaculate with a proteinase and dithioerythritol for purposes of decondensation, and with the DNA-specific fluorescent dye 4,6-diamino-2-phenyl-indole (DAPI), the degree of fluorescent intensity is measured for each individual cell. Sensitivity begins at $10^{-13}$ g DNA per cell. The measurements are presented in the form of histograms; the fluorescent intensity (= measure of DNA concentration) is shown on the x-axis and the number of impulses on the y-axis (Hartmann et al. 1982; Otto and Oldiges 1978; Hettwer et al. 1981).

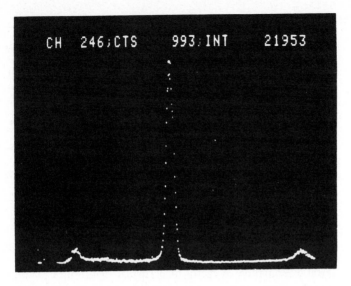

**Fig. 1.** Histogram for normospermia form impulse cytophotometric measurement of sperm stained with the fluorescent dye DAPI. The high peak in the middle of the figure represents cells with a DNA content of 1c. The smaller peak to the left represents diploid (2c) spermatozoa. The values to the right and to the left of the 1c peak are from destroyed cells

Figure 1 shows a typical histogram for normospermia. The peak in the 1c area corresponds to the DNA in the haploid gametes. The preparation technique used by Otto et al. (1979a) is such that only mature spermatozoa survive to combine with the fluorescent dye. Other cells are destroyed, and appear to the left or the right of the 1c peak, depending on their DNA content. The smaller peak in the 2c area indicates diploid spermatozoa which – according to an investigation by Carothers and Beatty (1975) – are likely to occur in the case of normospermia in approximately 0.5% of the spermatozoa.

Ejaculates containing a higher proportion of immature spermatozoa, spermatogenic cells or somatic cells produce stronger fluctuations in the background level, a less steeply sloping peak or additional peaks. This fluctuation indicates those cells which are included within the total cell count, but which are not registered in the separate peaks (continuous/discontinuous background). The only way to differentiate between spermatids and leukocytes, which are destroyed during the preparation procedure and the fragments of which are shown on the histogram in the hypohaploid area to the left of 1c, is by means of cytomorphological methods. Similarly, the only way of deciding whether the peak in the 2c area represents nonseparated double-headed forms or diploid spermatozoa is by analysis of a differential spermiogram (Otto et al. 1979b; Hofmann et al. 1980a; Hettwer et al. 1981; Hartmann et al. 1982). If analysis shows that double-headed forms are not present, the occurrence of diploid spermatozoa indicates a disturbance in the premeiotic/meiotic phases (Stolla et al. 1978). A disturbance of this type is more serious, both from a therapeutic and from a prognostic point of view, than a disturbance which occurs during the spermiohistogenetic phase.

Some ejaculates produce no obvious peak formations, even though the spermio-gram indicates no unusual characteristics. This is attributable to inadequate fluoro-chroming. However, by modifying the preparation technique (proteolysis using pepsin instead of papain), Hettwer et al. (1981) were able to achieve satisfactory staining results.

The reserves of energy required by the spermatozoa during their ascension are supplied by fructose. The quantity of fructose present and its catabolism (fructolysis) therefore have a decisive influence on spermatozoal motility. According to Schirren (1977) the level of fructose in the seminal plasma falls steadily between the ages of 20 and 70. The same author reports a negative correlation between the concentration of fructose and sperm density.

Spermatogenesis is regulated by the so-called hypothalamic-pituitary-testicular axis.

Testosterone (T), the principal androgen, is produced in the interstitial cells of the testes (Leydig cells). Production of androgens by the Leydig cells is stimulated by the luteinizing hormone (LH), which is secreted by the anterior lobe of the pituitary gland (adenohypophysis). Secretion is controlled by the hypothalamus by the gonadotropin-releasing hormone (GnRH). Hypothalamic production of GnRH is in turn regulated by the plasma testosterone concentration, which functions as a kind of feedback mechanism.

The pituitary gland or hypophysis also secretes the so-called follicle-stimulating hormone (FSH). This activates the function of the seminiferous tubules, and hence stimulates spermiogenesis. The feedback mechanism is not yet fully understood, but the inhibins (In) which have been found in rats are also thought to be present in the human organism. FSH is needed in order to transport a sufficient supply of testos-terone to the germinal epithelium. This cannot occur if the tubular membrane is sclerosed or hyalinized. The result is an increase in FSH secretion, which can be measured accordingly. An increased LH concentration indicates a Leydig cell insuf-ficiency. In the case of a low or normal LH concentration the degree to which the adenohypophysis can be stimulated may be ascertained by means of a GnRH test. If the testosterone level is reduced, it is necessary to distinguish between a primary (testicular) insufficiency and a secondary (hypophysal) insufficiency. The response of the Leydig cells to stimulation can be tested by the administration of HCG.

There are a number of factors which can have a toxic effect on spermatogenesis. In a report published in 1973 Schirren discussed the effect of various environmental factors on male fertility. Testicular infections are included in this category – in particular orchitis resulting from mumps, but not bacterial inflammations such as Neisseria gonorrhoeae – as well as allergic reactions, medications such as cytotoxic and anabolic drugs, pesticides, nicotine, and iatrogenic disorders.

The harmful effect on spermatogenesis of the cytotoxic drugs used in chemo-therapy of cancer (Meyhöfer 1973; Qureshi et al. 1972; Asbjornsen et al. 1976; Schilsky et al. 1980) is the basis for the recently developed "mouse sperm-abnormality assay" for mutagenic and/or carcinogenic substances (Wyrobek and Bruce 1978; Wyrobek 1982).

Many other forms of medication are known to have an anti-spermatogenic effect (Steinberger 1976; van Thiel et al. 1979; Babb 1980; Drife 1982). Drife's survey is mainly concerned with Sulphasalazine, which is used in the treatment of ulcerative colitis. In addition to oligozoospermia, some studies have also revealed morphological changes in the spermatozoa. Drife (1982) also discusses the question of chemical contraception for the male, pointing out that the antiandrogen cyproterone acetate has the effect of lowering the sperm count. Joffe (1979) also discusses the possible effect of medication on spermatogenesis. It is conceivable that the spermatozoa are directly affected, in which case the first effect likely to be observed would be a reduction in motility of the otherwise undamaged spermatozoa. The sperms may also be affected indirectly via damage to the germinal epithelium or via the system of hormonal regulation.

Various studies have also been conducted on the effect of chemicals in the environment. Some of these studies are concerned with the effects of chemicals in the working environment. Barlow and Sullivan (1981) identified lead, methyl mercury, beryllium, polychlorinated biphenyls, benzene, chlordecane, dibromochloropropane, hexachlorobenzene, carbon monoxide and carbon disulphide as substances which may have a damaging effect on both the male and female reproductive systems. Lancranjan et al. (1975) found that asthenospermia, hypospermia, and teratospermia – all associated with impaired fertility – occurred more frequently in workers exposed to lead. Whorton et al. (1977) reported an abnormally high incidence of azoospermia and oligozoospermia in pesticide workers exposed to the chemical 1,2-dibromo-3-chloropropane (DBCP). These workers also exhibited abnormally high levels of FSH and LH, although the testosterone level was found to be within normal limits. In 1981 Whorton reported upon other investigations (some of them unpublished) which confirmed his findings that a higher incidence of azoospermia and oligozoospermia is to be found in industrial workers exposed to DBCP.

Lantz and associates (1981) examined 71 workers in the petrochemical industry, and again found oligozoospermia in the case of 15 men who were exposed to the nematocide DBCP. Extratesticular abnormalities were also diagnosed in 7 workers (Group I), while exposure to DBCP was clearly the sole cause of oligozoospermia diagnosed in a further 7 of the 15 workers (Group II). FSH concentrations, both basal and stimulated, were higher in two patients from Group II. Testosterone concentrations were within normal limits for all patients, but significantly lower than the levels recorded for a control group. At 18 to 21 months after the last exposure to DBCP there was an improvement in sperm density. Cannon et al. (1978) examined workers engaged in the production of kepone (a polychlorinated-hydrocarbon insecticide), and found in addition to other symptoms a high incidence of oligozoospermia. Wyrobek et al. (1981a) found an abnormally high concentration of sperms with misshapen heads in workers exposed to carbaryl. This phenomenon was not dependent on age, smoking habits or other clinical symptoms. Workers who had been exposed 6.3 years previously (on average) exhibited a slightly but nonetheless significantly higher level of sperm abnormalities than the control group. On the basis of this result the authors tentatively suggest that the effects are non-reversible. No correlation was found with the degree or duration of exposure. The authors therefore suspect that other factors (as yet unknown) also play a role.

Hudec et al. (1981) examined 123 samples of semen and found that 34 contained tris(dichloropropyl)phosphate, a flame-retardant that causes mutations and sterility.

They believe that these and other chemicals in the environment may be responsible for the decline in male fertility in America over the last three decades.

Wyrobek et al. (1981b) have published the results of a study, in which they found that anesthesists who had been working in air-conditioned operating theaters for at least 1 year exhibited no spermatological differences compared with colleagues who were only just embarking on their careers. A comparison with a control group revealed that the proportion of subjects with sperm abnormalities was significantly higher in subjects who had varicoceles, a recent history of medication, a high tobacco consumption, a history of recent illness or who frequently visited the sauna.

Other data also indicates that spermatogenesis is adversely affected by prolonged or severe exposure to X-rays (Lushbaugh and Casarett 1976; Wyrobek 1979; Hahn et al. 1982; Thorsland and Paulsen 1972; Rowley et al. 1974).

The reversibility of the damage sustained is a question of major significance in any evaluation of toxic substances affecting spermatogenesis. In so far as the authors of the above-mentioned studies have addressed themselves to this problem, they found that in most cases spermatological findings improved within a few months if the subjects were no longer exposed to the substance. As regards the effects of mutagenic agents, further epidemiological data are needed before any conclusions can be drawn. In particular it will be necessary to examine the effects on one or two successive generations. Sperm anomalies can apparently be transmitted to the progeny, as demonstrated in mice by Wyrobek and Bruce (1978).

## 1.2 Paternal Smoking and Fertility

According to a study by Larson et al. (1961), the belief that tobacco smoking may have an effect on sexual activity dates back to the year 1622, when the Turkish Sultan Mourach imposed a ban on tobacco because of its alleged aphrodisiac effects. Attitudes had changed by the 19th century. Whereas Jacoby (1898) concluded that smoking had no effect at all on sexual activity, Schrottenbach (1933) refers to "nicotine impotence". A historical survey of the subject may be found in Sterling and Kobayashi (1975).

A number of recent studies have been concerned with the same question. In a retrospective study by Cendron and Vallery-Masson (1971), 70 men aged between 45 and 90 years were questioned on their sexual behaviour between the ages of 25 and 40 years and thereafter. While 77% of the smokers reported a decline in frequency of intercourse, the figure for non-smokers was only 33%. As a physiological basis for this allerged reduction in sexual activity, the findings of Briggs (1973) are often quoted. Briggs observed that testosterone concentration increases within normal limits when subjects give up smoking, but it seems doubtful whether this view is tenable. Money (1961) and Salmimies et al. (1982) have shown that while there is a link between individual sexual reactions and testosterone, further doses of testosterone which cause the plasma testosterone concentration to rise above normal minimum levels do not effect a further increase in sexual activity.

Animal experiments studying the influence of smoking on male fertility were carried out many years ago (Larson et al. 1961). The most common disorders observed were cases of testicular atrophy, the extent of which depended on the quantity of tobacco or nicotine administered. On the basis of these experiments one would expect to find that small – or at least smaller – testes occur more frequently in smokers than in non-smokers. However, this does not appear to be the case. Gey (1971) found small testes in only 19.5% of the smokers sampled, but in 27.4% of the non-smokers. If subjects with an enlarged epididymis were also taken into account, then small testes were found in a further 6.6% of smokers, and 9.4% of non-smokers. Nebe and Schirren (1980b,c) concluded that the toxic effect of smoking on testes of reduced size and consistency must be greater than the effect on testes exhibiting no palpable abnormalities.

Vogt et al. (1979) observed a significantly higher incidence of varicoceles in smokers as compared with non-smokers. In an initial group comprising 39 non-smokers and 30 smokers aged between 18 and 23 years, varicoceles were found in 10% of the non-smokers and 37% of the smokers. Investigation of a second group comprising 203 non-smokers and 202 smokers (of the same age and with the same smoking habits as the former group) confirmed this observation. The incidence of varicoceles among the non-smokers was 9%, and among the smokers 20%. If this higher incidence of varicoceles among smokers is actually due to smoking, then the mechanisms responsible are as yet totally unknown (Klaiber et al. 1980).

Medical literature contained few references to the problem of impaired fertility as a result of tobacco smoking before an exact andrological diagnosis was possible. Uchigaki (1927) reported that the addition of nicotine salts to spermatozoa samples reduced their life-span. In his animal experiments René (1878) found that doses of nicotine led to a reduction in sperm motility, a finding that was not subsequently confirmed by Rothschild (1953). However, Fürbringer (1926), Dörfel and Lutterberg (1937), and Stiasny (1944) were all of the opinion that smoking impaired sperm motility. As practical methods of sperm analysis were introduced, investigators began to focus their attention on three important variables related to fertility: sperm density (i.e. the number of spermatozoa per ml of semen), sperm motility and sperm morphology. Mellan (1967) was unable to confirm his initial suspicion that smoking is causally related to fertility failure. Nonetheless his investigations did reveal that up to 87% of the males found to be infertile were smokers, a percentage which is clearly higher than the percentage of smokers in the population.

Schirren and Gey (1969) carried out a study of 1,957 patients and found normozoospermia in 48.4% of the non-smokers, but only 43% of the smokers. No dose-response relationship was found between smoking and its effect on sperm density and motility. A further study of 4,372 andrological patients by Schirren (1972) revealed normozoospermia in 51.5% of the non-smokers and 52% of the smokers. Astenozoospermia was found in 9.5% of the non-smokers and 8% of the smokers. In a study carried out with 1,619 andrological patients in the year 1975 (Nebe and Schirren 1980a–c), no damaging effect of smoking on sperm density, motility or morphology was found. Indeed, it was found that the degree of sperm motility was actually higher among some smokers with a consumption of more than 30 cigarettes a day than among others who smoked only 10 a day (Nebe and Schirren 1980c).

In 1969 Viczian reported a significantly lower sperm count for smokers as compared with non-smokers, although by current standards all the values he obtained would be regarded as normal. Similarly, sperm motility was slightly impaired in the case of smokers, although there was no apparent correlation with the daily number of cigarettes smoked. The author particularly emphasized the significantly higher rate of pathological cells in smokers, and in the light of his earlier animal experiments (Viczian 1968) he attributed these findings to a nicotine-related impairment of the meiotic divisions.

Subsequent literature on this subject is likewise inconclusive. Campbell and Harrison (1979) found reduced sperm densities in 26% of the non-smokers they examined and in 41% of the smokers, while diminished sperm motility was observed in 26% of the non-smokers and 35% of the smokers. Of their 253 fertility patients, 53% were smokers. Stekhun (1980) investigated 247 smokers, ex-smokers and non-smokers and found oligospermia in 58.6% of the smokers, 41.6% of the ex-smokers, and 18.2% of the non-smokers. The decrease of sperm density and motility corresponded to the age at which smoking was started and the number of cigarettes smoked per day. Likewise, Vogel et al. (1979) reported a significant reduction in both sperm count and sperm motility in smokers. In a study by Evans et al. (1981), in which a group of 43 smokers was compared with a group of 43 non-smokers, the percentage of abnormal sperm forms was considerably higher among the smokers, although there was no discernable correlation between the degree of abnormality and the daily number of cigarettes smoked. An important feature of this study was the exclusion of subjects with possible previous damage unrelated to smoking. Godfrey (1981), who applied the same exclusion criteria as Evans, found no evidence that smoking affects spermatogenesis. A comparison of 74 non-smokers with 40 smokers who smoked 20 cigarettes or less a day and 35 smokers who smoked more than 20 cigarettes a day revealed no statistically significant differences in terms of sperm count, motility or sperm morphology. Recently Rodriguez-Rigau et al. (1982) reported the semen quality of 159 "normal men" (husbands of infertile women). They found no differences in sperm count, sperm motility, and percentage of normal spermatozoa between the groups of smokers (58 subjects) and non-smokers (101 subjects). In a group of 97 men with varicocele (59 non-smokers and 38 smokers) no significant differences related to smoking were found in these parameters. When compared with the group of 159 "normal men", a significantly higher percentage of men with varicocele had sperm counts of < 20 million/ml.

If cigarette smoking is thought to affect spermatogenesis adversely, it is only logical to consider the possible effects of smoking on the progeny. As far as maternal smoking is concerned, research has confirmed an association between lower birth weight of babies and their mothers' smoking habits during pregnancy (Hytten 1973; Mau 1977a; Murphy 1980). Whether smoking during pregnancy affects the development of the fetus in other ways is not yet known, however, the possibility of deficiencies in physical growth and in intellectual and psychological development cannot be excluded (Mau 1977b; U.S. Public Health Service 1979).

Far less appears to be known about the effects of paternal smoking. Borlee et al. (1978) found a reduction in the birth weight of newborn babies whose fathers were smokers. However, Yerushalmy (1962, 1972) and McMahon et al. (1966) were unable to confirm these findings. According to Mau and Netter (1974), there is a correlation between deformities in newborn babies and the father's cigarette consumption, but the correlation is only significant in the case of fissures of the lip, palate, and the alveolar process (cheilognathopalatoschisis).

Mau and Netter (1974) and Mau (1977b) also found an increase in perinatal mortality among the children of smokers (as compared with the children of non-smokers) if the father smoked more than 10 cigarettes a day. In another study by Comstock and Lundin (1967) the perinatal mortality rate among children of non-smoking mothers whose husbands were smokers was similarly found to be higher than among the children of non-smoking parents. Yerushalmy (1971) found that infants of low birth weight born to non-smoking mothers and smoking fathers were more vulnerable than those born to smoking mothers and non-smoking fathers.

The question that arises in connection with these retrospective studies is whether or not *all* the factors that can affect the course of pregnancy and child development have been excluded. Since this is improbable, and perhaps even impossible, certain doubts necessarily still remain as to the causal nature of the relationship between paternal smoking and the effects on progeny as described in these studies.

### 1.3 Objectives of the Present Study

The published findings on the influence of smoking on spermatogenesis present a picture which is full of contradictions. Our objective is to examine and clarify this issue in such a way that the results will stand up to rigorous scientific scrutiny as far as possible. We are particularly concerned to exclude all sources of error that are associated with the selection of subjects and with the subjective attitudes of investigators to the issue.

In contrast to all previous studies, in which presumably only male patients attending andrological/gynaecological infertility clinics took part, we selected only healthy male volunteers. Investigating subjects who report at a clinic for a special disorder will never reveal differences between smokers and non-smokers in their disease status although the risk for the disease might be different in both groups. This major disadvantage is circumvented with our procedure of selecting healthy volunteers. All subjects were asked to complete a comprehensive questionnaire and undergo a medical examination, thus enabling us to exclude from the sample any males in whom spermatogenesis may have been impaired by other factors apart from smoking. This study also differed from other studies in that it was a "blind study", i.e. the examining physician had no previous knowledge of the subject's medical history or habits.

In the following sections we compare the findings for smokers and non-smokers, and also take account of factors other than smoking which could have an influence on spermatogenesis. Finally, the scientific significance of the findings is discussed.

## 2 Subjects and Test Procedure

### 2.1 Subjects

By placing notices in Munich's various universities and colleges we obtained a group of subjects (cigarette smokers and non-smokers) aged between 20 and 40 who were prepared to give single specimens of ejaculate and blood, complete a questionnaire on their medical history and undergo a medical examination. The subjects were to receive a small payment upon completion of the examination, as well as information about the andrological findings if they so desired.

The clinical and andrological examinations were carried out in one institute, and the anamnestic data collected in another institute, by two independent investigators. The only personal information available to the andrologist throughout the entire examination was the subject's code number. In selecting the subjects care was taken to ensure that a sufficient number of smokers were included in the sample. Subjects were classified as "smokers" if they claimed to have smoked (and inhaled) at least 10 cigarettes a day for at least 1 year prior to their examination. The non-smoker group comprised those who claimed that they had never smoked, or not to have smoked for at least 1 year previously.

## 2.2 Medical History

The questionnaire, which contained a total of 80 questions, covered the following aspects:

- General illnesses
- Deformities and diseases in the genital region
- Operations performed in the genital region
- Exposure to radiation
- Medication
- History of sexual behavior
- Personal and social life
- Smoker and smoking behavior
- Consumption of alcohol and drugs

## 2.3 Clinical Examination

After a general inspection of the whole body the examining physician concentrated on a local examination of the genital organs, noting in particular any abnormalities in size or consistency, any deformities, vestiges of infection or traumata. Testicular volume was measured with an orchidometer developed by Hynie.

## 2.4 Spermatological Examination

The subjects were instructed to abstain from sexual activity for 5 days prior to giving their specimen of ejaculate. The ejaculate specimen was obtained by masturbation in the clinic. The volume of the ejaculate was measured on the graduated scale of the collection vessel. The pH value was measured using special indicator paper pH 6.6–8.0 (E. Merck, Darmstadt, West Germany). Liquefaction time and consistency were tested with a glass rod.

A sample of the ejaculate was removed with a sterile platinum loop and inoculated onto boiled blood agar and whole blood agar to test for pathogenic organisms, as well as onto a special culture medium (BAG, Lich/Main, West Germany), to test for myco-plasma. After appropriate incubation in a $CO_2$-enriched incubator, the cultures were analyzed.

Sperm motility was determined by putting liquefied ejaculate on a slide by means of a loop, and a cover glass placed on top. The percentage of motile sperm and the quality of motility were recorded within approximate limits. If less than 50% of the sperms were motile, an eosin test (vitality test) was carried out. Necrobiotic cells will absorb the dye, whereas the vital sperms do not stain.

An exact sperm count was conducted using Neubauer's hemocytometer. If no spermatozoa were found, the ejaculate was centrifuged and the sediment again sub-jected to microscopic examination. Cytomorphological differentiation of the cells was carried out by preparing a smear in the same way as a blood smear, fixing with methanol, and staining with hematoxylin-eosin. Cells were differentiated according to the following categories: spermiogenic cells, immature forms, degraded forms, microcephalous, bicephalous, mis-shapen head, mis-shapen middle piece, bicaudal, and other degenerate forms. The proportion of mis-shapen spermatozoa was express-ed as a percentage. In each case the total number of cells, including morphologically normal cells, added up to 100%.

The concentration of fructose in the seminal plasma was determined using the method developed by Nennstiel and Alich (1969).

## 2.5 Impulse Cytophotometry

An impulse cytophotometer (ICP 22) manufactured by Ortho Instruments (Westwood, Mass., USA) was used. The following functions were automated: sequential transport of the individual cells into the sharply focussed area of the microscope photometer; optical measurement of the DNA concentration in the cell; electronic sorting and storage of all measurements; depiction of results in the form of histograms and graphs. The ejaculate specimens were prepared and stained using the method developed by Otto et al. (1979a,b) (see Fig. 1).

The ejaculate obtained by masturbation was stored for approximately 2 h at room temperature until liquefaction was complete. One ml of ejaculate was then fixed by dilution with 20 ml of a 10% acetic acid solution and placed in a refrigerator until required. For the remaining tests the following solutions were needed:

*I. Tris buffer pH 7.5:*

| | |
|---|---:|
| 1. Aqua bidest | 330 ml |
| 2. Standard solution A [24.6 g tris(hydroxymethyl)aminomethane] | 250 ml |
| 3. Standard solution B (0.1 n HCl) | 420 ml |

*II. Decondensation solution:*

| | |
|---|---:|
| 1. Papain 60,000 U/g (Carl Roth, Karslruhe) | 0.4 g |
| 2. Walpole buffer pH 5.5 | 80 ml |
| consisting of: | |
| Standard solution A (13.6 g $CH_3COONa \times 3\ H_2O$ in 1,000 ml aqua bidest) | 88 ml |

| | |
|---|---|
| + Standard solution B (5.7 ml CH$_3$COOH 100% in 1,000 ml aqua bidest) | 12 ml |
| 3. Dithioerythritol (DTE) | 250 mg |
| 4. Dimethyl sulphoxide (DMSO) | 0.8 ml |

*III. Dye solution:*

| | |
|---|---|
| 1. Tris buffer pH 7.5 | 200 ml |
| 2. 4′,6-diamidino-2-phenylindole × 2 HCl (DAPI 1,000, Serva, Heidelberg) 0.05% aqueous solution | 2 ml |
| 3. MgCl$_2$ × 6 H$_2$O | 0.8 g |
| 4. NaCl | 1.0 g |

To separate the sperms from the fixing solution, 2 ml of the sample were diluted with 8 ml of the Tris buffer pH 7.5, centrifuged for 15 min at 1,500 rpm, and the supernatant liquid discarded. The sediment was then incubated for 10 min at room temperature in the decondensation solution, and centrifuged for 15 min at 1,500 rpm. The cells were resuspended in 5 ml of the dye solution. The suspension was subsequently passed through a nylon filter in order to remove any larger particles which might be present. The chicken erythrocytes used as an internal standard were subjected to the same treatment. The system was washed after examination of each sample.

## 2.6 Hormone Analysis

Blood samples for hormone determinations were performed between 08:30 and 10:00 h for all subjects. The testosterone concentration in the serum was determined by radioimmuno-assay using the technique developed by Pirke (1973) The variance in the series was 7% at a mean concentration of 579 ng/100 ml.

The concentration of LH and FSH in the serum were measured radioimmunologically in one step (Serono, Freiburg, West Germany). The double antibody method was used. The variance of LH in the series was 8.2% at a mean concentration of 4.7 mU/ml, while the variance of FSH was 9.0% at a mean concentration of 4.9 mU/ml.

## 2.7 Statistical Analysis

At first a number of plausibility tests were run to check the validity of the variable values. The analyses were based on the SPSS program, Statistical Package for the Social Sciences, Version 8.1 (Beutel et al. 1980) and on own computer programs and were conducted using standard statistical methods. Description of distributions are mainly given by the arithmetic mean ($\bar{x}$) and the standard deviation (s). Differences between smokers and non-smokers were checked for significance using an appropriate test: t-test, Mann-Whitney-test, $\chi^2$-tests [modified if necessary (Lienert 1978)].

In reading the results one should be aware that many significance tests are involved and so the problem of simultaneous inference arise (Jones and Rushton 1982). We have not adjusted our p-values for the effect of the many tests performed and therefore the reader should have in mind that some significant results could happened by chance alone.

# 3 Results

Out of a total of 350 males aged between 19 and 40 who volunteered to take part in the study, 333 were selected. Their occupations are listed in Table 2. The majority of the subjects were students.

**Table 2.** Occuaption of subjects

| Occupation | Number of subjects |
|---|---|
| Graduate | 14 |
| Salaried employee | 25 |
| Public servant | 6 |
| Businessman | 1 |
| Skilled trades | 12 |
| Arts/crafts | 14 |
| Medical professions | 5 |
| Armed forces/social services as alternative to compulsory military service | 3 |
| Students (University etc.) | 239 |
| Students (school) | 2 |
| No reply | 12 |

## 3.1 General Description of Groups

Of the 333 males, 140 were classified as non-smokers and 193 as smokers. Since it is known that particular diseases, certain forms of medication, alcohol, drugs, and radiation can have a damaging effect on the germinal epithelium, those men whose anamnesis revealed any exposure to these factors were excluded from both groups. Table 3 lists those factors which led to the exclusion of 58 non-smokers (reducing the group to a total of 82). The smoker group was reduced by 95 to 98 for the same reasons. The percentage of smokers excluded (49.2%) was larger than the corresponding percentage of non-smokers (41.4%), but the difference was not statistically significant.

In many cases it was found that several of the criteria for exclusion applied to one subject. Alcohol consumption was regarded as grounds for exclusion if the subject's consumption exceeded a certain limit (see Table 5). Subjects were also excluded if they admitted to using drugs (including the so-called "soft drugs" hashish, marijuana, and LSD, irrespective of the amount taken), if they had been exposed to radiation over and above the limits received in a routine X-ray examination, or if they had been prescribed certain forms of medication (tetracyclines, hormones, psychotropic drugs). Other criteria for exclusion were a testicular volume of less than 10 ml (even if only one testis was affected), and operations performed in the genital area. Three subjects (1 non-smoker, 2 smokers) were excluded because they gave no response to questions on their drug intake, while 1 smoker was excluded due to his failure to respond to questions on his consumption of drugs and alcohol. The group of respondents prior to exclusion was referred to as "Group I", and the group which remained after the above-mentioned procedure was referred to as "Group II".

**Table 3.** Causes of possible damage to the germinal epithelium which were used as criteria for exclusion. Distribution of excluded subjects among smokers and non-smokers

| Exclusion criteria | Number of subjects excluded | | |
|---|---|---|---|
| | Total group | Non-smokers | Smokers |
| Alcohol consumption (more than 1 l of beer or 1/2 l of wine per day) | 35 | 5 | 30 |
| Drugs | 37 | 5 | 32 |
| Medication | 6 | 3 | 3 |
| Exposure to radiation | 16 | 7 | 9 |
| Inflammatory disease of the genital organs | 3 | 2 | 1 |
| Undescended testis | 10 | 6 | 4 |
| Varicocele testis | 60 | 33 | 27 |
| Varicocele   right-sided | 1 | 0 | 1 |
| left-sided | 56 | 26 | 23 |
| bilateral | 3 | 2 | 1 |
| Operations performed in the genital area | 7 | 7 | – |
| Size of testes (less than 10 ml) | 4 | 2 | 2 |
| No reply | 4 | 1 | 3 |

The smoker and non-smoker subgroups within Group I and II were compared in terms of the following parameters (see Tables 4–6):

- age
- height
- weight
- previous illnesses
- material status and number of children
- smoking habits (smoker subgroups).

The smoker and non-smoker subgroups in Group I were fairly similar in terms of age. In Group II, however, the smokers were significantly older than the non-smokers – evidently the exclusion criteria applied more frequently to young smokers. As far as marital status is concerned, the number of divorcees in the smoker subgroup (prior to adjustment using the exclusion criteria) was significantly higher.

The number of children in smoker and non-smoker groups are not significantly different and relatively low. The latter is obviously due to age and social status of our subjects.

Table 5 shows the figures for alcohol and drug consumption and smoking behavior within the various subgroups. The proportion of subjects who do not drink alcohol was the same in both the smoker and non-smoker subgroups. The proportion of subjects with a low level of alcohol consumption was, however, greater among non-smokers than among the smokers. Similarly, the proportion of subjects with a moderate to heavy consumption of alcohol was significantly higher among the smokers. This also applied to drug intake, which was significantly higher among the smoker subgroup.

**Table 4.** A comparison of the smoker and non-smoker subgroups of Groups I and II in terms of age, height, weight, and marital status

| Variable | I | | | | | | II | | | | | |
|---|---|---|---|---|---|---|---|---|---|---|---|---|
| | Non-smokers N = 140 | | Smokers N = 193 | | Difference | | Non-smokers N = 82 | | Smokers N = 98 | | Difference |
| | x̄ | s | x̄ | s | | | x̄ | s | x̄ | s | |
| Age (years) | 25.3 | 3.7 | 25.7 | 3.8 | ns | | 25.0 | 3.4 | 26.2 | 4 | p<0.03 |
| Height (cm) | 179.9 | 6.6 | 180.5 | 6.6 | ns | | 180.4 | 6.6 | 181.0 | 6.6 | ns |
| Weight (kg) | 71.37 | 7.7 | 72.66 | 9.3 | ns | | 71.76 | 7.6 | 73.3 | 9.4 | ns |
| Marital status | | | | | | | | | | | |
| Married or similar relationship | 73 (52 %) | | 101 (52 %) | | ns | | 42 (51 %) | | 51 (52 %) | | ns |
| Divorced | 5 (3.6%) | | 18 (9.4%) | | p<0.05 | | 0 (0 %) | | 4 (4 %) | | ns |
| 1 child | 4 (2.9%) | | 9 (4.7%) | | ns | | 2 (2.4%) | | 5 (5.1%) | | ns |
| 2 children | 2 (1.4%) | | 4 (2.1%) | | ns | | 2 (2.4%) | | 2 (2.0%) | | ns |

x̄ = Arithmetic mean;   s = Standard deviation

**Table 5.** Smoking habits, drug, and alcohol consumption among smokers and non-smokers in Groups I and II

| Variable | I | | | II | | |
|---|---|---|---|---|---|---|
| | Non-smokers N = 140 | Smokers N = 193 | Difference | Non-smokers N = 82 | Smokers N = 98 | Difference |
| Daily alcohol consumption | | | | | | |
| None | 14 (10.1%) | 23 (12.1%) | ns | 10 (12%) | 16 (16.3%) | ns |
| Low[a] | 95 (68.4%) | 85 (44.7%) | p<0.01 | 59 (72%) | 53 (54.1%) | p<0.02 |
| Moderate[a] | 25 (18.0%) | 52 (27.4%) | p<0.02 | 13 (16%) | 29 (29.6%) | p<0.04 |
| Heavy[a] | 5 (3.6%) | 30 (15.8%) | p<0.01 | | | |
| Daily drug consumption | | | | | | |
| None | 134 (96.4%) | 157 (83.1%) | ns | | | |
| Hashish | 4 (2.9%) | 24 (12.7%) | p<0.02 | | | |
| Marijuana | 0 (0 %) | 4 (2.1%) | ns | Excluded | | |
| Other drugs (e.g. LSD) | 1 (0.7%) | 4 (2.1%) | ns | | | |
| Total drugs | 5 (3.6%) | 32 (16.9%) | p<0.01 | | | |
| Smoking habits | | | | | | |
| Length of time smoked (in years) | | 8.36 s = 4,2 | | | 8.76 s = 4.3 | |
| No. of cigarettes per day | | 20.4 s = 8.0 | | | 19.7 s = 6.8 | |
| 20 or fewer cigarettes per day | | 129 (67%) | | | 70 (71%) | |
| More than 20 cigarettes per day | | 63 (23%) | | | 28 (29%) | |

[a] Low    = 1/2 l beer or 1/4 l wine per day
Moderate = 1 l beer or 1/2 l wine per day
Heavy   = more than 1 l beer or 1/2 l wine per day

There was no difference in smoking habits between Groups I and II. In Group I, for example, the proportion of smokers who rolled their own cigarettes was 15%, while the corresponding figure for Group II was 17%. 74% of the smokers in Group I claimed to smoke regularly, compared with 75% in Group II. The proportion of smokers who smoked regularly throughout the day was 42% in the Group I smoker subgroup, and 44% in the Group II smoker subgroup. The remaining smokers smoked mainly in the evenings. Of the Group I non-smokers, 42% claimed to have smoked when they were younger (ex-smokers), while for Group II the figure was 41%. On average the ex-smokers had smoked for a period of 4.5 years (Group I) and 4.6 years (Group II), while their average daily consumption during that period had been 10.5 and 10.0 cigarettes respectively. All of the ex-smokers had given up smoking at least 1 year prior to the beginning of our study.

As spermatogenesis can be affected by past or present illness, and even by vegetative disorders, the incidence of such illnesses in smokers and non-smokers was compared in Groups I and II (Table 6), however no significant differences were found.

**Table 6.** Anamnestic data relating to past or present illnesses or vegetative disorders in Groups I and II

| Illness | I | | | II | | |
|---|---|---|---|---|---|---|
| | Non-smokers N = 140 | Smokers N = 193 | Differ-ence | Non-smokers N = 82 | Smokers N = 98 | Differ-ence |
| Diabetes mell. | 1  (0.7%) | 3  (1.6%) | ns | 0  (0  %) | 0  (0  %) | ns |
| Thyreopathy | 8  (5.8%) | 11  (5.7%) | ns | 4  (4.9%) | 4  (4.1%) | ns |
| Tuberculosis | 1  (0.7%) | 7  (3.7%) | ns | 1  (1.2%) | 4  (4.1%) | ns |
| Jaundice | 7  (5.0%) | 14  (7.3%) | ns | 5  (6.1%) | 9  (9.2%) | ns |
| Mumps | 66 (47.0%) | 72 (37.0%) | ns | 40 (49  %) | 37 (38  %) | ns |
| Hemorrhoids | 31 (22.1%) | 43 (22.5%) | ns | 18 (22  %) | 22 (22  %) | ns |
| Constipation | 14 (10.0%) | 15  (7.9%) | ns | 6  (7.3%) | 7  (7.1%) | ns |
| Somnipathy | 24 (17.1%) | 33 (17.3%) | ns | 12 (14.6%) | 18 (18.4%) | ns |

## 3.2  Results of Andrological Examination

The smokers and non-smokers subgroups within Groups I and II were compared with respect to

- sexual development
- general clinical findings
- spermatological findings
- endocrinological findings.

The results are discussed below.

Table 7 indicates that there are no substantial differences in sexual behaviour between smokers and non-smokers. The figures do show, however, that the Group I smokers were significantly younger than the non-smokers when they had their first experience of coitus (p < 0.01). An equal proportion of smokers and non-smokers experienced difficulties in coitus. It is worth noting that reports of several such problems occurring in conjunction with one another were more frequent among smokers than among non-smokers. The incidence of reported varicoceles was higher among the non-smokers. The fact that varicoceles were found in two non-smokers of Group II, although varicoceles was one of the criteria for excluding subjects from this group, is due to a discrepancy between the anamnestic data collected from the subjects and the clinical findings of the examining physician.

Table 8 shows the clinical-andrological findings. Once again we find that the differences between the smoker and non-smoker subgroups are either very slight or non-existent. Balanitis was the most common pathological change of the penis. The incidence of varicoceles actually found was significantly higher in non-smokers (p<0.03).

The spermatological findings are presented in Table 9. The percentages given for motility refer to the percentage of motile spermatozoa observed after a period of

**Table 7.** Anamnestic data from Groups I and II relating to sexual development, sexual activity, diseases and operations in the genital area

| Variable | I | | | II | | |
|---|---|---|---|---|---|---|
| | Non-smokers N = 140 | Smokers N = 193 | Differ- ence | Non-smokers N = 82 | Smokers N = 98 | Differ- ence |
| 1st experience of pollution/ masturbation (age) | 13.3 (s = 1.9) | 13.2 (s = 2.4) | ns | 13.3 (s = 2.0) | 13.0 (s = 2.6) | ns |
| 1st experience of coitus (age) | 18.8 (s = 2.4) | 17.4 (s = 2.0) | p<0.01 | 18.4 (s = 2.1) | 17.8 (s = 2.2) | ns |
| Frequency of masturbation (per month) | 8.7 (s = 9.7) | 7.5 (s = 9.3) | ns | 8.7 (s = 7.5) | 7.2 (s = 10.7) | ns |
| Frequency of coitus (per month) | 8.5 (s = 6.6) | 10.1 (s = 7.8) | ns | 8.8 (s = 6.7) | 9.9 (s = 7.3) | ns |
| Coital dif- ficulties associated with | | | | | | |
| Libido | 3 (2.1%) | 3 (4.7%) | ns | 2 (2.4%) | 4 (4.1%) | ns |
| Erection | 4 (2.9%) | 13 (6.7%) | ns | 2 (2.4%) | 9 (9.2%) | ns |
| Orgasm | 5 (3.6%) | 7 (3.6%) | ns | 1 (1.2%) | 4 (4.1%) | ns |
| Ejaculation | 3 (2.1%) | 5 (2.6%) | ns | 2 (2.4%) | 3 (3.1%) | ns |
| Ectopia of the testicle | 6 (4.3%) | 4 (2.1%) | ns | – | – | – |
| Herniotomy | 10 (7.1%) | 15 (7.8%) | ns | 5 (6.1%) | 8 (8.2%) | ns |
| Hydrocele | 2 (1.4%) | 1 (0.5%) | ns | 0 (0 %) | 1 (1 %) | ns |
| Varicocele | 7 (5 %) | 1 (0.5%) | p<0.01 | 2 (2.4%) | 0 (0 %) | ns |
| Orchitis | 1 (0.7%) | 1 (0.5%) | ns | – | – | – |
| Epididymitis | 1 (0.7%) | 5 (2.6%) | ns | 1 (1.2%) | 3 (3.1%) | ns |
| Prostatitis | 2 (1.4%) | 4 (2.1%) | ns | 1 (1.2%) | 3 (3.1%) | ns |
| Non-specific urethritis | 14 (10 %) | 17 (8.8%) | ns | 9 (11 %) | 10 (10.2%) | ns |
| Gonorrhoeal urethritis | 9 (6.4%) | 25 (13 %) | ns | 6 (7.1%) | 12 (12.2%) | ns |

20 min. With the single exception of the total sperm count none of the parameters studied revealed any significant differences between the various subgroups. Moreover, the significant difference in sperm density (total sperm count) between the smoker and non-smoker subgroups in Group I are no longer observable in Group II. The ejaculate specimens were bacteriologically normal in all but 16 cases (6 non-smokers, 10 smokers).

Similarly, the differential spermiocytogram (Table 10) revealed no differences be- tween the smoker and non-smoker subgroups except that in Group I a significantly

**Table 8.** Clinical-andrological findings in the genital area for Groups I and II

| Condition | I | | | II | | |
|---|---|---|---|---|---|---|
| | Non-smokers N = 140 | Smokers N = 193 | Differ- ence | Non smokers N = 82 | Smokers N = 98 | Differ- ence |
| Phimose | 2 (1.4%) | 2 (1.0%) | ns | 2 (2.4%) | 2 (2 %) | ns |
| Circumcision | 19 (13.6%) | 24 (12.4%) | ns | 11 (13.4%) | 10 (10.2%) | ns |
| Other changes of the penis | 8 (5.8%) | 21 (11.2%) | ns | 6 (7.3%) | 15 (15.3%) | ns |
| Testicular volume | | | | | | |
| Right (ml) | $\bar{x}$ = 25.3 | $\bar{x}$ = 25.0 | ns | $\bar{x}$ = 26.4 | $\bar{x}$ = 25.8 | ns |
| Left (ml) | $\bar{x}$ = 23.7 | $\bar{x}$ = 24.1 | ns | $\bar{x}$ = 25.2 | $\bar{x}$ = 25.1 | ns |
| Testicular consistency | | | | | | |
| Resilient | 136 (97.1%) | 193 (100 %) | | 81 (99 %) | 98 (100 %) | |
| Hard | 2 (1.4%) | 0 (0 %) | ns | 1 (1 %) | 0 (0 %) | ns |
| Soft | 2 (1.49%) | 0 (0 %) | | 0 (0 %) | 0 (0 %) | |
| Changes in epididymis | 7 (5 %) | 7 (3.6%) | ns | 3 (3.7%) | 1 (1.0%) | ns |
| Thickening of spermatic cord | 7 (5 %) | 5 (2.6%) | ns | 0 (0 %) | 3 (3.1%) | ns |
| Varicocele | 33 (23.7%) | 27 (14.1%) | p<0.03 | Excluded | | |

greater frequency of microcephalous forms was found in the non-smoker subgroup than in the smoker subgroup.

Figures 2–4 illustrate these various findings in diagrammatic form. Figure 5 shows the distribution of sperm density within the smoker and non-smoker subgroups of Groups I and II. The scale on the x-axis was determined by the criteria of the European Andrology Club. For Group I there was a significant difference in sperm density between the smoker and non-smoker subgroups (p < 0.04 – see Table 9). This difference is primarily due to a reduction in sperm density in smokers with normozoospermia (40–250 million/ml, Fig. 2a). Among subjects with pathological sperm densities the differences between the smoker and non-smoker subgroups are minimal, and contribute only insignificantly to the lowering of the average sperm count for smokers. In Group II the total sperm count is still an average 8 million lower than for non-smokers, yet the difference is again not significant (Table 9). This is clearly illustrated in Fig. 2b, where the non-smoker subgroup of Group II is compared with the smoker subgroup. From this data, one cannot exclude some effect of smoking on the sperm count.

Figures 3 and 4 show the distribution of morphologically normal spermatozoa and sperm motility within the smoker and non-smoker subgroups of Groups I and II. The curves show only slight differences in the distribution of the percentage of normal spermatozoa for the two subgroups (Fig. 3a,b). The distribution curves for sperm motility are almost identical for smokers and non-smokers (Fig. 4a,b).

**Table 9.** Ejaculate parameters for Groups I and II

| Variable | I | | | | | II | | | | | |
|---|---|---|---|---|---|---|---|---|---|---|---|
| | Non-smokers N = 140 | | Smokers N = 193 | | Differ-ence | Non-smokers N = 82 | | Smokers N = 98 | | Differ-ence |
| | x̄ | s | x̄ | s | | x̄ | s | x̄ | s | |
| Volume (ml) | 4.23 | 1.86 | 3.90 | 1.97 | ns | 4.06 | 1.90 | 3.93 | 2.06 | ns |
| pH value | 7.51 | 0.25 | 7.52 | 0.27 | ns | 7.52 | 0.24 | 7.50 | 0.27 | ns |
| Fructose (mg/ml) | 2.46 | 1.1 | 2.47 | 1.1 | ns | 2.37 | 1.07 | 2.26 | 0.97 | ns |
| Total sperm count (millions/ml) | 66.85 | 60.5 | 55.09 | 53.5 | $p < 0.04$ | 70.62 | 66.6 | 62.08 | 56.9 | ns |
| Motility after 20 min | 71.9% | 16.6 | 72.4% | 15.0 | ns | 71.3% | 17.4 | 72.3% | 15.6 | ns |
| Liquefaction | | | | | | | | | | |
| Under 30 min | 91 (65 %) | | 125 (65%) | | | 54 (66%) | | 60 (63%) | | |
| 30–99 min | 16 (11 %) | | 23 (12%) | | ns | 8 (10%) | | 10 (11%) | | ns |
| Over 99 min | 33 (24 %) | | 43 (23%) | | | 20 (24%) | | 25 (26%) | | |
| Epithelial cells (more than 5%) | 2 (1.4%) | | 7 (37%) | | ns | 2 (2%) | | 3 (3%) | | ns |
| Leucocytes | | | | | | | | | | |
| None | 16 (11%) | | 27 (14%) | | | 11 (13%) | | 13 (13%) | | |
| A few | 118 (84%) | | 155 (81%) | | ns | 59 (72%) | | 66 (61%) | | ns |
| Many | 6 (2%) | | 8 (4%) | | | 9 (11%) | | 15 (15%) | | |
| Very many | 0 (0%) | | 2 (1%) | | | 3 (4%) | | 3 (3%) | | |

Table 10. Spermatozoal morphology in Groups I and II

| Morphological forms | I Non-smokers N = 140 | | Smokers N = 193 | | Difference | II Non-smokers N = 82 | | Smokers N = 98 | | Difference |
|---|---|---|---|---|---|---|---|---|---|---|
| | $\bar{x}$ | s | $\bar{x}$ | s | | $\bar{x}$ | s | $\bar{x}$ | s | |
| Normal spermatozoa | 67.3 % | 15.5 | 67.0 % | 13.8 | ns | 67.1 % | 16.9 | 67.7 % | 13.1 | ns |
| Immature forms | 6.77 % | 6.0 | 6.32 % | 5.2 | ns | 7.94% | 6.2 | 6.35% | 5.0 | ns |
| Degraded forms | 7.71 % | 8.1 | 8.86 % | 8.1 | ns | 7.14% | 8.5 | 8.86% | 8.1 | ns |
| Degenerate forms | 18.2 % | 12.1 | 17.7 % | 9.8 | ns | 17.8 % | 13.2 | 17.1 % | 9.5 | ns |
| Microcephalous | 0.218% | 0.54 | 0.087% | 0.34 | $p < 0.01$ | 0.25% | 0.6 | 0.12% | 0.04 | ns |
| Bicephalous | 0.368% | 2.23 | 0.361% | 0.82 | ns | 0.48% | 2.8 | 0.39% | 0.91 | ns |
| Mis-shapen head | 9.77 % | 7.1 | 9.30 % | 7.3 | ns | 9.91% | 7.3 | 9.55% | 7.1 | ns |
| Deformities of middle piece | 4.90 % | 6.1 | 5.60 % | 6.3 | ns | 4.06% | 5.8 | 5.21% | 6.1 | ns |
| Bicaudal | 0.26 % | 0.61 | 0.30 % | 0.65 | ns | 0.25% | 0.58 | 0.24% | 0.62 | ns |
| Others | 2.64 % | 5.8 | 2.00 % | 2.8 | ns | 2.82% | 7.1 | 1.58% | 2.0 | ns |

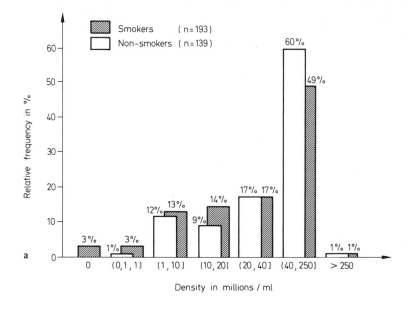

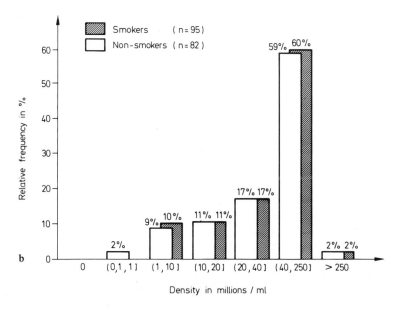

**Fig. 2a,b. a** Sperm density in smokers and non-smokers in Group I. **b** Sperm density in smokers and non-smokers in Group II

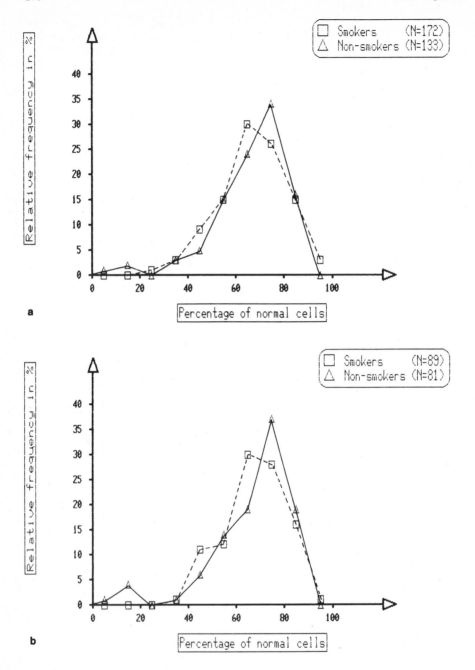

**Fig. 3a,b. a** Percentages of normal spermatozoa in smokers and non-smokers in Group I. **b** Percentages of normal spermatozoa in smokers and non-smokers in Group II

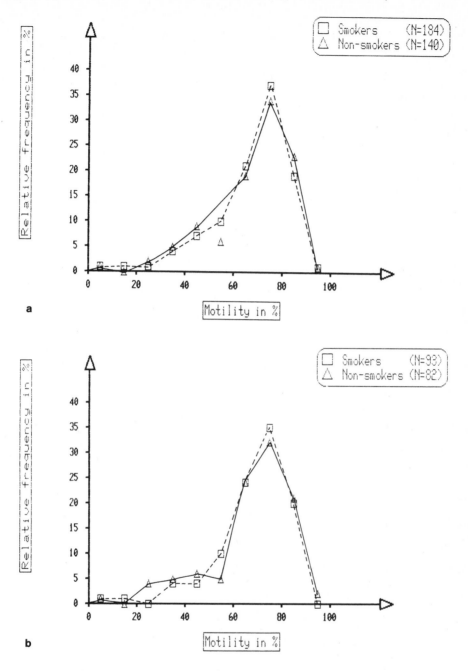

**Fig. 4a,b. a** Motility of spermatozoa in smokers and non-smokers in Group I. **b** Motility of spermatozoa in smokers and non-smokers in Group II

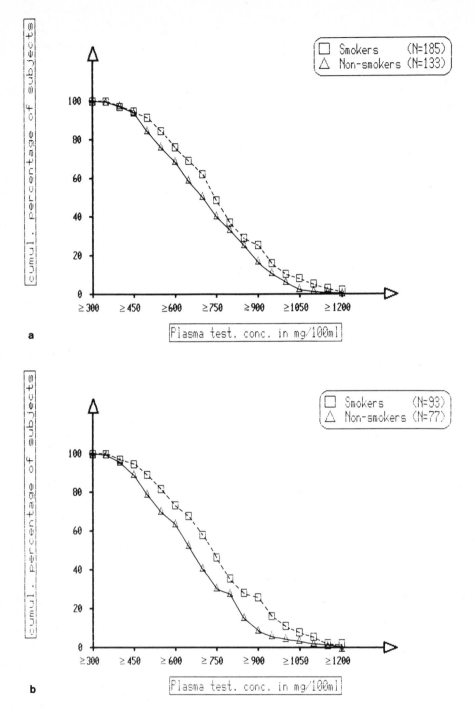

**Fig.5a,b. a** Cumulative testosterone concentrations in smokers and non-smokers in Group I.
**b** Cumulative testosterone concentrations in smokers and non-smokers in Group II

If the smoker and non-smoker subgroups are classified spermatologically according to the categories devised by the European Andrology Club, the differences between the two subgroups are again not significant (see Table 11). If the cases of
cryptozoospermia and azoospermia are added together, we find a significant difference ($p < 0.02$) between the smokers and non-smokers of Group I which is not apparent in Group II.

**Table 11.** Spermatological classification of smokers and non-smokers in Groups I and II

| Spermatological classification | I | | | II | | |
|---|---|---|---|---|---|---|
| | Non-smokers N = 140 | Smokers N = 193 | Differ-ence | Non-smokers N = 82 | Smokers N = 98 | Differ-ence |
| Polyzoospermia | 2 (1.5%) | 2 (1 %) | | 2 (2.4%) | 1 (1%) | |
| Normozoospermia | 61 (44 %) | 73 (38 %) | | 35 (43 %) | 44 (45%) | |
| Normo-asthenozoo-spermia | 5 (4 %) | 6 (3 %) | | 4 (5 %) | 4 (4%) | |
| Normo-teratozoo-spermia | 12 (9 %) | 15 (8 %) | ns | 8 (10 %) | 7 (7%) | ns |
| Normo-astheno-teratozoospermia | 5 (4 %) | 1 (0.5%) | | 2 (2.4%) | 1 (1%) | |
| Oligozoospermia | 53 (38 %) | 84 (43 %) | | 31 (38 %) | 35 (36%) | |
| Cryptozoospermia | 1 (0.5%) | 5 (2.5%) | | – | 3 (3%) | |
| Azoospermia | – | 6 (3 %) | | – | 2 (2%) | |

Table 12 demonstrates the influence of cigarette smoking on sperm motility in
relation to spermatological classification. Sperm motility differs from one category
to another, but for any given category there is no significant difference between the
smoker and non-smoker subgroups.

A comparison of the DNA concentrations in the spermatozoa of smokers and non-
smokers was made by means of impulse cytophotometry. It was not possible, however, to include all the specimens of ejaculate. Fifteen of the specimens (6 from
non-smokers, 9 from smokers) were rejected due to inadequate fluorochroming,
23 (10 non-smokers, 13 smokers) due to hypospermia, and 6 samples from smokers
due to azoospermia. Table 13 shows the findings of impulse cytophotometry applied
to 289 usable ejaculate specimens from both smokers and non-smokers of Group I
and 156 specimens from both subgroups in Group II. Analysis of the results reveals
no significant differences between smokers and non-smokers.

If the results of impulse cytophotometry are analyzed in terms of spermatological
classification (Table 14), the only significant differences between smokers and non-
smokers in Group I are to be found in the normospermia count for the hypohaploid
range ($p < 0.04$), as well as the number of normozoo-asthenospermia in the total cell
count ($p < 0.02$). The significantly higher impulse count in the hypohaploid range
for non-smokers with normospermia could, at least partly, be due to the higher number of microcephalous spermatozoa which are also associated with non-smokers. In
Group II we find no differences between smokers and non-smokers.

**Table 12.** Percentage of motile spermatozoa in smokers and non-smokers in Groups I and II according to the spermatological classification of the European Andrology Club (Eliasson 1970)

| Spermatological classification | I | | | | | | II | | | | | |
| --- | --- | --- | --- | --- | --- | --- | --- | --- | --- | --- | --- | --- |
| | Non-smokers N = 140 | | Smokers N = 193 | | Difference | | Non-smokers N = 82 | | Smokers N = 98 | | Difference | |
| | $\bar{x}$ | N | $\bar{x}$ | N | | | $\bar{x}$ | N | $\bar{x}$ | N | | |
| Polyzoospermia | 87.5% | 2 | 82.5% | 2 | ns | | 87.5% | 2 | 80.0% | 1 | ns | |
| Normozoospermia | 78.9% | 61 | 79.1% | 73 | ns | | 78.1% | 35 | 79.2% | 44 | ns | |
| Normo-asthenozoo-spermia | 46.0% | 5 | 37.5% | 6 | ns | | 45.0% | 4 | 36.3% | 4 | ns | |
| Normo-teratozoo-spermia | 78.8% | 12 | 75.7% | 15 | ns | | 78.1% | 8 | 72.9% | 7 | ns | |
| Normo-astheno-teratozoospermia | 42.0% | 5 | 55.0% | 1 | ns | | 40.0% | 2 | 55.0% | 1 | ns | |
| Oligozoospermia | 67.9% | 53 | 68.8% | 84 | ns | | 66.1% | 31 | 67.7% | 35 | ns | |
| Cryptozoospermia | 50.0% | 1 | 60.0% | 3 | ns | | – | – | 30.0% | 1 | – | |

**Table 13.** Results of impulse cytophotometry on ejaculate specimens from smokers and non-smokers in Groups I and II

| Impulse cytophoto-metric variable | I | | | | | | II | | | | | |
| --- | --- | --- | --- | --- | --- | --- | --- | --- | --- | --- | --- | --- |
| | Non-smokers N = 124 | | Smokers N = 165 | | Difference | | Non-smokers N = 71 | | Smokers N = 85 | | Difference | |
| | $\bar{x}$ | s | $\bar{x}$ | s | | | $\bar{x}$ | s | $\bar{x}$ | s | | |
| Total cell count | 27,800 | 7,217 | 27,100 | 8,673 | ns | | 27,763 | 6,369 | 27,407 | 9,910 | ns | |
| Hypohaploid range | 1,555 | 1,575 | 1,427 | 1,927 | ns | | 1,555 | 1,478 | 1,318 | 1,595 | ns | |
| Haploid peak (1 c) | 21,191 | 5,489 | 21,039 | 5,069 | ns | | 21,048 | 4,092 | 20,955 | 5,807 | ns | |
| Diploid peak (2 c) | 1,012 | 1,224 | 1,045 | 1,036 | ns | | 1,265 | 1,340 | 1,186 | 1,206 | ns | |
| $V_D$[a] | 4,041 | 3,517 | 3,746 | 4,121 | ns | | 3,896 | 2,333 | 4,261 | 5,082 | ns | |

[a] $V_D$ = total impulse count – (hypohaploid range + haploid peak + diploid peak)

**Table 14.** Results of impulse cytophotometry for ejaculate specimens from smokers (S) and non-smokers (NS) in Groups I and II according to spermatological classification

| Impulse cytophotometric variable | | Total cell count | | | | | Hypohaploid range | | | Haploid peak (1 c) | | | Diploid peak (2 c) | | |
|---|---|---|---|---|---|---|---|---|---|---|---|---|---|---|---|
| | | NS $\bar{x}$ | N | S $\bar{x}$ | N | D | NS $\bar{x}$ | S $\bar{x}$ | D | NS $\bar{x}$ | S $\bar{x}$ | D | NS $\bar{x}$ | S $\bar{x}$ | D |
| P | I | 25,854 | 2 | 25,655 | 1 | ns | 970 | – | – | 19,165 | 19,645 | ns | 571 | 1,642 | ns |
| | II | 25,854 | 2 | – | – | – | 970 | – | – | 19,165 | – | – | 571 | – | – |
| No | I | 27,096 | 58 | 25,393 | 71 | ns | 1,343 | 1,002 | p<0.04 | 20,767 | 20,337 | ns | 999 | 1,144 | ns |
| | II | 26,496 | 33 | 25,807 | 42 | ns | 1,075 | 1,057 | ns | 20,633 | 20,229 | ns | 1,250 | 1,323 | ns |
| N-A | I | 27,827 | 5 | 42,626 | 4 | p<0,02 | 1,419 | 6,230 | ns | 22,181 | 27,364 | ns | 1,310 | 749 | ns |
| | II | 27,486 | 4 | 43,246 | 3 | ns | 1,774 | 4,328 | ns | 21,647 | 28,662 | ns | 1,473 | 435 | ns |
| N-T | I | 26,349 | 11 | 25,332 | 15 | ns | 1,403 | 1,253 | ns | 20,582 | 20,342 | ns | 1,050 | 845 | ns |
| | II | 26,675 | 7 | 23,614 | 7 | ns | 1,741 | 1,426 | ns | 18,925 | 18,710 | ns | 1,411 | 715 | ns |
| N-A-T | I | 41,191 | 4 | 21,742 | 1 | ns | 3,985 | 658 | ns | 26,899 | 19,412 | ns | 569 | – | – |
| | II | 32,972 | 1 | 21,742 | 1 | ns | 1,150 | 658 | ns | 25,590 | 19,412 | ns | 1,230 | – | – |
| O | I | 28,517 | 43 | 28,874 | 69 | ns | 1,733 | 1,602 | ns | 21,861 | 21,909 | ns | 1,072 | 1,021 | ns |
| | II | 29,848 | 24 | 31,353 | 29 | ns | 2,189 | 1,442 | ns | 22,107 | 22,945 | ns | 1,268 | 1,302 | ns |
| C | I | 3,910 | 1 | 19,602 | 4 | ns | – | 2,505 | – | – | 13,709 | – | – | 785 | – |
| | II | 6,563 | – | 6,563 | 3 | – | – | 304 | – | – | 4,422 | – | – | – | – |

P = Polyzoospermia, No = Normozoospermia, N-A = Normo-asthenozoospermia, N-T = Normo-teratozoospermia, N-A-T = Normo-astheno-teratozoospermia, O = Oligozoospermia, C = Cryptozoospermia, N = Number of usable ejaculate specimens, D = Difference

Similarly, no differences between smokers and non-smokers are apparent in the shape of the haploid peak of the cytophotogram (Table 15).

**Table 15.** Shape of haploid peak based on cytophotometric investigation of smokers and non-smokers in Groups I and II

| Characteristic form | I | | | II | | |
|---|---|---|---|---|---|---|
| | Non-smokers N = 135 | Smokers N = 178 | Differ-ence | Non-smokers N = 81 | Smokers N = 93 | Differ-ence |
| Symmetrical | 62 (50%) | 84 (51%) | | 39 (56%) | 43 (52%) | |
| Gradual slope/right | 5 (4%) | 10 (6%) | | 5 (7%) | 4 (5%) | |
| Gradual slope/left | 42 (34%) | 49 (30%) | ns | 22 (31%) | 26 (31%) | ns |
| Steep slope/right | 2 (2%) | 6 (4%) | | 1 (1%) | 4 (5%) | |
| Steep slope/left | 12 (10%) | 15 (9%) | | 3 (4%) | 6 (7%) | |

Spermatogenesis is a hormone-dependent process. Part of our investigation was therefore concerned with measuring the plasma concentrations of those hormones which control the process of spermatogenesis. In Table 16 the concentrations of testosterone, LH, and FSH in the blood plasma are shown for the smoker and non-smoker subgroups. After adjustment of the subgroups it was found that plasma testosterone concentrations were significantly higher among the smokers than in the non-smoker subgroups. No significant differences were observed for the other parameters.

**Table 16.** Plasma concentrations of testosterone, LH and FSH in smokers and non-smokers in Groups I and II (figures for normal range in brackets)

| Hormone | I | | | | Dif-fer-ence | II | | | | Differ-ence |
|---|---|---|---|---|---|---|---|---|---|---|
| | Non-smokers N = 140 | | Smokers N = 193 | | | Non-smokers N = 82 | | Smokers N = 98 | | |
| | $\bar{x}$ | s | $\bar{x}$ | s | | $\bar{x}$ | s | $\bar{x}$ | s | |
| Testosterone (300–1,000 mg%) | 714 | 192 | 756 | 206 | ns | 672 | 178 | 749 | 213 | $p < 0.03$ |
| LH (2–20 mU/ml) | 6.9 | 3.2 | 7.0 | 2.4 | ns | 6.42 | 2.7 | 6.61 | 2.2 | ns |
| FSH (1–10 mU/ml) | 3.3 | 2.0 | 3.6 | 2.1 | ns | 2.89 | 1.3 | 3.38 | 1.8 | ns |

Figure 5 illustrates the cumulative testosterone concentrations in the blood plasma, while Fig. 6 shows the cumulative LH concentrations and Figure 7 the cumulative

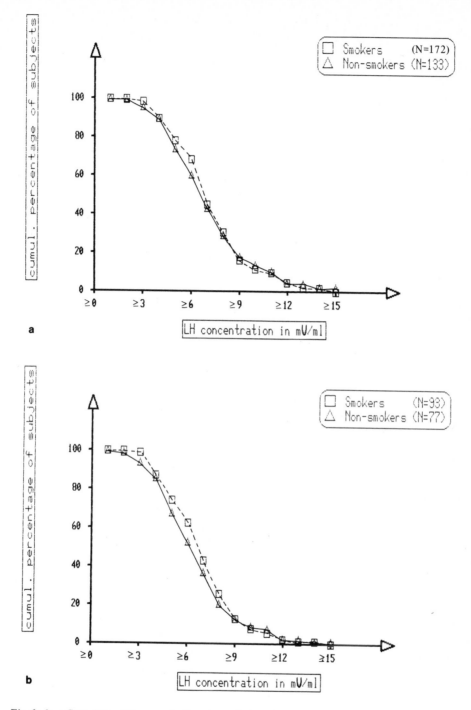

**Fig. 6a,b.** a Cumulative LH concentrations in smokers and non-smokers in Group I. b Cumulative LH concentrations in smokers and non-smokers in Group II

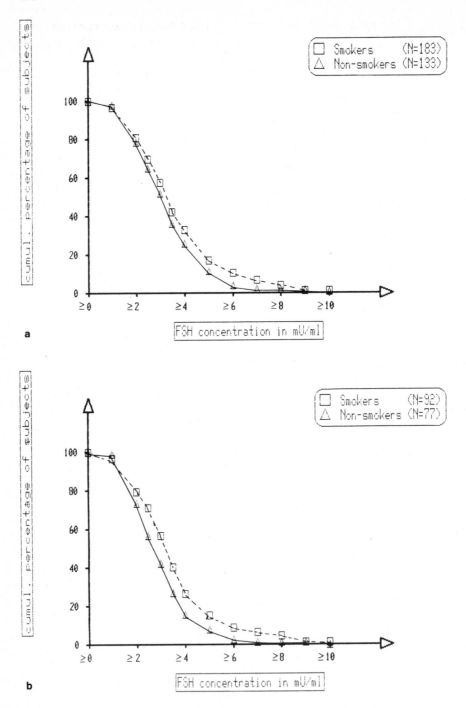

**Fig. 7a,b. a** Cumulative FSH concentrations in smokers and non-smokers in Group I. **b** Cumulative FSH concentrations in smokers and non-smokers in Group II

FSH concentrations reveal no significant differences between smokers and non-smokers, whereas a shift to the left is observable for testosterone in non-smokers compared to smokers.

# 4  Discussion

For some hundreds of years the merits and disadvantages of tobacco smoking were debated and argued on a purely emotional level. Only since the second half of the 19th century have experimental studies been conducted with the objective of proving or disproving that the smoking of tobacco is a cause of ill health or behavioral disorders. The published results of research into sexual behavior and fertility, and into the effects of paternal smoking on child development, now constitute an extensive scientific literature, but one which nonetheless remains full of contradictions.

The present study is concerned with the process of spermatogenesis in smokers and non-smokers. It examines and compares the anamnestic data furnished by the subjects, together with the results of clinical, spermatological, cytophotometric, and endrocrinological analysis.

The existing literature contains virtually no reference to genital abnormalities in smokers as compared with non-smokers. The only exceptions are Vogel et al. (1979) and Klaiber et al. (1980), who report a higher incidence of varicoceles in smokers. However, there is considerable disagreement on the question of fertility, expecially sperm density, sperm motility, and sperm morphology. Scientists whose investigations have led them to the conclusion that smoking has a damaging influence on sperm density, sperm motility, and sperm morphology (Viczian 1969; Campbell and Harrison 1979; Stekhun 1980; Evans et al. 1981; Vogel et al. 1979) disagree with others who have found no evidence that smoking has any such effect (Mellan 1967; Nebe and Schirren 1980c; Godfrey 1981; Rodriguez-Rigau 1982). The endocrinological findings for smokers and non-smokers are also contradictory. The results obtained by Briggs (1973), for example, indicate a lowering of testosterone concentrations in the serum, while other authors either found no such differences between smokers and non-smokers (Winternitz and Quillen 1977; Persky et al. 1977) or have arrived at the opposite conclusion (Gutai et al. 1981; Dotson et al. 1975; Dai et al. 1981).

If all the scientific evidence produced so far is taken into consideration, it is fair to say that on balance there is no conclusive proof that smoking is harmful to spermatogenesis. Nevertheless, it is true that a number of toxic or mutagenic substances which are absorbed into the body along with cigarette smoke could theoretically lead to an impairment of the spermatogenic process. Since a discrepancy between the actual findings and these theoretical possibilities exists, further investigation is undoubtedly warranted. It is essential, however, that the planning and organization of such a study, and the analysis of the data, should be carried out to the most stringent scientific standards. In order to define those standards it is first of all necessary to identify possible sources of error when conducting such a study, and to devise means of eliminating them.

It appears that all previous studies have concentrated mainly on patients attending infertility clinics. In other words, this selection method was biased and resulted in groups of subjects for whom smoking was probably only one of a number of risk factors – or even a habit generally in keeping with the patient's overall behavioral pattern (Eysenck 1968, 1973). Even where comparable anamnestic and somatic abnormalities have resulted in the exclusion of certain individuals from the groups (cf. Evans et al. 1981; Godfrey 1981), it is still quite possible that spermatogenesis has been impaired in other subjects as a result of some psychic disturbance. Finally, we must consider the likelihood that many males attending infertility clinics are in a state of great emotional stress due to their wounded self-esteem, particularly given the fact that in 40%–50% of cases the cause of infertility is to be found in the male (Stauber 1979; Vogt 1977; Vogt and Mayer 1980). In the light of these considerations the subjects for a study of male fertility should be selected exclusively from healthy males whose general state of physical and mental health is at least equivalent to the national norm. The only factor differentiating the smoker group from the non-smoker group should be the smoking habit itself.

It is difficult to think of another field of research which is so influenced by emotion and prejudice as the study of smoking and its effects on human health. So much is already known about the negative consequences of smoking that people are naturally predisposed to believe the worst. In order to ensure that the investigator's own personal opinions can have no influence whatsoever on the outcome of a research project of this nature, it is therefore absolutely essential to conduct a "blind" investigation. This means that the investigator himself must not know from which subject any given data have been collected, or to which test group a particular subject belongs. With the exception of the investigations by Evans et al. (1981) and by Godfrey (1981), which produced contradictory results, this condition has not been satisfactorily fulfilled by any of the studies hitherto published. In the case of those retrospective studies involving comparisons between patients attending an andrological outpatient clinic, no information is generally supplied about the group selection criteria.

Comparative studies of this kind are frequently based on groups containing very small numbers. The less homogeneous such small groups are in their composition, the greater the probability of chance differences appearing to be statistically significant. Indeed, it is interesting to note that most of the investigators who concluded that smoking is harmful based their research on small groups. Moreover, the smokers and non-smokers who made up these groups differ from one another by many other parameters apart from smoking. Apart from the studies of Evans et al. (1981) and Godfrey (1981), this was not taken into account in any of the other investigations we have seen.

Our own study was based exclusively on subjects who were in good health and aged between 20 and 40 years. Most of those who volunteered to take part were students (see Table 2). In terms of sociological structure, therefore, the group of subjects certainly cannot be regarded as representative of this age group in the male population as a whole. But as far as the spermatological findings for smokers and non-smokers are concerned, there is no reason at all why they should not be considered applicable to the population in general.

We selected our subjects from this age group because these years correspond to the main reproductive phase in the life of the male, during which the calculated parameters remain constant (MacLeod and Gold 1950, 1951; Heinke and Doepfmer 1960; Schirren 1971; Nebe and Schirren 1980b). At the same time this is the period in a smoker's life when cigarette consumption is at its highest (Statistisches Bundesamt 1978). On the basis of this sample, it is not possible to draw any conclusions about the long-term effects of smoking on the germinal epithelium. The main focus of our study, however, is not on the long-term effects of smoking, but on the influence of smoking on spermatogenesis during the years of maximum reproductive potential.

In the initial group (Group I) of 333 subjects (193 smokers and 140 non-smokers), a number of test parameters were checked to establish differences between smokers and non-smokers (see Table 3). In order to isolate the influence of smoking on spermatogenesis as far as possible from other factors, subjects were excluded from our study if they were found to exhibit one or more characteristics which are known to have a potentially negative effect on the germinal epithelium or on the maturation of the spermatozoa.

According to Stekhun (1979), excessive consumption of alcohol results in a lower sperm count and reduced sperm motility. Amelar et al. (1980) have discussed the effect of alcohol on spermatogenesis via a reduction in testosterone levels. Five non-smokers (3.6%) and 30 smokers (15.5%) were excluded from our study on the basis of this criterion.

A link between drug intake (specifically hashish and marijuana) and spermatogenesis has been suggested by Kolodny et al. (1974), Nahas (1979), Blevins and Regan (1976), and Hembree et al. (1978). Whereas only five non-smokers (3.6%) stated that they took drugs, 32 smokers (16.6%) admitted to taking drugs. All these subjects were excluded from the study.

Various forms of medication are also known to have an influence on spermatogenesis (Qureshi et al. 1972; Meyhöfer 1973; Asbjornsen et al. 1976; Steinberger 1976; Wyrobek and Bruce 1978; Schilsky et al. 1980; Drife 1982; Wyrobek 1982). Six subjects – three smokers (1.6%) and three non-smokers (2.1%) – were excluded on the grounds that they had been taking such medication.

The harmful effects of ionizing irradiation on spermatogenesis are generally acknowledged (Thorsland and Paulsen 1972; Rowley et al. 1974; Lushbaugh and Casarett 1976; Wyrobek 1979; Hahn et al. 1982). Excessive exposure to radiation during the year prior to the study (beyond a routine chest X-ray) resulted in the exclusion of nine smokers (4.7%) and seven non-smokers (5.0%).

Subjects who have suffered from orchitis, as well as those who have an undescended testis or a condition following an operation in the genital region, are likely to find their fertility impaired as a result. These criteria applied to 5 smokers (2.6%) and 15 non-smokers (10.7%) in our sample. The incidence of varicoceles may also be associated with a deterioration in spermatological findings (Russell 1954; Jecht 1977; Pontonnier et al. 1979). Varicoceles were diagnosed in 18.0% of our subjects – 27 smokers (14.1%) and 33 non-smokers (23.7%). The difference is statistically significant.

It should be mentioned that significantly more non-smokers (36%) than smokers (18%) were excluded for inflammatory disease of the genital organs, undescend-

testis, varicosed testis, operations performed in the genital area or small testes (Table 3). We cannot offer any plausible reason for this highly significant ($p < 0.001$) difference. A testicular volume of less than 10 ml (unilateral or bilateral) generally correlates with impaired sperm production (Schirren 1977; Heite and Wokalek 1980). Two smokers (1.0%) and two non-smokers (1.4%) who fell into this category were also excluded.

The application of these exclusion criteria to the initial subject group produced a new group of subjects (Group II) comprising 180 males (98 smokers and 82 non-smokers). Groups I and II were listed separately in all tables, and the results for the two groups compared. Whereas the subjects in Group I were fairly comparable in terms of age, height, and weight, analysis of Group II shows that the non-smokers are significantly younger than the smokers ($p < 0.03$), the average difference being 1.2 years. The exclusion criteria evidently applied more often to younger smokers (see Table 4).

Numerous studies have now established beyond all doubt that smokers and non-smokers differ in many other respects apart from their smoking behavior (Faust and Mensen 1974). For example, it is generally acknowledged that smokers are more extroverted (Eysenck 1968, 1973). This is associated with the greater willingness of smokers to take risks, as exemplified by their higher intake of alcohol, drugs, and medication (Arnold-Krüger 1971; Goode 1972; Wynder 1977; U.S. Public Health Services 1979; Stransky 1981), their driving behavior (Eiser et al. 1982), and their general attitude concerning health (Greitemeyer 1981a, 1981b).

Our own data support these findings. We found that the proportion of alcohol drinkers among smokers becomes significantly larger as the consumption of alcohol increases. In the case of "heavy drinking" – one of the criteria for excluding subjects – the difference is particularly striking ($p < 0.01$; see Table 5). As far as drug intake is concerned, we found that smokers take hashish with significantly greater frequency ($p < 0.02$) than non-smokers, and that their total intake of all drugs is significantly higher ($p < 0.01$).

The smokers in Group I had their first experience of sexual intercourse significantly earlier than non-smokers. In Group II the difference is no longer significant, but the trend is still discernible (Table 7). The same trend was observed by Koponen (1960) and Raboch and Mellan (1974) in their studies of Czechoslovakian adolescents. Faust (1982) reports that English boys and girls who had experienced coitus smoked more than those whose previous sexual contact had been confined to petting, while the latter group themselves smoked more than their peers who had had no sexual relations whatsoever. In a study of smoking behavior in young adults Malcolm and Shephard (1978) found that smokers were more liberal than non-smokers both in their attitude towards sexuality and in their own sexual behavior.

As far as our subjects are concerned, no differences were found between smokers and non-smokers in terms of the frequency of coition or coital difficulties. In Group I the number of divorced smokers is significantly larger ($p < 0.05$) than the number of divorced non-smokers, while in Group II we are left with only four divorces, all of them smokers. This correlates with the findings of Haenszel et al. (1956), who found a higher percentage of smokers among those who were divorced or widowed – regardless of age or sex – than among those who were married or single. Eysenck (1968) similarly found that the divorce rate was higher among smokers than non-smokers.

The differences between smokers and non-smokers which are revealed by an analysis of andrological and spermatological findings are such that it is not possible to attribute them solely to the biological effects of compounds in tobacco smoke. Clearly the entire psychological and social environment in which these persons are placed must also be considered. In applying the exclusion criteria we could only allow for the most important factors known to contribute to fertility disorders. It should, however, be assumed that even after eliminating subjects for the reasons stated, certain differences between smokers and non-smokers remain whose effects on spermatogenesis are not known.

The only notable difference revealed by the clinical examination was a significantly higher incidence of varicoceles in non-smokers as compared to smokers. This was evident in both the anamnestic data furnished by the subjects (Table 7) and in the results of the clinical-andrological examination (Table 8). Ducot et al. (1981) examined 576 males and found varicoceles in 16.9% of smokers and 18.4% of non-smokers, i.e. they found that smoking had no influence on this parameter. These findings conflict with those of Vogel et al. (1979) and Klaiber et al. (1980), who reported varicoceles in 37% of 30 smokers and 10% of 39 non-smokers, and 20% of 202 smokers and 9% of 203 non-smokers respectively. The theory that tobacco smoking contributes in any way to a congenital disposition towards varicosity can probably be excluded. We are inclined to regard our own findings – which would tend to suggest that smoking actually protects against varicoceles – and those of Vogel et al. (1979) and Klaiber et al. (1980), from which the opposite may be inferred, as purely chance results.

No differences in testicular size between smokers and non-smokers were found either in Group I or in Group II (Table 8). This indicates that the results of animal experiments in which testicular atrophy was reported following the administration of nicotine (Larson et al. 1961) cannot be extrapolated to humans. Our findings do not allow comment on the hypothesis advanced by Gey (1971), Nebe and Schirren (1980c), Viczian (1969), and Campbell and Harrison (1979), who suggest that smoking only has an effect in cases where the testes have already been damaged in other ways.

The results of the spermatological analysis indicate that sperm densities in Group I are significantly lower for smokers than for non-smokers. In Group II, however, the difference is no longer significant, although the trend is still apparent (Fig. 3 and Table 9). Sperm densities for both smoker subgroups are well within normal limits. At first glance our own findings in respect of sperm density would appear to confirm the results reported by Viczian (1969), Vogel et al (1979), and Stekhun (1980). The fact that no significant difference in sperm density was evident in Group II suggests that the reduction in sperm density may be partially due to the exclusion criteria we have described, which were *not* applied in the aforementioned analyses. A further contributory factor – and in our opinion probably the most significant one – is the fact that the smoker and non-smoker subgroups differ in other behavioural patterns apart from smoking. For example, an increased level of sexual activity in smokers – which our findings did not strongly confirm, but which was reported by Malcolm and Shephard (1978) – could account for a reduction in sperm density. Whether such differences between smokers and non-smokers in sperm density have anything to do with smoking is a question which our study cannot answer.

Sperm motility is another important index of fertility. No differences in sperm motility were found between smokers and non-smokers either in Group I or in Group II (Fig. 4, Table 9). This is contrary to the findings of Viczian (1969), Vogel et al. (1979), and Stekhun (1980). We can offer no explanation for this discrepancy. It may be of some significance that we selected only healthy volunteers for our study, whereas the above-mentioned authors based their research on patients attending an infertility clinic. In this connection it is worth mentioning the hypothesis put forward by Campbell and Harrison (1979), who suggest that smoking may have a negative influence on spermatogenesis in connection with damage already caused by other factors. This has led to the theory, based on the concept of the "individual factor" in medicine propounded by Gottron (1939), that there may exist such a thing as an individual sensitivity to nicotine (Eyband 1949). This theory is discussed at length by Schirren and Gey (1969) and Schirren (1972) and illustrated by reference to specific case histories. Our own findings are certainly consistent with such an explanation. If we look at the distribution of cryptozoospermia and azoospermia in Group I and Group II, we find that both these conditions occur only in the smoker subgroups, apart from a single exception in Group I (see Table 11).

After sperm density and motility, sperm morphology is the third major criterion of fertility. Apart from microcephalous forms, which occur with significantly greater frequency in both non-smoker subgroups (Group I and Group II), there are no differences in sperm morphology between smokers and non-smokers (Table 10). Since this particular degenerative form accounts for only 0.2% of the total sperm count, we do not attribute any biological significance to this finding, and are inclined to regard the one significant difference as a chance result.

The absence of any difference in sperm morphology between smokers and non-smokers is in accordance with the findings of Nebe and Schirren (1980c), Rodriguez-Rigau et al. (1982), but conflicts with those of Viczian (1969) and Evans et al. (1981). The study by Viczian (1969) is hardly suitable, in our view, as a basis for comparing smokers and non-smokers, as the smokers were selected from a group of patients with fertility disorders, while the non-smokers were without exception the fathers of healthy children. Both Viczian (1969) and Evans et al. (1981) found little or no correlation between sperm damage and smoke dosage. It should also be noted that the paper published by Evans et al. (1981) does not state the age of the smokers and non-smokers studied, nor does it contain sufficient information about the control group.

In our study none of the other ejaculate parameters analyzed revealed any significant differences between the smoker and non-smoker subgroups. In the absence of any other than marginal differences between smokers and non-smokers with regard to sperm density, sperm motility, and sperm morphology, no differences are to be expected in terms of the classification devised by the European Andrology Club (see Tables 11 and 12).

Information about the DNA content of the spermatozoa can be obtained by means of impulse cytophotometry, a method which is still relatively new to andrological studies. Its chief value at the present time is as an instrument for supplementing the differential spermiogram in individual diagnosis (Otto et al. 1980; Hofmann et al. 1980a; Hettwer et al. 1981; Hartmann et al. 1982). We were the first to apply the technique of impulse cytophotometry to a relatively large, healthy group of subjects. Analysis of 289 usable ejaculate specimens did not reveal any differences between smokers and non-smokers in Groups I and II with respect to total impulse count, as well as impulse counts for the haploid, hypohaploid, and diploid ranges (Table 13). The same result is obtained when the ejaculates are first classified in accordance with the Andrology Club nomenclature (Table 14). However, it should be pointed out that the subgroups in question (with the exception of those with normozoospermia and oligozoospermia) are too small to be regarded as representative.

A broadening of the haploid peak (1 c), particularly where the peak is asymmetrical, indicates that the DNA content of the sperm sample analyzed is not homogeneous (Otto and Oldiges 1978; Hofmann et al. 1980b). Our examination of the peak showed no differences at all between smokers and non-smokers (Table 15). Incipient DNA damage or abnormal DNA distribution would have caused a broadening of the 1 c peak flanks (especially a displacement towards the left), a rise in the impulse count in the hypohaploid region, or a general rise in the background level. However, calculation of the difference ($V_D$) between the total number of cells and the number of cells represented in the peaks also showed no differences between the smoker and non-smoker groups. The results of the cytophotometric analysis therefore do *not* show anomalies in the DNA content of spermatozoa.

The hormonal regulation of spermatogenesis is controlled essentially by FSH, LH, and testosterone. It is conceivable that smoking could have an effect on spermatogenesis via some change in these hormones, and this possibility has been discussed (Vogel et al. 1979). While there were no differences between our smoker and non-smoker subgroups as far as FSH and LH levels were concerned, the smokers in Group II exhibited significantly higher testosterone concentrations than non-smokers ($p < 0.03$) (see Figs. 5–7, Table 16). All the results are within normal limits. We cannot explain the significant difference in testosterone concentrations between smokers and non-smokers in Group II. It should be noted, however, that Persky et al. (1977) discovered a decline in testosterone levels associated with alcohol consumption.

Our own findings agree with most of those reported in the published literature. They confirm those of Persky et al. (1977) and Winternitz and Quillen (1977), who found no differences in FSH and LH levels between smokers and non-smokers. While these authors also found no difference in the testosterone levels recorded for smokers and non-smokers, our finding that smokers exhibited higher concentrations of testosterone was confirmed by Dotson et al. (1975), Gutai et al. (1981), and Dai et al. (1981). On the other hand, however, Briggs (1973) reports that testosterone concentrations in smokers who refrained from smoking for one week rose from an initially depleted level to match those recorded for non-smokers. As all the recorded testosterone concentrations are within normal limits, and secretions of gonadotropin are likewise normal, there is no reason to suppose that smoking causes any hormonal dysfunction or that it therefore has a negative effect on spermatogenesis, nor is the capacity for coitus in any way impaired. Although individual response is testosterone-related, additional doses of testosterone which raise the plasma testosterone concentration above the lower end of the normal range have no effect in terms of increased sexual activity (Money 1961; Salmimies et al. 1982). In the light of our own findings and those reported in more recent literature, we believe that the results obtained by Briggs (1973) - which have never been repeated - should be disregarded. We cannot comment at present on the causes or the significance of the higher testosterone concentrations which we found in smokers (and which other researchers have also reported). Possibly we are dealing here with yet another phenomenon that is linked in some way with those personality differences between smokers and non-smokers to which we previously referred.

## 5 Conclusion

Form our results it can be concluded that cigarette smoking has no effect on spermatogenesis in healthy adult males between ages of 20 and 40. We attribute the marginal spermatological differences we found between the smokers and non-smokers mainly to the fact that the behavioral characteristics of smokers differ from non-smokers in a number of aspects which seem not to be connected with smoking. The effect of such behavioral patterns on spermatogenesis must be considered quite separately in any study of this nature.

*Acknowledgements.* We are grateful to Prof. Dr. Helmut Schievelbein (Deutsches Herzzentrum München) and Mrs. von Ziegesar, who recruited the subjects and thus enabled the study to be conducted as a "blind study". The manuscript was translated by Brian Rasmussen.

## References

Amelar RD, Schoenfeld C (1980) Sperm motility. Fertil Steril 34:197−215
Arnold-Krüger MA (1971) Beziehungen zwischen Rauchen, Unfallquote und Persönlichkeitsaspekten. Inauguraldissertation, Freiburg, pp 100−102
Asbjornsen G, Molne K, Klepp O, Aarvaag A (1976) Testicular function after combination chemotherapy for Hodgkin's disease. Scand J Haematol 16:66−69
Babb RR (1980) Cimetidine: clinical uses and possible side effects. Postgrad Med 68/6:87−93
Barlow SM, Sullivan FM (1981) Reproductive hazards and industrial chemicals. Ann Occup Hyg 24:359−361
Beutel P, Küffner H, Schubo W (1980) SPSS 8-Statistik-Programm − System für die Sozialwissenschaften. Fischer, Stuttgart New York
Blevins RD, Regan JD (1976) Delta-9-tetrahydrocannabinol: Effect on macromolecular synthesis in human and other mammalian cells. Arch Toxikol 35:127−135
Borlee I, Bouckaert A, Lechat MF, Misson CB (1978) Smoking patterns during and before pregnancy: Weight, length, and head circumference of progeny. Eur J Obstet Gynaecol Reprod Biol 8:171−177
Briggs MH (1973) Cigarette smoking and infertility in men. Med J Aust (I):616−617
Campbell JM, Harrison KL (1979) Smoking and infertility. Med J Aust (I):342−343
Cannon SB, Veazey JM Jr, Jackson R et al. (1978) Epidemic kepone poisoning in chemical workers. Am J Epidemiol 107:529−537
Carothers AD, Beatty RA (1975) The recognition and incidence of haploid and polyploid spermatozoa in man, rabbit, and mouse. J Reprod Fertil 44:487−500
Cendron H, Vallery-Masson J (1971) Tabac et comportement sexuel chez l'homme. Vie Méd 25: 3027−3030
Comstock GW, Lundin EE (1967) Parental smoking and perinatal mortality. Am J Obstet Gynecol 98:708−718

Dai WS, Kuller LH, LaPorte RE, Gutai JP, Falvo-Gerard L, Caggiula A (1981) The epidemiology of plasma testosterone levels in middle-aged men. Am J Epidemiol 114:804–816

Dörfel H, Lutterberg W (1937) Unfruchtbarkeit des Mannes. Dermatol Wochenschr 104:1–14

Dotson LE, Robertson LS, Tuchfeld B (1975) Plasma alcohol, smoking, hormone concentrations and selfreported aggression. J Stud Alcohol 36:578–586

Drife JO (1982) Drugs and sperm. Br Med J 284:844–845

Ducot B, Mayaux MJ, Spira A (1981) Testicular varicoceles and tobacco consumption. Fertil Steril 36:686–687

Eiser JR, Sutton SR, Wober M (1982) Smoking, seatbelts, and beliefs about health. World Smoking and Health 7/1:12–17

Eliasson R, Hellinga G, Lübcke F, Meyhöfer W, Niermann H, Steeno O, Schirren C (1970) Empfehlungen zur Nomenklatur in der Andrologie. Andrologia 2:186–187

Evans HJ, Fletcher J, Torrance M, Hargreave TB (1981) Sperm abnormalities and cigarette smoking. Lancet I:627–629

Eyband M (1949) Nikotinempfindlichkeit und vegetative Tonuslage. Schweiz Arch Neurol Psychiat 64:55–82

Eysenck HJ (1968) Rauchen, Gesundheit und Persönlichkeit. Rau, Düsseldorf

Eysenck HJ (1973) Personality and the maintenance of the smoking habit. In: Dunn WL Jr (ed) Smoking behavior: motives and incentives. Winston, Washington D.C., pp 113–146

Faust V, Mensen H (1974) Zur Psychologie des Rauchens. Hippokrates 45:210–225

Faust V (1982) Drogen-ABC III/7: Rauschgift, Alkohol, Nikotin. Notabene Medici 12:413

Fürbringer P (1926) Sterilität des Mannes. In: Marcuses Handbuch der Sexualwissenschaft, 2. Aufl. Marcus and Weber, Bonn

Gey GH (1971) Einfluß des Rauchens von Zigaretten auf die Spermienzahl bei andrologischen Patienten. Inauguraldissertation, Hamburg

Godfrey B (1981) Sperm morphology in smokers. Lancet I:948

Goode E (1972) Cigarette smoking and drug use on a college campus. Int J Addict 7:133–140

Gottron HA (1939) Der personale Faktor bei Hautkrankheiten. In: Adam-Curtius, Individualpathologie, Berlin

Greitemeyer M (1981a) Zur Risikobereitschaft von Rauchern und Nichtrauchern. Med Welt 32:1730–1733

Greitemeyer M (1981b) Dissertation

Gutai L, LaPorte R, Kuller L, Dai W, Falvo-Gerard L, Caggiula A (1981) Plasma testosterone, high density lipoprotein cholesterol and other lipoprotein fractions. Am J Cardiol 48:897–902

Haenszel W, Shimkin MB, Miller HP (1956) Tobacco smoking patterns in the United States. Public Health Monog No 45:1–111

Hahn EW, Feingold SM, Simpson L, Batata M (1982) Recovery from aspermia induced by low-dose radiation in seminoma patients. Cancer 50:337–340

Hartmann W, Hettwer H, Hofmann N, Freundl G, Otto FJ (1982) Die Pepsinvorbehandlung menschlichen Spermas als Standardmethode in der impulszytophotometrischen Analyse. Andrologia 14:135–142

Heinke E, Doepfner R (1960) Fertilitätsstörungen beim Manne. In: Schürmann H, Doepfner R (eds) Handbuch der Haut- und Geschlechtskrankheiten. J Jadassohn, Erg-Werk, Bd. VI-3. Springer, Berlin Heidelberg New York

Heite HJ, Wokalek H (eds) (1980) Männerheilkunde. Fischer, Stuttgart New York

Hembree W, Huang J, Nahas G (1978) Altérations morphologiques des cellules spermatiques chez les fumeurs de chanvre (marihuana). Ann Méd NANCY 17:417–419

Hettwer H, Hartmann W, Otto FJ, Hofmann N (1981) Die Anwendung der Impulszytophotometrie in der Andrologie. Lab Med 5:183–186

Hofmann N, Otto FJ, Hettwer H, Oldiges H (1980a) Die klinische Anwendung der impulszytophotometrischen Spermaanalyse. FDF 8:185–187

Hofmann N, Otto FJ, Freundl G, Hettwer H, Oldiges H (1980b) Desintegration von Spermatozoen bei der impulszytophotometrischen Analyse des menschlichen Spermas. Andrologia 12:534–539

Hudec T, Thean J, Kuehl D, Dougherty RC (1981) Tris(dichloropropyl)phosphate, a mutagenic flame retardant: frequent occurrence in human seminal plasma. Science 211:951–952

Hytten FE (1973) Smoking in pregnancy. Dev Med Child Neurol 15:355–357

Jacoby GW (1898) Die chronische Tabaksintoxication, speciell in aetiologischer und neurologischer Hinsicht. Berl Klinik 126:1–30

Jecht E (1977) Varikozele und Fertilität. Zentralbl Haut- u. Geschl Krh 138:177

Joffe MJ (1979) Influence of drug exposure of the father on perinatal outcome. Clin Perinatol 6:21–36

Jones DR, Rushton L (1982) Simultaneous inference in epidemiologic studies. Int J Epidemiol 11:276–282

Klaiber EL, Broverman DM, Vogel W (1980) Increased incidence of testicular varicoceles in cigarette smokers. Fertil Steril 34:64–65

Kolodny RC, Masters WH, Kolodner RM, Toro G (1974) Depression of plasma testosterone levels after chronic intensive marihuana use. New Engl J Med 290:872–874

Koponen A (1960) Personality characteristics of purchasers. J Adv Res 1:6–13

Lancranjan J, Popescu HI, Gavanescu O, Klepsch I, Serbanescu M (1975) Reproductive ability of workmen occupationally exposed to lead. Arch Environ Health 30:396–401

Lantz GD, Cunningham GR, Huckins C, Lipshultz LI (1981) Recovery from severe oligospermia after exposure to dibromochloropropane. Fertil Steril 35:46–53

Larson PS, Haag HB, Silvette H (1961) Tobacco. Experimental and clinical studies. A comprehensive account of the world literature. Williams and Wilkins, Baltimore

Leuchtenberger C, Weir DR, Schrader F, Leuchtenberger R (1956) Decreased amounts of desoxyribose nucleic acid (DNA) in male germ cells as a possible cause of human male infertility. Acta Genet (Basel) 6:272–278

Lienert GA (1978) Verteilungsfreie Methoden in der Biostatistik. Hain, Meisenheim a.G.

Lushbaugh CC, Casarett GW (1976) The effects of gonadal irradiation in clinical radiation therapy: A review. Cancer 37:1111–1120

Macleod J, Gold RZ (1950) The male factor in fertility and infertility. An analysis of ejaculate volume in 800 fertile men and in 600 men in infertile marriage. Fertil Steril 1:347–361

Macleod J, Gold RZ (1951) The male factor in fertility and infertility, IV. Sperm morphology in fertile and infertile marriage. Fertil Steril 2:394–414

McMahon B, Alpert M, Salber EJ (1966) Infant weight and parental smoking habits. Am J Epidemiol 82:247–261

Malcolm S, Shephard RJ (1978) Personality and sexual behavior of the adolescent smoker. Am J Drug Alcohol Abuse 5:87–96

Mau G (1977a) Rauchen in der Schwangerschaft. Dtsch Ärzteblatt 74:2115–2116

Mau G (1977b) In: Deutsche Forschungsgemeinschaft: Schwangerschaftsverlauf und Kindesentwicklung. Forschungsbericht. Boldt KG, Boppard, p 81

Mau G and Netter P (1974) Die Auswirkungen des väterlichen Zigarettenkonsums auf die perinatale Sterblichkeit und die Mißbildungshäufigkeit. Dtsch Med Wochenschr 99:1113–1118

Mellan J (1967) Smoking and male fertility. Prakt Lek (Praha) 47:890–892

Meyhöfer W (1963) Mikrospektrophotometrische Messungen des Nucleinsäuregehaltes von Spermien fertiler und infertiler Männer. Arch Klin Exp Derm 216:556–614

Meyhöfer W (1973) Auswirkungen von Cytostatika auf die Spermiogenese Andrologia 5:107–108

Meyhöfer W, Herrmann R, Knoth W (1960) Über Ultraviolett-Absorptionsmessungen an Spermatozoen. Arch Clin Exp Derm 209:637–642

Miescher F (1897) In: Die histochemischen und physiologischen Arbeiten von Friedrich Miescher. Vogel, Leipzig

Money JW (1961) Sex hormones and other variables in human eroticism. In: Young WC (ed) Sex and internal secretions, vol II. Williams and Wilkins, Baltimore

Murphy JF (1980) Der Einfluß des mütterlichen Rauchens auf Geburtsgewicht und Wachstum des Schädeldurchmessers. Gynaecologia (Basel) 87:462–466

Nahas GG (1979) Current status of marijuana research. JAMA 242:2775–2778

Nebe KH, Schirren C (1980a) Statistische Untersuchungen bei andrologischen Patienten. I. Grundhäufigkeiten. Andrologia 12:360–372

Nebe KH, Schirren C (1980b) Statistische Untersuchungen bei andrologischen Patienten. II. Befunde an den Genitalorganen. Andrologia 12:417–425

Nebe KH, Schirren C (1980c) Statistische Untersuchungen bei andrologischen Patienten. III. Nikotin und Ejakulatparameter. Andrologia 12:493–502

Nennstiel HJ, Alich R (1969) Fuktosebestimmung im Ejakulat mittels Mikrotitersystem. Ärztl Lab 15:197–198

Otto FJ, Hacker U, Zante J, Schumann J, Goehde W, Meistrich ML (1979a) Flow cytometry of human spermatozoa. Histochemistry 61:249–254

Otto FJ, Hofmann N, Hettwer H, Oldiges H (1979b) Die Analyse von Spermaproben mit Hilfe impulszytophotometrischer DNS-Bestimmungen. Andrologia 11:279–286

Otto FJ, Hofmann N, Schumann J, Hacker U, Hettwer H (1980) Präparation menschlicher Spermatozoen für die impulszytophotometrische DNS-Messung. FDF 8:181–184

Otto FJ, Oldiges H (1978) Requirements and procedures for chromosomal DNA measurements for rapid karyotype analysis in mammalian cells. Pulse-cytophotometry, part 3, pp 393–400

Persky H, O'Brien CP, Fine E, Howard WJ, Khan MA, Beck RW (1977) The effect of alcohol and smoking on testosterone function and aggression in chronic alcoholics. Am J Psychiatry 134: 621–625

Pirke KM (1973) A comparison of three methods of measuring testosterone in plasma: competitive protein binding, radioimmunoassay without and radioimmunoassay including thin-layer chromatography. Acta Endocrinol (Kbh) 74:168–176

Pontonnier F, Plante P, Mansat A et al. (1979) Etude informatisée de 124 cas de varicoceles operés pour hypofertilité. J Urol Nephrol 85:639–647

Qureshi MSA, Pennington JH, Goldsmith HJ, Cox PE (1972) Cyclophosphamide therapy and sterility. Lancet II:1290–1291

Raboch J, Mellan J (1974) Parameter Zigarette. Sexualmedizin 4:175–176

René A (1878) A propos de l'immunité des bêtes a cornes pour la nicotine. Quatre cas d'empoisonnement. Gas Hosp 51:806

Rodriguez-Rigau LJ, Smith KD, Steinberger E (1982) Cigarette smoking and semen quality. Fertil Steril 38:115–116

Rothschild Lord (1953) The fertilisation reaction in the sea urchin. The induction of polyspermy by nicotine. J Exp Biol 30:57–67

Rowley MJ, Leach DR, Warner GA, Heller CG (1974) Effect of graded doses of ionizing radiation on the human testis. Radiat Res 59:665–678

Russell K (1954) Varicocele in groups of fertile and subfertile males. Br Med J 1:1231–1233

Salmimies P, Kockott G, Pirke KM, Vogt HJ, Schill WB (1982) Effects of testosterone replacement on sexual behavior in hypogonadal men. Arch Sex Behav 11:345–354

Sandritter W (1958) Ultraviolett-Mikrospektrophotometrie. Handbuch der Histochemie, Bd I, 1. Teil. Fischer, Stuttgart

Sandritter W, Müller D, Schiemer G (1958) Über den Nukleinsäuregehalt und das Trockengewicht von haploiden und diploiden Zellen. 55. Verh Anat Ges in Frankfurt. Fischer, Jena

Schilsky RL, Lewis BJ, Sherins RJ, Young RC (1980) Gonadal dysfunction in patients receiving chemotherapy for cancer. Ann Intern Med 93:109–114

Schirren C (1971) Praktische Andrologie. Brüder Hartmann, Berlin

Schirren C (1972) Die Wirkung des Nikotin auf die Zeugungsfähigkeit des Mannes. Rehabil 25: 23–24

Schirren C (1973) Umweltschäden und Fertilität des Mannes. Schädigende exogene Einflüsse. Andrologia 5:91–104

Schirren C (1977) Einführung in die Andrologie. Wissenschaftliche Buchgesellschaft, Darmstadt

Schirren C, Gey G (1969) Der Einfluß des Rauchens auf die Fortpflanzungsfähigkeit bei Mann und Frau. Z Haut- u Geschl Kr 44:175–182

Schrottenbach H (1933) Nervenkrankheiten und Nikotin. Wien Med Wochenschr 83:850–851

Statistisches Bundesamt Wiesbaden (1978) Fragen zur Gesundheit 1978. Gesundheitswesen, Fachserie 12, Reihe 3:1–80

Stauber M (1979) Psychosomatik der sterilen Ehe. In: Schirren C, Semm K (eds) Fortschr Fertilitätsforschung, vol 7. Grosse, Berlin

Steinberger E (1976) Recent advances in regulation of male fertility. In: Moghisse KS, Evans TN (eds) Regulation of human fertility. Wayne State University Press, Detroit, pp 274–290

Stekhun FI (1979) Alcohol and tobacco smoking as possible causes of male sterility. Vestnik Dermatologii i Venerologii, 61–65 (in Russian); English summary in Abstr Hyg 55:563–564 (1980)

Stekhun FI (1980) Effect of tobacco smoking on spermatogenesis indices. Vrachebnoe Delo 7: 93–94. English summary in Bibliography on Smoking and Health, U.S. Dept HHS (1981), no 81-0535, p 122

Sterling TD, Kobayashi D (1975) An critical review of reports on the effect of smoking on sex and fertility. J Sex Res 11:201–217

Stiasny H (1944) Unfruchtbarkeit beim Mann. Enke, Stuttgart

Stolla RA, Gropp A, Leidl W, Hofmann N (1978) Teratozoospermie aus human- und tierärztlicher Sicht. Hautarzt 29:518–525

Stransky M (1981) Tabak- und Alkoholkonsum bei jungen Männern. Soz Praeventiv Med 26: 248–251

Thorsland TW, Paulsen CA (1972) Effects of X-ray irradiation on human spermatogenesis. Proc Nat Symp Natural and Mammade Radiation in Space. NASA Document TM X-2440, pp 229 to 232

Uchigaki S (1927) Biological research of sterility. Part III. Influence of drugs and medicines on spermatozoa. Kinki fujinka Gakkai Zass 10:1109–1129. English Abstract in Jpn J Med Sci, Pharmacol 3:155 (1929)

U.S. Public Health Service (1979) The health consequences of smoking. A report of the Surgeon General. U.S. Dept of Health, Education and Welfare, DHEW Publication No 79-50066, pp 18-13–18-15

Van Thiel DH, Gavaler JS, Smith WI JR, Paul G (1979) Hypothalamic-pituitary-gonadal dysfunction in men using cimetidine. New Eng J Med 300:1012–1015

Viczián M (1968) The effect of cigarette smoke inhalation on spermatogenesis in rats. Experientia 24:511–513

Viczian M (1969) Ergebnisse von Spermauntersuchungen bei Zigarettenrauchern. Z Haut- u Geschl Kr 44:183–187

Vogel W, Broverman DM, Klaiber EL (1979) Gonadal, behavioral, and electroencephalographic correlations of smoking. In: Remond A, Izard C (eds) Electrophysiological effects of nicotine. Elsevier/North-Holland Biomedical, Amsterdam, pp 201–215

Vogt HJ (1977) Infertilität als Folge psychosexueller Störungen des Mannes. Gynäkol Prax 1: 697–702

Vogt HJ, Mayer G (1980) Psychische Barrieren bei ungewollt kinderloser Ehe. In: Schirren C, Mettler L, Semm K (eds) Fortschr Fertilitätsforschung, vol 8. Grosse, Berlin, pp 107–110

Whorton MD (1981) The effects of the occupation on male reproductive function. In: Spira A, Jouannet P (eds) Human fertility factors. INSERM 103:339–350

Whorton MD, Krauss RM, Marshall S, Milby TH (1977) Lancet II:1259–1261

Winternitz WW, Quillen D (1977) Acute hormonal response to cigarette smoking. J Clin Pharmacol 17:389–397

Wynder EL (1977) Interrelationship of smoking to other variables and preventive approaches. In: Jarvik ME, Cullen JW, Gritz ER, Vogt TM, West LJ (eds) Research on smoking behavior. NIDA Research Monograph 17. U.S. Dept of Health, Education and Welfare, Washington, pp 67–97

Wyrobek AJ (1979) Changes in mammalian sperm morphology after x-ray and chemical exposures. Genetics 92:105–119

Wyrobek AJ (1982) Sperm assays as indicators of chemically induced germ-cell damage in man. In: Heddle JA (ed) Mutagenicity: New horizons in genetic toxicology. Academic, New York, pp 337–350

Wyrobek AJ, Bruce WR (1978) The induction of sperm-shape abnormalities in mice and humans. In: Hollaender A, de Serres FJ (eds) Chemical mutagens. Principles and methods for their detection, vol 5. Plenum, New York, pp 257–285

Wyrobek AJ, Watchmaker G, Gordon L, Wong K, Moore D, Whorton D (1981a) Sperm shape abnormalities in carbaryl-exposed employees. Environ Health Perspect 40:225–265

Wyrobek AJ, Brodsky J, Gordon L, Moore DH, Watchmaker G, Cohen EN (1981b) Sperm studies in anesthesiologists. Anesthesiology 55:527–532

Yerushalmy J (1962) Statistical considerations and evaluation of epidemiological evidence. In: James G, Rosenthal T (eds) Tobacco and Health. Thomas, Springfield, pp 208–230
Yerushalmy J (1971) The relationship of parents' cigarette smoking to outcome of pregnancy. Implications as to the problem of interferring causation from observed associations. Am J Epidemiol 93:443–456
Yerushalmy J (1972) Infants with low birth weight born before their mothers started to smoke cigarettes. Am J Obstet Gynecol 112:277–284

# Methods to Estimate the Genetic Risk

U.H. EHLING[1]

## 1 Introduction

In this chapter those effects are considered that may be inherited through radiation- or chemically-induced injury to the genes or chromosomes of germ cells. Mutational changes represent a broad spectrum of alterations in the deoxyribonucleotide structure of the genes. At one end of the spectrum this change can be represented by a single nucleotide base substitution, base addition or deletion. At the other end of the mutation spectrum is a complete deletion of the entire gene and/or adjacent genes. For higher organisms these submicroscopic changes are not further resolvable in many instances, and they are simply classified as mutations.

All organisms are the products of a long evolutionary history during which favorable genes have been preserved and deleterious genes eliminated by natural selection. Beneficial mutations of the past are part of the present population. A random change is much more likely to be deleterious than to improve the fitness. In the framework of genetic possibility, a mutation is essentially a random change. Therefore, the general harmfulness of mutation is both an empirical fact and an expectation based on evolutionary reasoning.

The Mendelian mechanism is remarkly effective in maintaining variability. Crow (1981) emphasized that "if any way of reducing the spontaneous mutation rate can be found, it could have enormous humanitarian benefits". Based on this reasoning we have to avoid an increase of the spontaneous human mutation rate due to the exposure to ionizing radiations and chemical mutagens.

To prevent an increase in the mutation rate, for chemical mutagens, the first step is the *Hazard Identification*, the determination of whether a substance is mutagenic. The second step is the *Hazard Evaluation*, the investigation of the chemical, physical, and biological factors that affect mutagenesis. The third step is the *Risk Estimation*, the estimation of effect per unit of exposure. The risk estimation can be divided into two components: estimation of mutagenic effects on germ cells (damage) and estimation of effects on health and welfare of future generations (impact). The Hazard Evaluation and the Risk Estimation are also relevant for the exposure to ionizing radiation. Both aspects will be discussed in detail in this chapter.

---

1  Institut für Genetik, Gesellschaft für Strahlen- und Umweltforschung (GSF), 8042 Neuherberg, FRG

Mutations in Man, ed. by G. Obe
© Springer-Verlag Berlin Heidelberg 1984

The aim for the strategies of risk analysis is the prevention of induced irreparable genetic damage of the human genome. This aim requires the development of animal model systems for the study of mutation induction by radiation and chemical mutagens. The relevance to man of the experimental studies results from the following factors: the genetic material is of the same chemical nature in all higher organisms and its arrangement in the form of chromosomes in animals and plants implies that the response to environmental agents such as radiation and chemical mutagens will be very similar.

## 2 Hazard Evaluation

The model system for hazard evaluation is the specific locus method of the mouse developed by W.L. Russell (1951). With the specific locus method, mutations to recessive alleles of a small number of selected loci can be detected in the first generation. The multiple recessive tester stock that has been extensively used in radiation and chemical mutagenesis studies has the following 7 markers: a/a (non-agouti); b/b (brown); $c^{ch}p/c^{ch}p$ (chinchilla, pink-eyed dilution); d se/d se (dilute, short-ear); s/s (piebald). The phenotypes of the mutants are shown in Fig. 1. The 7 loci are distributed among five autosomes. The $c^{ch}$- and p-loci are linked on chromosome 7, with an average recombination of about 14%, and the d- and se-loci are closely linked, being only 0.16 centimorgans apart on chromosome 9 (Ehling 1978; Green 1966).

A detailed description of the other multiple tester stocks was published by Ehling (1978). A mutation in the germ cells of wild-type animals at any of the 7 loci represented by the recessives in the test stock will be detected in the first generation offspring. Different types of intragenic mutations, small deficiencies involving the

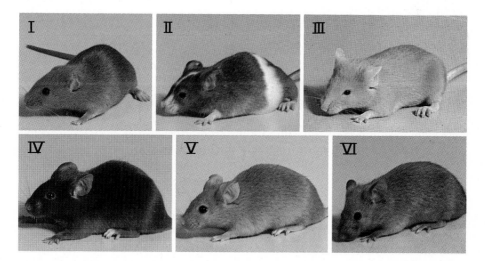

**Fig. 1.** Phenotypes of specific locus mutations recovered with the T-stock: *I* d se (double mutant), d dilute; se short-ear. *II* s piebald spotting. *III* p pink-eyed dilution. *IV* a non-agouti. *V* b brown. *VI* $c^{ch}$ chinchilla

marked loci and, rarely, events of a grosser nature, like nondisjunction, can be detected with the specific locus mutation assay. In addition, it is possible to detect mosaics and clusters.

Estimates of spontaneous mutation frequencies at specific loci obtained by different workers and laboratories are in good agreement and there have been no reports so far of heterogeneity associated with seasonal or other factors in normal environment (Searle 1975). Based on these observations Bateman (1979) suggested the use of a historical control rate for mutagenicity testing. The advantage of a concurrent control is the check on the genetic and mutational behavior of the stocks in absence of treatment.

Generally, the induction of specific locus mutations was tested in $(101 \times C3H)F_1$ male mice, 10–14 weeks old. Immediately after irradiation or application of the test compound, each male was caged separately with an untreated test stock female of the same age. The offspring were counted, sexed and carefully examined externally at birth. The litters were examined again when cages were changed, the final examination being at weaning age. Classification by phenotype was generally confirmed by an allelism test. A detailed description of the method has been published elsewhere (Cattanach 1971; Ehling 1978; Russell et al. 1981; Searle 1975).

The technique is simple and no special skills are needed apart from those of animal husbandry and good powers of observation. Furthermore, the mutations can be studied in detail and the test can easily be combined with other methods, for example, the detection of induced dominant mutations (Sect. 3.2.1). The advantage of a combined investigation of dominant cataract mutations and specific locus mutations in mice is at least threefold:

1. The number of scorable mutations is increased by a factor of four.
2. The combined investigation allows the comparison of the mutation frequency of selected and unselected loci.
3. In the same experiment the frequency of mutations with a dominant and a recessive mode of expression can be compared.

The specific locus method is especially useful to investigate the effects that various biological factors and treatment conditions might have on mutation frequency.

## 2.1 Radiation-Induced Specific Locus Mutations in Male Mice

The mutagenic effects of radiation vary markedly depending on the developmental stage of the germ cell at time of exposure (Ehling 1966; Ehling et al. 1982). In the mature testis, certain cells, called $A_s$ (stem cell) spermatogonia, continue throughout the reproductive life span to bud off, by division. Cells which go through meiotic divisions and through successive differentiation stages are known as spermatocytes, spermatids, and spermatozoa. The production of a mature sperm from a spermatogonial stem cell takes about 5 weeks in the mouse (Oakberg 1956) and almost twice as long in man (Heller and Clermont 1964). The spermatozoa, spermatids, and spermatocytes do not persist for long in the body, less than 0.7% of the human generation time. Therefore, these germ cell stages are less important than spermatogonia for the hazard evaluation.

The results of the induction of specific locus mutations in spermatogonial stem cells of the mouse from three different laboratories are summarized in Table 1.

## 2.2 Radiation-Induced Specific Locus Mutations in Female Mice

In the female mouse, germ cells enter meiosis on or soon after the 12th day of embryonic life and reach the dictyate stage shortly after birth. They are arrested in this stage until 12 h before ovulation. For this reason, almost all the studies on the genetic response of the female mouse to irradiation have concentrated on the dictyate stage (Searle 1974).

It is of interest to compare the specific locus mutation rates obtained for mature and maturing oocytes with the mutation rate in the male mouse for chronic irradiation of the germ cell stage primarily at risk in man, the spermatogonia. The mutation rate, for the same 7 loci, in mouse spermatogonia irradiated at dose rates of 0.009 R/min and below, was calculated by Searle (1974) to be $6.6 \times 10^{-8}$ per locus per R. The rates for low-level irradiation of mature and maturing oocytes were 0.17 to 0.44 times as effective (Russell 1977).

In adult male mice, no effect of the interval between irradiation and fertilization has ever been observed on the induced specific locus mutation frequency in spermatogonia (Russell 1972). In mouse oocytes, on the other hand, the interval between irradiation and conception has turned out to be one of the most effective factors affecting the mutation rate. The distribution of offspring and specific locus mutations in conceptions occurring during successive weeks after irradiation of female mice with 50 or 200 R at 90 R/min and with 400 R at 0.8 R/min are summarized in Table 2.

The factors affecting the mutation rate described in Sects. 2.1 and 2.2 have to be taken into account for the estimation of the radiation-induced genetic risk.

## 2.3 Chemically Induced Specific Locus Mutations in Male Mice

The toxicological literature is replete with thresholded phenomena and traditional thinking about dose-response relationships generally assumes a threshold until data indicate otherwise. In contrast, the experience with radiation mutagenesis has introduced theories and supporting data for the absence of threshold with this particular class of toxic agents. Chemical mutagens share the pharmacokinetics and metabolism of traditional toxins, as well as the potential for the stochastic inherently non-thresholded mechanism of radiation induction of mutations. Not surprisingly there are strong differences of opinion as to whether, in the absence of data, one should assume a threshold or a non-thresholded response for genotoxic effects.

After reviewing the literature for the presence or absence of thresholds for the induction of mutations it was concluded by Ehling et al. (1983) that "for the assessment of genotoxic risks, there is no alternative at present to a case by case analysis based on the laborious acquisition of good dose-response data. For those many situations where good information is lacking, we would reaffirm the principle enumerated in the BEIR-Report (1980): 'Use simple linear interpolation between the

**Table 1.** Radiation-induced specific locus mutations in $A_s$-spermatogonia of mice

| Category | Source of radiation | Exposure (R) | Dose rate (R/min) | Mutations No. | Offspring No. | Frequencies per locus × 10⁵ | Ref. |
|---|---|---|---|---|---|---|---|
| Control | $^{60}$Co | | | 28 | 531,500 | 0.8 | Russell and Kelly (1982) |
| | $^{137}$Cs | | | 11 | 157,421 | 1.0 | Searle (1974) |
| | | | | 13[a] | 205,794 | 0.9 | Ehling and Neuhäuser-Klaus (1982) |
| | | | | 52 | 894,715 | 0.8 | |
| Chronic | $^{60}$Co | 37.5 | 0.001 | 7 | 79,364 | 1.3 | Searle (1974) |
| | $^{137}$Cs | 86 | 0.001 | 6 | 59,810 | 1.4 | |
| | $^{137}$Cs | 300 | 0.0007 | 11 | 48,358 | 3.2 | |
| | $^{137}$Cs | 300 | 0.001 | 15 | 49,569 | 4.3 | |
| | $^{137}$Cs | 300 | 0.009 | 10 | 58,457 | 2.4 | Russell and Kelly (1982) |
| | $^{137}$Cs | 300 | 0.005 | 24 | 84,831 | 4.0 | |
| | $^{137}$Cs | 516 | 0.009 | 5 | 26,325 | 2.7 | |
| | $^{137}$Cs | 600 | 0.001 | 22 | 53,380 | 5.9 | |
| | $^{137}$Cs | 600 | 0.8 | 10 | 28,059 | 5.1 | |
| | $^{60}$Co | 618 | 0.008 | 5 | 22,682 | 3.1 | Searle (1974) |
| | $^{60}$Co | 671 rem | 0.005 | 20 | 58,795 | 4.9 | |
| | $^{137}$Cs | 861 | 0.009 | 12 | 24,281 | 7.1 | Russell and Kelly (1982) |
| Acute | X-ray | 300 | 90 | 40 | 65,548 | 8.7 | Russell and Kelly (1982) |
| | $^{137}$Cs | 400 | 60 | 10[b] | 9,406 | 15.2 | Neuhäuser-Klaus (1983) |
| | $^{137}$Cs | 534 | 53 | 7 | 10,212 | 9.8 | Ehling et al. (1982) |
| | $^{137}$Cs | 600 | 60 | 10 | 10,453 | 13.7 | Neuhäuser-Klaus (1983) |
| | $^{137}$Cs | 600 | 53 | 14 | 11,095 | 18.0 | Ehling et al. (1982) |
| | X-ray | 600 | 90 | 111 | 119,326 | 13.3 | Russell and Kelly (1982) |
| | X-ray | 670 | 72 | 12 | 11,138 | 15.4 | Searle (1974) |

[a] 13 independent mutational events, one of which was a cluster of 6 presumed s-mutants and one was a cluster of 2 se-mutants. The total number of mutations observed was 19

[b] One presumed double non-disjunction included

**Table 2.** Distribution of offspring and mutations at seven specific loci in conceptions occurring during successive weeks after irradiation of female mice with various doses and dose rates. (Russell 1977)

| Week | 200 R, 90 R/min | | 50 R, 90 R/min | | 400 R, 0.8 R/min | |
|---|---|---|---|---|---|---|
| | No. of offspring | No. of mutations | No. of offspring | No. of mutations | No. of offspring | No. of mutations |
| 1 | 28,547 | 13 | 84,614 | 6 | 49,039 | 14 |
| 2 | 3,604 | 2 | 8,016 | 0 | 7,928 | 3 |
| 3 | 454 | 1 | 2,904 | 0 | 737 | 1 |
| 4 | 1,138 | 0 | 9,587 | 1 | 844 | 4 |
| 5 | 8,395 | 12 | 45,179 | 3 | 7,342 | 6 |
| 6 | 3,327 | 5 | 16,304 | 3 | 4,990 | 2 |
| 7 | 328 | 0 | 13,868 | 0 | 170 | 0 |
| 8 | 17 | 0 | 9,092 | 0 | 2 | 0 |
| 9 | 5 | 0 | 11,096 | 0 | 4 | 0 |
| 10 | 1 | 0 | 7,433 | 0 | – | – |
| 11 | – | – | 50,570 | 0 | – | – |

lowest reliable dose data ... In order to get any kind of precision from experiments of manageable size, it is necessary to use dosages much higher than expected for the human population. Some mathematical assumption is necessary and the linear model, if not always correct, is likely to err on the safe side'".

The effectiveness of the blood-testis barrier (Setchell 1970) was sometimes wrongly used as an argument to conclude that mammalian assays are insensitive. On the contrary, the tests for induction of gene mutations in mammals belong to the few methods, which provide information about compounds, which are able to cross the blood-testis barrier. Such information is of great importance for the evaluation of a chemical mutagen.

The complexities of the reproductive system in mammals can also be demonstrated indirectly by factors affecting the yield or the quality of mutations in mammals.

## 2.3.1 Differential Spermatogenic Response

Ionizing radiation induces specific locus mutations in all spermatogenic stages of mice. Only the yield of the induced mutation rate is different in the various germ cell stages. In contrast to radiation, chemical mutagens may induce mutations only in certain spermatogenic stages. Examples of specific locus mutations induced by different chemicals primarily in postspermatogonial germ cell stages are summarized in Table 3.

The induction of specific locus mutations after i.p. injection of methyl methanesulfonate (MMS) and cyclophosphamide demonstrates the differential spermatogenic response (Table 3). A dose of 20 mg/kg of MMS increased significantly the mutation frequency in the mating interval 5-12 days posttreatment (P = 0.009). The highest mutation rates induced by MMS and cyclophosphamide are 28-30 times higher than the control rate. For cyclophosphamide the induced specific locus mutation rate in spermatozoa (1-7 days posttreatment) is 85 times higher than the mutation frequen-

**Table 3.** Highly effective compounds for the induction of specific locus mutations in spermatozoa and spermatids of mice[a]

| Compound | Dose (mg/kg) | Mating intervals (days) | Average litter size | No. of offspring | No. of mutations at 7 loci | Frequencies per locus × 10⁵ | Ref. |
|---|---|---|---|---|---|---|---|
| Methyl-methane-sulfonate (MMS) | 20 | 1– 4 | 6.6 | 5,638 | 1 | 2.5 | |
| | | 5– 8 | 6.6 | 5,593 | 2 | 5.1 | |
| | | 9–12 | 6.7 | 5,518 | 2 | 5.2 | |
| | | 13–16 | 7.0 | 5,785 | 1 | 2.5 | |
| | | 17–20 | 6.9 | 5,759 | 0 | – | Ehling and Neuhäuser-Klaus (1982) |
| | 40 | 1– 4 | 6.0 | 2,902 | 2 | 9.8 | |
| | | 5– 8 | 4.0 | 1,716 | 3 | 25.0 | |
| | | 9–12 | 4.1 | 1,750 | 3 | 24.5 | |
| | | 13–16 | 6.0 | 2,799 | 1 | 5.1 | |
| | | 17–20 | 6.6 | 3,072 | 0 | – | |
| | 4 × 10[b] | 1– 4 | 6.6 | 1,426 | 0 | – | |
| | | 5– 8 | 4.4 | 898 | 1 | 15.9 | |
| | | 9–12 | 5.6 | 1,186 | 1 | 12.0 | |
| | | 13–16 | 7.7 | 1,638 | 0 | – | |
| | | 17–20 | 7.6 | 1,692 | 0 | – | |
| Cyclophos-phamide | 120 | 1– 7 | 4.4 | 1,627 | 3 | 26.3 | |
| | | 8–14 | 4.8 | 1,794 | 3 | 23.9 | |
| | | 15–21 | 4.9 | 1,744 | 1 | 8.2 | Ehling (1982) |
| | | 22–42 | 7.1 | 2,678 | 1 | 5.3 | |
| | | >43 | 7.4 | 12,573 | 1 | 1.1 | |

a   See Table 5 for control rate
b   24 h apart

cy in spermatogonia. The mutation frequency in spermatogonia is not significantly different from the control frequency (P = 0.56). The reduction of average litter size indicates an induction of dominant lethal mutations. The data prove that the peak sensitivity to the induction of specific locus mutations corresponds well with the sensitivity pattern for the induction of dominant lethal mutations (Ehling 1974, 1978).

In contrast to MMS and cyclophosphamide other compounds, like procarbazine (Ehling and Neuhäuser 1979), and triethylenemelamine (TEM) (Cattanach 1966), induce mutations in postspermatogonial germ cell stages and spermatogonia, whereas ethylnitrosourea (ENU) and mitomycin C induce specific locus mutations primarily or exclusively in spermatogonia (Ehling 1978; Ehling et al. 1982; Russell et al. 1979).

### 2.3.2 Changes of the Mutation Spectrum with Different Noxa

It is interesting to compare, under the same laboratory conditions, the mutation spectra of chemically and radiation-induced specific locus mutations. Ionizing radiation induced in spermatogonia more mutations at the s-locus (36%) than procarbazine (13%) and ENU (8%). In addition, among 87 ENU- and 39 procarbazine-induced specific locus mutations in spermatogonia no double mutants were observed of the closely linked d/se-loci. In a sample of 42 $\gamma$-ray-induced specific locus mutations in spermatogonia of mice we observed 4 double mutants, in another sample of 30 $\gamma$-ray-induced specific locus mutations 1 double mutant was observed in spermatogonia (Ehling and Neuhäuser 1979; Ehling in press).

### 2.3.3 Differences in Viability of Mutants

For a total of 135 mutants, 15 from control groups and 120 from different experiments with chemical mutagens, the viability tests are complete. All mutations that as homozygotes cause death before maturity have been classified as lethals. The overall control frequency so far shows 13% (2 out of 15) homozygous lethals. In the procarbazine group, 2 of 3 mutations (67%), and in the MMS group, 9 of 12 mutations (75%), induced in postspermatogonial stages were lethal. In contrast to the high frequency of lethals in postspermatogonial germ cell stages, only 7 of 29 mutations in the procarbazine group (24%), 3 of 12 mutations in the mitomycin C group (25%), and 17 of 59 in the ENU group (29%) induced in spermatogonia were lethal in homozygous condition. TEM-induced mutations have a similar high frequency of lethal as radiation-induced specific locus mutations (Table 4).

The high frequency of homozygous lethal mutations induced in postspermatogonia suggests that small deficiencies are the main cause for mutations induced in these germ cell stages. This observation explains likewise the correlation between the induction of dominant lethal and specific locus mutations. The absence of double mutants and the relatively low frequency of homozygous lethal mutations suggest that mutations induced by ENU, mitomycin C, and procarbazine in spermatogonia may be mainly due to base-pair changes.

When comparing the results of chemically-induced specific locus mutations with the results of radiation-induced mutations, two points are of special interest (Table 4).

**Table 4.** Lethality of specific locus mutations after treatment of male mice with different noxa

| Treatment | Germ cell stage treated | Number lethal/number tested per locus | | | | | | | | | Percent lethal | Ref. |
|---|---|---|---|---|---|---|---|---|---|---|---|---|
| | | a | b | c | d | se | d+se[a] | p | s | Total | | |
| Control[b] | – | – | 0/ 7 | 0/ 3 | 0/ 1 | – | – | 2/ 4 | – | 2/15 | 13 | Neuhäuser-Klaus (1983) |
| Acute X-rays | | 2/2 | 10/18 | 2/ 6 | 12/12 | 1/2 | – | 6/14 | 38/38 | 71/92 | 77 | Searle (1974) |
| X- and γ-rays | | 0/2 | 1/10 | 5/11 | 14/14 | 0/1 | 1/1 | 1/ 9 | 8/ 9 | 30/57 | 53 | |
| Fission neutrons | g[c] | 0/2 | 4/ 4 | 2/ 6 | 6/ 6 | 0/4 | 2/2 | 2/ 9 | 19/24 | 35/57 | 56 | Neuhäuser-Klaus (1983) |
| γ-rays | | – | 0/ 6 | 6/ 9 | 8/ 8 | 1/2 | 2/2 | 0/ 5 | 12/13 | 29/45 | 64 | |
| Lethality (%) | | 33 | 39 | 47 | 100 | 22 | 100 | 24 | 92 | 165/251 | 66 | |
| Procarbazine | pg[d] | – | – | – | – | – | – | 0/ 1 | 2/ 2 | 2/ 3 | 67 | Ehling (1980a) |
| MMS | | – | 3[e]/ 3 | – | – | 1/1 | 1/1 | 3/ 4 | 1[f]/ 3 | 9/12 | 75 | Ehling and Neuhäuser-Klaus (1982) |
| Lethality (%) | | – | 100 | – | – | 100 | 100 | 60 | 60 | 11/15 | 73 | |
| TEM | | 1/1 | – | – | 3/ 3 | – | – | 0/ 1 | – | 4/ 5 | 80 | Cattanach (1966) |
| Procarbazine | g[c] | 0/1 | 0/ 7 | 0/ 3 | 5/ 6 | 0/3 | – | 0/ 7 | 2/ 2 | 7/29 | 24 | Ehling (1980a) |
| Mitomycin C | | 0/1 | 0/ 1 | 0/ 1 | 3/ 3 | – | – | 0/ 5 | 0/ 1 | 3/12 | 25 | |
| ENU | | 0/2 | 0/ 6 | 0/ 5 | 13/16 | 0/5 | – | 0/21 | 4/ 4 | 17/59 | 29 | Ehling and Neuhäuser-Klaus (1982) |
| Lethality (%) | | 20 | 0 | 0 | 86 | 0 | – | 0 | 86 | 31/105 | 30 | |

a  Presumptive deficiencies only
b  Includes 5 mutations from conventionally bred controls
c  g = spermatogonia
d  pg = postspermatogonia
e  Includes one mutation with dominant deleterious effect
f  Includes one semisterile mutation (translocation carrier)

The high percentage of lethality of radiation-induced mutations and the relatively high frequency of dilute/short-ear double mutants. These observations suggest that radiation, in contrast to some chemical mutagens, induces primarily small deletions. Therefore, it seems necessary to reconsider the point of view by Wolff (1967) about the nature of the induced specific locus mutations in mice and the explanation of the dose-rate effect.

### 2.3.4 Comparison of Mutation Rates After Fractionated Injection

A single dose of 40 mg/kg of MMS induces in the mating interval 5–12 days post-treatment $23.9 \times 10^{-5}$ specific locus mutations per gamete, which is 27 times the spontaneous level. Dividing the total dose of 40 mg/kg into 4 equal parts of 10 mg/kg given 24 h apart induces in the same mating interval only 14 times the mutation rate as the control frequency ($12.9 \times 10^{-5}$ mutations per locus per gamete). However, the sample sizes of these experiments are small and the mutation frequencies for the same mating intervals are not significantly different ($P = 0.21$). The results of these experiments are summarized in Table 3.

In contrast to the induction of mutations in spermatozoa and spermatids the yield of specific locus mutations induced by fractionated doses of procarbazine or ENU in $A_s$-spermatogonia was reduced. The results of these experiments are summarized in Table 5. For procarbazine the difference between the mutation frequency after single injection of 600 mg/kg and the treatment $6 \times 100$ mg/kg in 24 h intervals was at the borderline of significance ($P = 0.048$). In dose fractionation experiments with $10 \times 10$ mg/kg of ENU injected in weekly intervals a pronounced reduction in the yield of mutations was observed.

The fractionation of the dose affects the yield of mutations differently in specific locus and dominant lethal experiments. In dominant lethal experiments with MMS the effectiveness of a dose fractionation was strictly additive (Ehling 1981). The following assumptions could explain these results. (1) The distinct genetic endpoints have different repair mechanisms. (2) The differences between germ cell stages (differentiating spermatids and spermatozoa versus the dividing spermatogonia).

### 2.4 Chemically Induced Specific Locus Mutations in Female Mice

The results of induced specific locus mutations in female mice are summarized in Table 6. The limited data base indicates differences in the mutagenic response of mature and early oocytes for different chemicals. Similarly, the sensitivity between male and female germ cells is different for the various chemicals tested.

The mature oocytes are much more sensitive than the immature oocytes for the induction of mutations by radiation, ENU, TEM, and mitomycin C. The only exception known until now is procarbazine. The mutants in the procarbazine group were born 70, 74, and 243 days posttreatment.

If spermatogonia and oocytes were equally sensitive to the induction of specific locus mutations by procarbazine, one would expect 26 mutants in a total of 81,089 offspring. The observed 3 mutations are significantly different from the ex-

**Table 5.** Induction of specific locus mutations in spermatogonia of mice after fractionation of the dose

| Compound | Dose (mg/kg) | Interval between fractions | No. of $F_1$-offspring | No. of mutations at 7 loci | No. of minimum of independent mutational events | Frequencies per locus $\times 10^5$ [a] | Ref. |
|---|---|---|---|---|---|---|---|
| Control | – | – | 205,794 | 19 | 13 | 0.9 | Ehling and Neuhäuser Klaus (1982) |
| Procarbazine | 1 × 600 | – | 45,413 | 16 | 16 | 5.0 | Ehling (1980a) |
| | 2 × 300 | Day | 13,908 | 3 | 3 | 3.1 | |
| | 6 × 100 | Day | 20,621 | 2 | 2 | 1.4 | |
| | 5 × 200 | Week | 18,393 | 4 | 4 | 3.1 | |
| Control | – | – | 531,500 | 28 | 28 | 0.8 | Russell et al. (1982) |
| ENU | 1 × 100 | – | 21,235 | 64 | 53 | 35.7 | |
| | 1 × 100 | – | 3,679 | 12 | 12 | 46.6 | |
| | 10 × 10 | Week | 19,991 | 9 | 8 | 5.7 | |

[a] Calculation based on the minimum number of independent mutational events

**Table 6.** Induction of specific locus mutations in female mice

| Treatment | Dose[a] | Interval between treatment and conception | | | | Ref. |
| | | Up to 7 weeks | | More than 7 weeks | | |
| | | No. of offspring | No. of mutations at 7 loci | No. of offspring | No. of mutations at 7 loci | |
| --- | --- | --- | --- | --- | --- | --- |
| Radiation | 30–400 rad | 319,399 | 84 | 259,683 | 3 | Russell (1972) |
| Procarbazine | 400 mg/kg | 9,369 | 0 | 23,233 | 2 | Ehling (in press) |
| | 600 mg/kg | 16,888 | 0 | 31,599 | 1 | |
| | Total | 26,257 | 0 | 54,832 | 3 | |
| Mito-mycin C | 2 mg/kg | 2,847 | 1 | 4,956 | 0 | Ehling (1982) |
| | 4 mg/kg | 1,515 | 0 | 1,204 | 0 | |
| | Total | 4.362 | 1 | 6,160 | 0 | |
| ENU | 160 mg/kg | 2,728 | 2 | 3,325 | 0 | Ehling and Neuhäuser-Klaus (1982) |
| | 250 mg/kg | 390 | 0 | 1,040 | 0 | |
| | Total | 3,118 | 2 | 4,365 | 0 | |
| TEM | 2 mg/kg | 10,812 | 2 | – | – | Cattanach (1982) |

[a] The total number of mutations observed was 8 in 204,639 control offspring. Three independent mutational events, one of which was a cluster of 6. (Russell 1977)

pected frequency of 26 ($P < 10^{-4}$). This result indicates that oocytes are less sensitive than spermatogonia for the induction of specific locus mutations (Tables 2 and 6). Similarly, ENU is in oocytes less effective than in spermatogonia. The results of mitomycin C are inconclusive. However, the mutation rate of TEM is similar in spermatogonia and oocytes (Cattanach 1982).

# 3  Risk Estimation

In using the data from the mouse to arrive at quantitative estimates of the radiation genetic risks for humans, three general assumptions are made, according to Sankaranarayanan (1982), unless there is evidence to the contrary:

1. The amount of genetic damage induced by a given type of radiation under a given set of conditions is the same in the germ cells of humans and in those of the test species which serves as a model.
2. The various biological and physical factors affect the magnitude of the damage in similar ways and to similar extents in the mouse and in humans.

3. At low doses and at low dose rates of low LET irradiation, there is a linear relationship between dose and frequency of genetic effects studied.

The first and second condition is also valid for chemical mutagens. The validity of the extrapolation from one species to another species has to be proven by mutagenicity testing. The problems of the uptake, transport, metabolism, and excretion of chemicals have to be considered for the extrapolation. The third assumption, linear dose effect relationship (Sect. 2.3), has to be proven for each chemical mutagen.

There are two main approaches in making genetic risk estimates. One of these, termed the direct method, expresses risks in terms of expected frequencies of genetic changes induced per unit dose, the other, referred to as the doubling dose method or the indirect method, expresses risks in relation to the observed incidence of genetic disorders now present in man. Both approaches will be discussed in the following sections.

## 3.1 Doubling Dose Method

The doubling dose can be defined as the dose necessary to induce as many mutations as occur spontaneously in one generation. One underlying assumption for the calculation of the doubling dose is a linear dose response relationship. Another assumption is the similarity between spontaneous and induced mutations. If these conditions are fulfilled the doubling dose can be used to calculate the individual risk and the population risk.

### 3.1.1 Individual Risk

Procarbazine is used in combination treatment of Hodgkin's disease. The induction of specific locus mutations by procarbazine fulfills the requirements for the doubling dose approach (Ehling and Neuhäuser 1979). The doubling dose for procarbazine based on the regression coefficient for the induction of mutations in $A_s$-spermatogonia is 110 mg/kg (Ehling 1982). By comparing the therapeutic dose of 215 mg/kg with the doubling dose of 110 mg/kg, one may conclude that the procarbazine treatment of a patient with Hodgkin's disease would induce two times as many mutations as arise spontaneously, provided man and mouse are equally sensitive. The different application schedules of man and mouse would reduce the calculated genetic risk. In addition, the calculation is based on the sensitivity of a male patient, the genetic risk of a female patient would be drastically lower.

The calculation of the individual risk is of importance for the treatment of patients with drugs or for the exposure to ionizing radiation. Similarly the determination of the individual risk is necessary for the safety of people working with chemicals or radiation. The information about the individual risk is the basis for human genetic counseling.

## 3.1.2 Population Risk

For the estimation of the population risk the UNSCEAR-Report (1982) used a doubling dose of 1 Gy for low dose rate, low LET irradiation on a population. This estimate was based on a direct analysis of γ-ray data of specific locus mutations in the mouse (Searle 1974) and indirectly from the values estimated by Lüning and Searle (1971) for different kinds of genetic endpoints. The former analysis gave a value of 127 R and the later gave a range between 16–51 R.

Schull et al. (1981) presented data on four indicators of genetic effects from studies of children born to survivors of the atomic bombings of Hirsohima and Nagasaki. The indicators are frequency of untoward pregnancy outcomes (stillbirth, major congenital defect, death during first postnatal week); occurrence of death in liveborn children, through an average life expectancy of 17 years; frequency of children with sex chromosome aneuploidy; and frequency of children with mutation resulting in an electrophoretic variant. In no instance is there a statistically significant effect of parental exposure; but for all indicators the observed effect is in the direction suggested by the hypothesis that genetic damage resulted from the exposure. On the basis of assumptions concerning the contribution that spontaneous mutation in the preceding generation makes to the indicators in question, it is possible to estimate the genetic doubling dose for radiation for the first three indicators (the data base is still too small for the fourth). The average of these estimates is 156 rems. They argue that "in general, human exposure to radiation will not be acute and of the magnitude experienced by the inhabitants of Hiroshima and Nagasaki, but either interrupted or chronic, and at much lower levels. Under these circumstances, the genetic yield of chronic radiation in mice is approximately one-third that of acute radiation. If mice and people are similar in this respect, the doubling dose for human chronic exposures suggested by these data becomes 468 rems, in contrast to the estimate of 100 rems for low-LET, low-dose, low-dose-rate exposure". However, this reasoning is not acceptable because the three indicators used for the calculation of the doubling dose do not fulfill the criteria for the use of this method mentioned in Sect. 3.1. Therefore, the following quantification is based on a doubling dose of 1 Gy accepted by the UNSCEAR-Committee (1982). In addition, the Proceedings of the Symposium on the Effects of Radiation on the Hereditary Fitness of Mammalian Populations (Roderick 1964) already demonstrated that it is not possible to use the hereditary fitness for the demonstration of radiation-induced mutations.

If the current incidence of autosomal dominant and X-linked diseases is 10,000 per 1 million liveborn then an exposure of 1 Gy per generation would induce the same number of diseases per 1 million liveborn at equilibrium. The first generation incidence is assumed to be 15%, therefore, the incidence in the first generation is 1,500 (Table 7).

The quality of the risk estimation depends on assumptions of the persistence of the induced mutations and the ability to determine the current incidence of the genetic diseases. The following examples should demonstrate the complexities of these problems.

The number of generations required to reach equilibrium will depend on the rate of elimination of the induced mutations from the population. If, for autosomal dominants, we were to take the mean persistence to be about five generations there

**Table 7.** Estimated effect of 1 Gy per generation of low dose or low dose rate, low-LET irradiation on a population of one million liveborn according to the doubling dose method. Assumed doubling dose: 1 Gy (UNSCEAR-Report 1982)

| Disease classification[a] | Current incidence[b] | Effect of 1 Gy per generation | |
|---|---|---|---|
| | | First generation[c] | Equilibrium |
| Autosomal dominant and X-linked diseases | 10,000[d] | 1,500 | 10,000 |
| Recessive diseases | 2,500[e] | Slight | Slow increase |
| Chromosomal diseases | | | |
| Structural | 400[f] | 240 | 400 |
| Numerical | 3,000[g] | Probably very small | Probably very small |
| Congenital anomalies, anomalies expressed later and constitutional and degenerative diseases | 90,000[h] | 450 | 4,500[i] |
| Total | 105,900 | 2,190 | 14,900 |

[a]  Follows that given in the BEIR-Report 1972, except that chromosomal diseases are divided into those with a structural and those with a numerical basis

[b]  Based on the results of Trimble and Doughty (1974) and other studies

[c]  The first generation incidence is assumed to be about 15% of the equilibrium incidence for autosomal dominant and X-linked diseases, about 3/5 of the equilibrium incidence for structural anomalies and about 10% of the equilibrium incidence for diseases of complex inheritance

[d]  Includes diseases with both early and late onset

[e]  Also includes diseases maintained by heterozygous advantage

[f]  Based on the pooled values of Table 2 of the UNSCEAR-Report 1982 but excluding euploid structural rearrangements, Robertsonian translocations and "others" (mainly mosaics)

[g]  Excluding mosaics

[h]  Includes an unknown proportion of numerical (other than Down's syndrome) and structural chromosomal anomalies

[i]  Based on the assumption of a 5% mutational component

would be about a 20% probability that the mutant would be eliminated in any given generation. Equilibrium would be reached when the rate of elimination was exactly balanced by the rate of addition of new mutants to the population. For all practical purposes, this would be achieved in some 10–20 generations in the example chosen. If the persistence is five generations, then the amount of first-generation expression would be one-fifth of the equilibrium expression; if it were 10 generations, the first-generation expression would be one-tenth of the equilibrium expression (BEIR-Report 1980). For the more complex situation involving irregularly inherited diseases (congenital anomalies, anomalies expressed later and constitutional and degenerative diseases) the UNSCEAR-Report 1982 assumed a mutational component of 5%, the BEIR-Report (1980) a mutational component of 5%–50%. The estimation for the increased number of irregularly inherited diseases at equilibrium is estimated by the UNSCEAR-Report 1982 after population exposure of 1 Gy per generation per 1 million liveborn to be 4,500 cases and by the BEIR-Report (1980) between 2,000 to 90,000 cases. Using the same assumption of a 5%–50% mutational component for

UNSCEAR-Report would give a range of 4,500–45,000. The higher value of the BEIR-Report is based on the assumption that the doubling dose may have a range between 0.5–2.5 Gy.

The same approach can be used for the estimation of the genetic risk of a population due to the exposure with chemicals if mouse and man are equally sensitive. The doubling dose for the induction of specific locus mutations in spermatogonia of mice is $\leqslant$ 1 mg/kg for mitomycin C, and 110 mg/kg for procarbazine. The calculation of the doubling dose is based on the regression coefficient of the dose response data (Ehling 1982). If a population is exposed to the doubling dose one would expect the same induced mutation frequencies as after exposure with 1 Gy (Table 7).

Based on the incidence of patients with Hodgkin's disease Ehling and Neuhäuser (1979) estimated the population exposure for procarbazine to be 5.12 $\mu$g/kg body weight. Comparing the expected impact of a total of 14,900 new cases with hereditary diseases in 1 million liveborn for the doubling dose (110 mg/kg) with the estimated population exposure of 5.12 $\mu$g/kg indicates that the mutation load due to procarbazine is negligible.

The expected number of hereditary diseases for the doubling dose for 1 million liveborn is tabulated in Table 7. From the known sales figures of those compounds where the doubling dose is established, it would be relatively easy to calculate for a given population the expected genetic hazard. Unfortunately until now it is not possible to get the sales figures for drugs. An amendment of the existing law could rectify this situation. In the future it is necessary to determine for known chemical mutagens the exposure level similar to the well-known figures of radiation exposure (BMI 1979).

It is interesting to compare the calculation of the genetic hazard of procarbazine based on the doubling dose approach with the results of the direct estimation based on the number of treated patients.

Calculation of the population dose would be the preferable way for estimating the genetic risk of a pollutant. For a drug it is not necessary to base the calculation on a hypothetical population dose. In the above calculation we made the assumption that patients with Hodgkin's disease are at no disadvantage in contributing to the next generation. It follows that of the 2 $\times$ 10$^6$ gametes needed to produce 10$^6$ zygotes, 1 in 42,000 would come from a patient with Hodgkin's disease, i.e. about 48 gametes. Assuming an overall spontaneous mutation rate of 0.8 $\times$ 10$^{-5}$ per locus per gamete based on specific locus results, we can calculate, for 30,000 mutable loci, a total spontaneous mutation frequency of 0.24 per gamete. The spontaneous mutation rate per gamete of 0.24 is enhanced by a factor of 2 owing to the procarbazine treatment for a total of 48 gametes (0.24 $\times$ 2 $\times$ 48 = 23). It follows that 23 new mutations can be expected in 1 million births owing to the procarbazine treatment (Ehling and Neuhäuser 1979). The great uncertainty in this calculation is the assumption about the number of mutable loci.

The difficulty in the estimation of the current incidence per million liveborn offspring can be illustrated by a comparison of the two BEIR-Reports from 1972 and 1980 (Table 8). The changes in the estimation of the frequencies are due to the population survey in British Columbia. Trimble and Doughty (1974) reported that at least 9.4% of all liveborn humans will be seriously handicapped at some time during

**Table 8.** Comparison of the estimation of the current incidence per million liveborn by the BEIR-Committee

| Disease classification | Current incidence | |
|---|---|---|
| | (1980) | (1972) |
| Autosomal dominant and X-linked disease | 10,000 | 10,000 |
| Irregularly inherited diseases | 90,000 | 40,000 |
| Recessive diseases | 1,100 | 10,000 |
| Chromosomal diseases | 6,000 | |
| Total | 107,100 | 60,000 |

their lifetimes by genetic disorders of complex etiology, manifested as congenital malformations, anomalies expressed later, or constitutional and degenerative diseases. However, they also reported that the incidence of simple, dominant conditions is probably 800 in 1 million liveborn, or about 12 times lower than the estimation of Stevenson (1959).

The difficulties of improving the estimation of the current incidence of genetic diseases or the persistence of the genes in the population led us to the development of a method for the direct estimation of the genetic risk (Ehling 1974, 1976).

## 3.2 Direct Method

The concept of the direct estimation of the genetic risk is based on the induction of mutations having dominant effects, the basic data are those from mouse studies on the induction of dominant cataract mutations and dominant skeletal mutations.

### 3.2.1 Dominant Cataract Mutations in Mice

A cataract is an opacity of the lens causing a reduction of visual function. The organogenesis of the lens in various mammals is similar. Therefore, a gene, which disturbs in different species the same process in the normal development of the lens, leads to the manifestation of the same cataract type in these different species. Ehling (1963) pointed out that morphologically comparable cataracts in humans and other mammals have very often the same mode of inheritance. The comparability of the genetic endpoint in mice and man is one advantage of this system.

The systematic investigation of induced dominant cataracts in mice was initiated by Ehling in 1977. In a series of papers the induction of radiation-induced dominant cataracts (Ehling 1980b; Kratochvilova and Ehling 1979) and ethylnitrosourea-induced dominant cataracts (Ehling et al. 1982; Favor 1982b; Favor 1983) were reported.

To detect dominant cataract mutations, the $F_1$-offspring were examined with a slit lamp at 4–6 weeks of age. Mydriasis was achieved with 1% atropin applied at least 10 min before examination. The lenses of mice were observed with a narrow beam slit lamp illumination which was at a 25°–30° angle from the direction of observation.

The normal variability of small lens opacities was described in detail by Kratochvilova (1981) and Favor (1983). Presumed mutant individuals exhibiting a lens opacity were outcrossed to normal mice, and at least 20 offspring were examined to confirm the genetic nature of the cataract. When, among 20 offspring, no individual exhibited the phenotype, it was concluded that the lens opacity was not due to a dominant mutation with a penetrance value equal or greater than 0.32 (Favor 1982a). For those $F_1$ individuals that produced offspring with the lens opacity phenotype, the presumed mutant was considered confirmed, and mutant lines were established to determine the penetrance and the expressivity of the gene. Furthermore the viability of the mutant and the effect of the mutation in the homozygous condition were studied.

A detailed description of radiation-induced dominant cataracts in mice was published by Kratochvilova (1981). She compared the 11 radiation-induced dominant cataracts with 8 dominant cataracts described earlier for the mouse and concluded that "all 11 cataract mutations differ from each other in respect to their morphological characteristics, degree of severity, and the associated lesions ... No similarities were found between the cataracts described" in her publication and the dominant cataracts reported in the literature. In humans at least 20 well defined cataracts with dominant inheritance are known (McKusick 1978).

For the development of a data base for dominant cataracts and for the comparison of induced recessive and dominant mutations the scoring of specific locus mutations was combined with the screening for dominant cataracts. The results of the radiation experiments were discussed recently in detail. A summary of the data is given in Table 9 (Ehling et al. 1982). A total of 15 dominant cataract mutations were observed in 29,396 offspring. Comparing the overall frequency of induced cataract mutations and specific locus mutations in postspermatogonia and spermatogonia, there were 2.5–2.7 times more recessive mutations than dominant mutations induced by $\gamma$-radiation. Taking into account that in humans 20 well-established dominant cataracts are known, according to McKusick (1978), then it is very likely that approximately three times as many loci coding for dominant cataracts are scored in this experiment than for recessive mutations. Therefore, on a per locus rate, radiation induced approximately eight times more recessive mutations than cataract mutations. In this regard it is interesting to note that H.J. Muller (1950) reported that the ratio of radiation-induced recessive visibles to dominant mutations in spermatogonia of *Drosophila* is 5:1.

Two characteristic features of radiation-induced specific locus mutations and of dominant skeletal mutations can also be demonstrated for the induction of dominant cataracts (Table 9). The mutation frequency per gamete for dominant cataracts due to the fractionation of the dose is 2.6 times higher than the combined frequency from two independent experiments after single exposure of spermatogonia. Similarly the frequency of dominant mutations induced in postspermatogonial stages, is 2–4 times higher than in spermatogonia (Ehling 1966).

## 3.2.2 Dominant Skeletal Mutations in Mice

The systematic investigation of induced dominant mutations affecting the skeleton of mice was initiated by Ehling and Randolph (1962). The skeleton was chosen be-

**Table 9.** Recessive and dominant mutations induced in mice by γ-rays. (Ehling et al. 1982)

| Dose (R) | Dose rate (R/min) | Germ cell stage treated | Number of F₁ offspring | Number of mutations at 7 specific loci | Frequencies per locus × 10⁵ | Number of dominant cataract mutations | Mutations per gamete × 10⁵ |
|---|---|---|---|---|---|---|---|
| 0 | – | – | 103,218 | 6[a] | 0.8 | – | – |
| 0 | – | – | 8,174 | 2 | 3.5 | 0 | 0 |
| 534 | 53 | Postspermatogonia | 1,721 | 3 | 24.9 | 1 | 58.1 |
| 600 | 53 | Postspermatogonia | 865 | 3 | 49.5 | 1 | 115.6 |
| 455 + 455 | 55 | Postspermatogonia | 272 | 2 | 105.0 | 1 | 367.6 |
| 534 | 53 | Spermatogonia | 10,212 | 7 | 9.8 | 3 | 29.4 |
| 600 | 53 | Spermatogonia | 11,095 | 14 | 18.0 | 3 | 27.0 |
| 455 + 455 | 55 | Spermatogonia | 5,231 | 9[b] | 24.6 | 6 | 114.7 |

[a] Untreated historical control of the laboratory
[b] A simultaneous d-se mutation included, which is caused by double non-disjunction

**Table 10.** Frequency of dominant mutations affecting the skeleton of mice

| Classification | Dose (R) | Interval between dose fractions | Germ cell stage treated | No. of F₁ skeletons examined | Mutations n | Mutations (%) | Ref. |
|---|---|---|---|---|---|---|---|
| | 0 | | | 1,739 | 1 | 0.06 | |
| Presumed Mutations | 600 | 0 | Postspermatogonia | 569 | 10 | 1.8 | Ehling (1966) |
| | 600 | 0 | Spermatogonia | 754 | 5 | 0.7 | |
| | 100 + 500 | 24 h | Spermatogonia | 277 | 5 | 1.8 | |
| | 500 + 500 | 10 weeks | Spermatogonia | 131 | 2 | 1.5 | |
| Mutations | 100 + 500 | 24 h | Spermatogonia | 2,646 | 31–37 | 1.2–1.4 | Selby and Selby (1977) |

cause it is formed over an extended period of development and is, therefore, presumably subject to modification by gene action falling within a wide range of time.

The classification used to separate, as far as possible, the existing natural variation from that caused by newly occurring genetic changes is based on the following considerations. Knowing the frequencies of mutation for specific loci (Table 1) one can assume that in an experiment, where the sample size is such that not more than one mutation would be expected for any particular gene locus, the screening for mutations could best be done by dividing the abnormalities according to whether they occur only once in the whole experiment (Class 1) or more frequently (Class 2). Some of the Class-1 abnormalities might be determined environmentally rather than mutationally, and some of the Class-2 abnormalities might result from mutations at different loci or from an unstable locus. One possibility for removing some of the undesired (i.e., presumed nonmutant) portion of Class-1 abnormalities is to increase the sample size; another is to examine statistically the frequency of abnormalities of the different subgroups of Class 1. Mutational events in Class 2 might be recognized by a statistically significant increase of the frequency in any subgroup of Class-2 abnormalities as a result of treatment.

Animals having Class-1 abnormalities were subdivided as having multiple or single abnormalities. In order to be classified as a multiple abnormality, an animal had to have at least one Class-1 abnormality plus others, regardless of whether Class 1 or Class 2. The Class-1 single category was further subdivided into abnormalities of the appendicular skeleton and abnormalities of the axial skeleton. The abnormalities of the appendicular skeleton were subdivided according to whether they were bilateral or unilateral. The most sensitive indication of a mutation is probably the statistically significant increase over the control value of Class-1 multiple abnormalities, and Class-1 abnormalities of the bilateral type in the appendicular skeleton (Ehling 1965, 1966). The frequency of these two subgroups of Class-1 abnormalities are summarized in Table 10 as presumed mutations.

A pilot experiment was designed specifically to permit breeding tests on a sample of presumed mutations. Three dominant mutations affecting the skeleton were found to be transmitted to the second and later generations. One of the mutants was found in an experiment where the offspring derived from sperm irradiated as spermatogonia. Two of these mutants were found in the pilot experiment where the offspring derived from sperm irradiated as spermatozoa or spermatids. One mutant was found in a group of 108 offspring which were produced by males irradiated with 600 R. The other mutant was found in a group of 86 offspring produced by males which were irradiated with 100 R followed by 500 R 24 h later (Ehling 1970). One of these mutations, a disproportionate micromelia (Dmm), was described in detail recently by Brown et al. (1981). In addition, the frequency of Class-1 abnormalities in mouse embryos after irradiation of the sire was determined by Bartsch-Sandhoff (1974).

In a subsequent experiment the inheritance of the presumed mutations was tested. Since the main purpose of the experiment was to determine whether the types of skeletal defects classified as presumed mutations were, in fact, true mutations, it was decided to concentrate on studying in a large sample exclusively the radiation-induced mutations (Selby and Selby 1977). The results of these experiments are summarized in Table 10. The mutation frequency determined by statistical considerations are in good agreement with the breeding test (Ehling 1979).

### 3.2.3 Quantification of First Generation Risk

Based on the induction of dominant mutations in mice, Ehling (1974/76) developed a concept for the direct estimation of the risk of radiation-induced genetic damage to the human population expressed in the first generation. The quantification of the genetic risk is based on the following assumptions:

1. The dose-effect-curve for the induction of dominant cataract mutations is linear.
2. Dominant cataract mutation rates are representative for all dominant mutations.
3. The ratio of dominant cataract mutations (20) to the total number of well established dominant mutations (736) in man (McKusick 1978) is the same as in the mouse. This ratio gives a multiplication factor of 36.8. The multiplication factor is used to convert the induced mutation rate of dominant cataracts to the estimation of the overall dominant mutation rate.

The radiation-induced frequency of dominant cataracts for single exposure (Table 9) is $0.45-0.55 \times 10^{-4}$ mutations/gamete/Gy. The mutation rate has to be multiplied by 36.8 for the calculation of the overall frequency for dominant mutations ($17-20 \times 10^{-4}$ mutations/gamete/Gy). For an acute exposure with high-intensity radiation of spermatogonia of man with 1 Gy per generation, one can expect 1,700-2,000 induced dominant mutations in the first generation in 1 million liveborn individuals (Table 11).

**Table 11.** Estimated effect of 1 Gy per generation of high intensity exposure of spermatogonia for 1 million liveborn individuals

|                                                          | Cataracts                    | Skeletal defects         |
|----------------------------------------------------------|------------------------------|--------------------------|
| Mutations/gamete/Gy                                      | $0.45-0.55 \times 10^{-4}$   | $10.1 \times 10^{-4}$    |
| Multiplication factor for the overall dominant mutation rate | 36.8                     | 4.6                      |
| Expected cases of dominant diseases                      | 1,700-2,000                  | 4,600                    |

No experimental data are available for the induction of dominant cataract mutations by low-intensity radiation in male mice or for the mutation rate in female mice. Therefore, an estimation of the population risk is only possible with reservations. Such an estimation can only be based on the generalization of results obtained with the specific locus method (Ehling 1980b). In general, for exposure with low-intensity we expect only one-third of the frequency with high-intensity exposure (UNSCEAR-Report 1982).

For the risk due to total population exposure, a sensitivity factor for female mice of 1.4 (Russell 1977) or 2 (Ehling 1980b) is used. Because of the higher DNA content in the germ cells of man in comparison to those of mice (Abrahamson et al. 1973) some authors use a factor of 1.2 for the extrapolation from mice to man (Ehling 1980b; UNSCEAR-Report 1966). Other authors claim that mice and man are equally sensitive to the induction of mutations by radiation (Sankaranarayanan 1982; UNSCEAR-Report 1982). Using these additional assumptions, it is possible to calculate the genetic risk of a population after exposure with 1 Gy per 1 million liveborn individuals to be 800 to 1,600 expected cases of dominant diseases in the first generation.

One advantage of the direct estimation of the genetic risk is that the results based on the induction of dominant cataracts can be compared with the data based on dominant hereditary disorders of another system, for example, on the induction of dominant skeletal mutations. For the quantification of the genetic risk based on the induction of dominant skeletal mutations similar assumptions are made as for the estimation based on dominant cataract mutations in mice:

1. The dose-effect-curve for the induction of dominant mutations affecting the skeleton of mice is linear.
2. Dominant skeletal mutation rates are representative for all dominant mutations.
3. The multiplication factor of 4.6 is used to convert the induced mutation rate of dominant skeletal defects to the estimation of the overall dominant mutation rate (Ehling 1980b).

The radiation-induced frequency of dominant skeletal mutations for single exposure is $10.1 \times 10^{-4}$ mutations/gamete/Gy (Table 10). This rate can be converted to the overall frequency for dominant mutations by multiplication with 4.6. It follows that for acute exposure with high-intensity radiation of spermatogonia of man with 1 Gy per generation, we can expect 4,600 induced dominant mutations in the first generation in 1 million liveborn individuals (Table 11).

In contrast to the determination of the multiplication factor for cataracts, it is difficult to determine the multiplication factor for the conversion of the mutation rate for dominant skeletal defects into the estimation of the overall dominant mutation rate (Jacobi et al. 1981). These difficulties are also well documented by the UNSCEAR-Report (1977) and the BEIR-III-Report (1980). In addition, in the experiments to determine the mutation rate of dominant skeletal defects by Ehling (1966) there was for the great majority of cases no proof of inheritance of the defect. The evaluation of the mutations was based on statistical arguments. The transmission experiments by Selby (1979) were evaluated on the basis of Ehling's presumed mutations without a control group.

Considering these differences between the quantification of the genetic risk based on skeletal and cataract mutations and the sample size of these experiments the estimations of Table 11 are in good agreement. The problems of the quantification of the genetic risk based on these two systems were discussed in detail recently (Jacobi et al. 1981).

Also for the induction of dominant skeletal mutations in mouse spermatogonia we have no information about the effectiveness of exposure with low-intensity. Additionally, no experiments were done until now with female mice.

Using similar suppositions as in the UNSCEAR-Report (1977) one can calculate the population risk. This calculation is based on the following assumption: $10.1 \times 10^{-4}$ dominant skeletal mutations/gamete/Gy $\times$ 4.6 (multiplication factor for conversion to the overall mutation rate) $\times$ 0.3 (correction factor for dose rate) $\times$ 1.4 (exposure of female mice) $\times 10^{6}$ (offspring) equals 2,000. This estimation of the effect of 1 Gy per generation of low dose, low dose rate, low LET irradiation on a population of one million liveborn individuals of 2,000 cases with autosomal dominant diseases is based on a generalization of the results obtained with the specific locus method. This figure of 2,000 cases is identical with the estimation of Ehling (1976) and the UNSCEAR-Report (1977).

For an improvement of the risk estimation it is necessary to extend the data base for the induction of dominant mutations in mice, especially information about the mutation frequency in female mice and in the low-intensity range of exposure. In addition, it is necessary to emphasize that a mutation with high penetrance has a better chance to be recovered in these experiments than a mutation with low penetrance. Furthermore, it is very likely that the multiplication factors for the determination of the overall estimates will increase as more genes with dominant inheritance are described. Therefore, these quantifications underestimate the radiation-induced genetic damage of the first generation.

Similar to the quantification of the radiation-induced genetic damage of the first generation the data of Ehling et al. (1982) can be used for the estimation of the expected number of dominant mutations in the first generation after ENU-exposure.

## 4 Problems and Perspectives

Results and conclusions presented in the preceding chapters indicate that the basis for estimating the genetic risks due to radiation exposure or environmental mutagens are based on results of mammalian genetic studies. The specific locus method is used for exploring the effect of various biological and physical factors on the induced mutation frequency at a sample of seven genes. However, this exploration is limited by a very specific genetic background of the cross (101 × C3H)$F_1$ × Test Stock mice. Knowing that the mutation induction may depend on the genetic background (Laskowski 1981) it is essential to explore these factors with other strains of mice. The development of techniques to study the induction of dominant mutations makes this exploration possible.

Multifactorial diseases, diseases of complex aetiology and congenital malformations account for 85% of the spontaneously occurring genetic diseases in humans (Table 7). The extent to which radiation exposure or chemical mutagens will increase their incidence is largely unknown. Nomura (1982) observed a significant increase in tumors among offspring of X-ray or urethane-treated ICR parent mice. About 90% of the tumors were in the lung, and were inherited as if they were dominant mutations with about 40% penetrance. Kirk and Lyon (1982) irradiated female mice with 108–504 rad of X-rays. The exposed females were mated at different intervals after irradiation (1–7, 8–14, 15–21, and 22–28 days). Uterine contents were examined at late pregnancy in order to detect fetal death and malformations in the live fetuses. Two trends were apparent from data on abnormal fetuses. At each weekly interval, the incidence of abnormalities tended to rise with increase in dose, and, at any given dose, the incidence tended to increase with time after irradiation. Dwarfism and exencephaly were the two most common malformations observed. The data base of these studies should be extended. The genetic component of these congenital anomalies should be carefully analyzed. Additional model systems should be developed.

The BEIR (1980), the UNSCEAR (1982), and the CCEM-Report (1983) emphasize the importance of dominant mutations for the quantification of the genetic risk. Because of the importance of these test systems, Sankaranarayanan (1982) pointed out correctly, that the data base for these estimations are very limited. It is essential to establish dose-response curves for the induction of dominant mutations. In addition, "there are no data on the induction of skeletal or cataract mutations in female mice" and "there is no experimentation so far to verify whether the correction factors used to estimate effects at low doses and at low dose rates are in fact valid for these kinds of mutational events". Similarly, it is necessary to establish a data base for the induction of dominant mutations by chemical mutagens. Because it is until now not possible to consider the total burden of chemical mutagens, it is therefore necessary to establish standards for the exposure limits of specific chemical mutagens. This limit can be expressed as a fraction of an increase of the spontaneous mutation rate or as a number of accepted mutations. For example, for a single compound the allowable level of risk for all dominant mutations could be $10^{-6}$. This figure is debatable and depends on the number of chemical mutagens a society will permit. Independent of the details of these exposure limits for a chemical mutagen it is necessary to discuss these problems and find an acceptable and responsible solution.

The most important aspect of risk estimation is the possible interaction of different chemical mutagens and of ionizing radiation with chemical mutagens. Some of the main problems involved in studying interaction between radiation and chemicals have been recently reviewed by Glubrecht et al. (1979) and Kada et al. (1979). It is likely that the action of a mutagenic agent is not direct and that cellular functions, such as mutator or repair systems, are involved in the mutagenesis initiated by an agent. Such cellular functions can be affected by a second agent. In sexually reproducing organisms, the two agents can also act on separate cells (male and female germ cells) which subsequently fuse. However, it is at present not possible to estimate the impact of the interaction between radiation and chemicals for the risk assessment. These problems have to be solved in the near future. This task will be extremely difficult, but to quote CP Snow: Scientists "are inclined to be impatient to see if something can be done: and inclined to think that it can be done, until it's proved otherwise. That is their real optimism".

*Acknowledgements.* Specific locus experiments were supported by Contract ENV-637-D(B) and the dominant cataract studies by BIO-E-395-81-D of the Commission of the European Communities.

# References

Abrahamson S, Bender MA, Conger AD, Wolff S (1973) Uniformity of radiation-induced mutation rates among different species. Nature 245:460–462

Bartsch-Sandhoff M (1974) Skeletal abnormalities in mouse embryos after irradiation of the sire. Humangenetik 25:93–100

Bateman AJ (1979) Significance of specific-locus germ-cell mutations in mice. Mutat Res 64: 345–351

BEIR-Report (1972) (Biological Effects of Ionizing Radiations) The effects on populations of exposure to low levels of ionizing radiation. National Academy, Washington, D.C.

BEIR-Report (1980) (Biological Effects of Ionizing Radiations) The effects on populations of exposure to low levels of ionizing radiation: 1980. National Academy, Washington, D.C.

BMI (Bundesministerium des Innern) Umweltradioaktivität und Strahlenbelastung. Jahresbericht 1979

Brown KS, Cranley RE, Greene R, Kleinman HK, Pennypacker JP (1981) Disproportionate micromelia (Dmm): An incomplete dominant mouse dwarfism with abnormal cartilage matrix. J Embryol Exp Morphol 62:165–182

Cattanach BM (1966) Chemically induced mutations in mice. Mutat Res 3:346–353

Cattanach BM (1971) Specific locus mutation in mice. In: Hollaender A (ed) Chemical mutagens; principles and methods for their detection, vol 2. Plenum, New York, pp 535–539

Cattanach BM (1982) Induction of specific-locus mutations in female mice by triethylenemelamine (TEM). Mutat Res 104:173–176

CCEM-Report (1983) (Committee on Chemical Environmental Mutagens) Identifying and estimating the genetic impact of chemical mutagens. National Academy, Washington, D.C.

Crow JF (1981) How well can we assess genetic risks? Not very. In: Lauriston S Taylor lectures in radiation protection and measurements, lecture no. 5. National Council on Radiation Protection and Measurements, Washington, D.C., pp 7–31 (quotation p 13)

Ehling U (1963) Vererbung von Augenleiden im Tierreich. Bericht über die 65. Zusammenkunft der Deutschen Ophthalmologischen Gesellschaft in Heidelberg. Bergmann, München, 228–238

Ehling UH (1965) The frequency of X-ray induced dominant mutations affecting the skeleton of mice. Genetics 51:723–732

Ehling UH (1966) Dominant mutations affecting the skeleton in offspring of X-irradiated male mice. Genetics 54:1381–1389

Ehling UH (1970) Evaluation of presumed dominant skeletal mutations. In: Vogel F, Röhrborn G (eds) Chemical mutagenesis in mammals and man. Springer, Berlin Heidelberg New York, pp 162–166

Ehling UH (1974) Die Gefährdung der menschlichen Erbanlagen im technischen Zeitalter (Vortrag beim Deutschen Röntgenkongreß 1974 in Baden-Baden). Fortschr Röntgenstr 124:166 to 171 (1976)

Ehling UH (1976) Estimation of the frequency of radiation-induced dominant mutations. ICRP, CI-TG 14, Task Group on Genetically Determined Ill-Health

Ehling UH (1977) Dominant lethal mutations in male mice. Arch Toxicol 38:1–11

Ehling UH (1978) Specific-locus mutations in mice. In: Hollaender A, de Serres FJ (eds) Chemical mutagens, principles and methods for their detection, vol 5. Plenum, New York, pp 233 to 256

Ehling UH (1979) Criteri di stima del rischio genetico. In: Enciclopedia della scienza e della tecnica. Mondadori, Milano, pp 125–134

Ehling UH (1980a) Induction of gene mutations in germ cells of the mouse. Arch Toxicol 46: 123–138

Ehling UH (1980b) Strahlengenetisches Risiko des Menschen. Umschau 80:754–759

Ehling UH (1981) Genetische Risiken durch Umweltchemikalien. In: Glöbel B, Gerber G, Grillmaier R, Kunkel R, Leetz H-K, Oberhausen E (Hrsg) Umweltrisiko 80. Thieme, Stuttgart New York, pp 400–411

Ehling UH (1982) Risk estimations based on germ-cell mutations in mice. In: Sugimura T, Kondo S, Takebe H (eds) Environmental mutagens and carcinogens (Proceedings of the 3rd International Conference on Environmental Mutagens). University of Tokyo Press, Tokyo New York, pp 709–719

Ehling UH (in press) In vivo gene mutations in mammals. Proceedings of the symposium "Critical evaluation of mutagenicity tests". Bundesgesundheitsamt, Berlin

Ehling UH, Neuhäuser A (1979) Procarbazine-induced specific-locus mutations in male mice. Mutat Res 59:245–256

Ehling UH, Neuhäuser-Klaus A (1982) Chemically-induced mutations in mice, Progress report: Mai 1982–November 1982. Commission of the European Communities

Ehling UH, Randolph ML (1962) Skeletal abnormalities in the $F_1$ generation of mice exposed to ionizing radiations. Genetics 47:1543–1555

Ehling UH, Favor J, Kratochvilova J, Neuhäuser-Klaus A (1982) Dominant cataract mutations and specific-locus mutations in mice induced by radiation or ethylnitrosourea. Mutat Res 92: 181–192

Ehling UH, Averbeck D, Cerutti PA, Friedman J, Greim H, Kolbye Jr AC, Mendelsohn ML (1983) Review of the evidence for the presence or absence of thresholds in the induction of genetic effects by genotoxic chemicals. Mutat Res 123:281–341

Favor J (1982a) The penetrance value tested of a presumed dominant mutation heterozygote in a genetic confirmation test for a given number of offspring observed. Mutat Res 92:192

Favor J (1982b) The dominant cataract mutation test in mice. Mutat Res 97:186–187

Favor J (1983) A comparison of the dominant cataract and recessive specific-locus mutation rates induced by treatment of male mice with ethylnitrosourea. Mutat Res 110:367–382

Glubrecht H, Gopal-Ayengar AR, Ehrenberg L (1979) Interactions of ionizing radiation and chemicals and mechanisms of action – summary. In: Okada S, Imamura M, Terashima T, Yamaguchi H (eds) Radiation research. Proceedings of the 6th International Congress of Radiation Research. Japanese Association for Radiation Research, Tokyo, Japan, pp 708–710

Green MC (1966) Mutant genes and linkages. In: Green EL (ed) Biology of the laboratory mouse, 2nd edn. McGraw Hill, New York, pp 87–150

Heller CG, Clermont Y (1964) Kinetics of the germinal epithelium in man. In: Pincus G (ed) Recent progress in hormone research. Academic, New York London, pp 545–575

Jacobi W, Paretzke HG, Ehling UH (1981) Strahlenexposition und Strahlenrisiko der Bevölkerung. GSF-Bericht, S-710

Kada T, Inoue T, Yokoiyama A, Russell LB (1979) Combined genetic effects of chemicals and radiation. In: Okada S, Imamura M, Terashima T, Yamaguchi H (eds) Radiation research. Proceedings of the 6th International Congress of Radiation Research. Japanese Association for Radiation Research, Tokyo, Japan, pp 711–720

Kirk M, Lyon MF (1982) Induction of congenital anomalies in offspring of female mice exposed to varying doses of X-rays. Mutat Res 106:73–83

Kratochvilova J (1981) Dominant cataract mutations detected in offspring of gamma-irradiated male mice. J Hered 72:302–307

Kratochvilova J, Ehling UH (1979) Dominant cataract mutations induced by $\gamma$-irradiation of male mice. Mutat Res 63:221–223

Laskowski W (1981) Biologische Strahlenschäden und ihre Reparatur. De Gruyter, Berlin New York

Lüning KG, Searle AG (1971) Estimates of the genetic risks from ionizing irradiation. Mutat Res 12:291–304

McKusick VA (1978) Mendelian inheritance in man, 5th edn. Johns Hopkins University Press, Baltimore

Muller HJ (1950) Radiation damage to the genetic material. Am Sci 38:33–59

Neuhäuser-Klaus A (1983) Personal communication

Nomura T (1982) Parental exposure to X rays and chemicals induces heritable tumours and anomalies in mice. Nature 296:575–577

Oakberg EF (1956) Duration of spermatogenesis in the mouse and timing of stages of the cycle of the seminiferous epithelium. Am J Anat 99:507–516

Roderick TH (1964) Proceedings of the symposium: The effects of radiation on the hereditary fitness of mammalian populations. The Jackson Laboratory, Bar Harbor, Maine, June 29 to July 1, 1964. Genetics 50:1023–1217

Russell LB, Selby PB, von Halle E, Sheridan W, Valcovic L (1981) The mouse specific-locus test with agents other than radiations. Interpretation of data and recommendations for future work. Mutat Res 86:329–354

Russell WL (1951) X-ray-induced mutations in mice. Cold Spring Harbor Symp Quant Biol 16: 327–336

Russell WL (1972) The genetic effects of radiation. In: Peaceful uses of atomic energy, vol 13. International Atomic Energy Agency, Vienna, pp 487–500

Russell WL (1977) Mutation frequencies in female mice and the estimation of genetic hazards of radiation in women. Proc Natl Acad Sci USA 74:3523–3527

Russell WL, Kelly EM (1982) Mutation frequencies in male mice and the estimation of genetic hazards of radiation in men. Proc Natl Acad Sci USA 79:542–544

Russell WL, Hunsicker PR, Carpenter DA, Cornett CV, Guinn GM (1982) Effect of dose fractionation on the ethylnitrosourea induction of specific-locus mutations in mouse spermatogonia. Proc Natl Acad Sci USA 79:3592–3593

Russell WL, Kelly EM, Hunsicker PR, Bangham JW, Maddux SC, Phipps EL (1979) Specific-locus test shows ethylnitrosourea to be the most potent mutagen in the mouse. Proc Natl Acad Sci USA 76:5818–5819

Sankaranarayanan K (1982) Genetic effects of ionizing radiation in multicellular eukaryotes and the assessment of genetic radiation hazards in man. Elsevier Biomedical, Amsterdam (quotations p 276, p 311)

Schull WJ, Otake M, Neel JV (1981) Genetic effects of the atomic bombs: a reappraisal. Science 213:1220–1227 (quototation p 1227)

Searle AG (1974) Mutation induction in mice. In: Lett JT, Adler HI, Zelle M (eds) Advances in radiation biology, vol 4. Academic, New York London, pp 131–207

Searle AG (1975) The specific locus test in the mouse. Mutat Res 31:277–290

Selby PB (1979) Induced skeletal mutations. Genetics 92:S127–S133

Selby PB, Selby PR (1977) Gamma-ray-induced dominant mutations that cause skeletal abnormalities in mice. I. Plan, summary of results and discussion. Mutat Res 43:357–375

Setchell BP (1970) Testicular blood supply, lymphatic drainage, and secretion of fluid. In: Johnson AD, Gomes WR, VanDemark NL (eds) The testis, vol 1. Academic, New York, pp 101 to 239

Snow CP (1964) The two cultures: And a second look. University Press, Cambridge (quotation p 7)

Stevenson AC (1959) The load of hereditary defects in human populations. Radiat Res Suppl 1:306–325

Trimble BK, Doughty JH (1974) The amount of hereditary disease in human populations. Ann Hum Genet 38:199–223

UNSCEAR-Report (1966) (United Nations Scientific Committee on the Effects of Atomic Radiation) Supplement no 14. United Nations, New York

UNSCEAR-Report (1977) (United Nations Scientific Committee on the Effects of Atomic Radiation) Sources and effects of ionizing radiation. United Nations, New York

UNSCEAR-Report (1982) (United Nations Scientific Committee on the Effects of Atomic Radiation) Ionizing radiation: Sources and biological effects. United Nations, New York

Wolff S (1967) Radiation genetics. In: Roman HL, Sandler LM, Stent GS (eds) Annual review of genetics, vol 1. Annual Reviews, Inc, Palo Alto, California, pp 221–244

# Subject Index

F. Vogel, A. G. Motulsky

# Human Genetics

## Problems and Approaches

2nd printing with corrections. 1982. 420 figures, 210 tables.
XXVIII, 700 pages
ISBN 3-540-09459-8

**Contents:** Introduction. – History of Human Genetics. –
Human Chromosomes. – Formal Genetics of Man. – Gene
Action. – Mutation. – Population Genetics. – Human
Evolution. – Genetics and Human Behavior. – Practical
Applications of Human Genetics and the Biological Future
of Mankind. – Appendices 1–9. – References. – Author
Index. – Subject Index.

This comprehensive text examines the principles and
problems of human genetics in the light of biological,
medical and social sciences. In addition to tracing its
historical development, the authors discuss the various
branches of human genetics and their relation to the field as
a whole. They critically assess research approaches and
methods, highlighting controversial problems in the application of genetic principles to medicine and the behavioral
sciences. An extensive bibliography is also included.
The book is directed at: a) students in genetics, medicine,
biology, anthropology, psychology, and statistics, especially
those seeking a critical assessment of modern human genetics; b) researchers and university teachers in the same
fields; c) physicians and scientists interested in the impact
on human genetics on medicine and society.

*From the reviews:*
"This book should be read by all who work in human genetics and who teach it to others. It should be read by all
advanced students in genetics, human or otherwise. And it
should be read by those outside the field who would like
insight into the way human genetics and human geneticists
work and into the accomplishments of the past and the
challenges for the future."
*American Journal of Human Genetics*

Springer-Verlag
Berlin
Heidelberg
New York
Tokyo

# Human Genetics

ISSN 0340-6717          Title No. 439

Recent years have witnessed significant advances in human genetics
and an ever-increasing realization of their importance for those
engaged in the life sciences. New concepts of the structure and function of the gene have been of major importance in understanding
the genetics of man. Simultaneously, methodological advances have
led to new insights concerning the genetic basis of health and
disease. Research in cytogenetics, biochemical genetics, population
genetics, immunogenetics, and pharmacogenetics now strongly
supplements studies in formal genetics.

All these fields are covered by original contributions which appear
in the journal **Human Genetics.** New observations in the fields of
medical genetics and cytogenetics are helping to improve genetic
diagnosis, prognosis, and counselling. In recent years, the journal
has increasingly become a forum in which such observations are
published and then discussed in Europe and all around the world.

Special emphasis is on **review articles** within the scope of clinical
and theoretical genetics. The editors consider the dialogue between
theoretically and clinically oriented human geneticists vital and
therefore give a very high priority to covering the broad range of this
field.

**Subscription information** and/or **sample copies** are available from
your bookseller or directly from
Springer-Verlag, Journal Promotion Dept.,
P. O. Box 105 280, D-6900 Heidelberg, FRG

Orders from North America should be addressed to:
Springer-Verlag New York Inc., Journal Sales Dept.,
44 Hartz Way, Secaucus, NJ 07094, USA

**Springer
International**